Advances in Experimental Medicine and Biology

Volume 1345

Advances in Experimental Medicine and Biology provides a platform for scientific contributions in the main disciplines of the biomedicine and the life sciences. This series publishes thematic volumes on contemporary research in the areas of microbiology, immunology, neurosciences, biochemistry, biomedical engineering, genetics, physiology, and cancer research. Covering emerging topics and techniques in basic and clinical science, it brings together clinicians and researchers from various fields.

Advances in Experimental Medicine and Biology has been publishing exceptional works in the field for over 40 years, and is indexed in SCOPUS, Medline (PubMed), Journal Citation Reports/Science Edition, Science Citation Index Expanded (SciSearch, Web of Science), EMBASE, BIOSIS, Reaxys, EMBiology, the Chemical Abstracts Service (CAS), and Pathway Studio.

2020 Impact Factor: 2.622

More information about this series at http://www.springer.com/series/5584

Abdol-Mohammad Kajbafzadeh
Editor

Decellularization Methods of Tissue and Whole Organ in Tissue Engineering

Springer

Editor
Abdol-Mohammad Kajbafzadeh
Children's Medical Center
Tehran University of Medical Sciences
Tehran, Iran

ISSN 0065-2598 ISSN 2214-8019 (electronic)
Advances in Experimental Medicine and Biology
ISBN 978-3-030-82734-2 ISBN 978-3-030-82735-9 (eBook)
https://doi.org/10.1007/978-3-030-82735-9

This Springer imprint is published by the registered company Springer Nature Switzerland AG
The registered company address is: Gewerbestrasse 11, 6330 Cham, Switzerland

Introduction

This book contains a comprehensive reference and teaching aid on decellularization and recellularization methods of tissues and organs in **Tissue Engineering and Regenerative Medicine**. The chapter of the current book will cover from the basics of regenerative medicine to more advanced topics for translational applications of tissue-engineered scaffolds. This edition will also provide an update on general understanding of tissue growth and development, the materials and methods needed to design tissues/organs, as well as a presentation by the world's professionals of what is currently known about tissue engineering of specific tissues/organs.

Tissue engineering is a scientific field concerned with the application of cells, biomaterials, biochemical (e.g., growth factors), and suitable scaffolds to improve or replace biological tissues. "**Tissue Engineering & Regenerative Medicine**" provides a platform for the advancement of research and technologies for the formation of new viable tissues/organs for a medical purpose. This interdisciplinary field may be able to help with the shortage of life-saving organs available through donation for transplantation in the near future. It can also help with the limited donor availability and rejection of the grafts by the immune system.

Introduced chapters have been dedicated to provide in-depth principles for many of the supporting and enabling knowledge during the tissue production process and also to expand focus on stem cells, including adult and embryonic stem cells and progenitor populations. The tissue fabrication process is illustrated with specific examples for more than 25 organs, which may soon lead to new tissue engineering therapies for several diseases that afflict humanity. Section-coverage **includes an overall introduction of tissue engineering; Materials & Methods; Cell seeding process; Clinical Applications; Limitations; Conclusion; and future challenges**.

The readers may turn to this up-to-date coverage for a widespread understanding of regenerative medicine that we believe will be useful to students and experts alike.

Readership

The readership consists of basic and clinical researchers in fields of medicine, biology, materials science, and engineering with an interest in tissue engineering.

Contents

About the Editor

Affiliations and Expertise

Professor/Founder and Chairman of the Department of Pediatric Urology/Founder and President of Pediatric Urology and Regenerative Medicine Research Center, Section of Tissue Engineering and Stem Cells Therapy, Children's Hospital Medical Center, Tehran University of Medical Sciences, Tehran, Iran (IRI)/President of the Iranian Fetal Diagnosis and Treatment Association.

Professor Abdol-Mohammad Kajbafzadeh is currently Head of Pediatric Urology and Regenerative Medicine Research Center, Section of Tissue Engineering and Stem Cells Therapy. His research specialty includes genitourinary cancer, physiology, nephrology, urology, radiology, stem cell and regenerative medicine. He has hundreds of scientific publications including definitive references in the fields of tissue engineering and regenerative medicine. His extensive experience in clinical, surgical and basic science research mainly tissue engineering and stem cell therapy as well as bio-scaffolds and invention of non toxic techniques for more than 40 tissues and whole organ decellularization paved the road to extrapolate his knowledge to clinical settings and use his expertise to solve patients' problems [1–53]. Over 28 years (1984–2012), he designed hundreds of studies with main focus on the stem cell therapy and tissue engineering tissue repair renal transplantation. Pulsating heart tissue was produced by seeding various heart cells in his laboratory, where live images can be viewed on site [54]. The human-sized kidney has been also recellularized and transplanted to the large laboratory animals [55]. Since 2007, the works done in his Center has been featured seven times on the cover of the US and UK urology journals. The first cell therapy for urinary incontinence was also performed 12 years ago at his center [56]. Renal damage was also repaired by autologous cells for the first time at his center [42]. His research center has focused primarily on designing cell therapy approaches in animal models, new clinical trials, and prospective clinical trials. He developed hundreds of new ideas in various aspects of tissue engineering, stem cell culture and differentiation therapy, bone marrow stem cells, adipose-derived stem cells, dental pulp stem cells, endometrial stem cells, muscle derived stem cells, and kidney stem cells. This clearly helps to understand the mechanism of the disease and to improve the therapeutic methods. These all together substantially add to his clinical and experimental expertise in translational medicine and from the bench to the bedside. It clearly helps to understanding of the mechanism of disease and improvement of treatment options.

References

1. Sabetkish S, Kajbafzadeh AM, Sabetkish N, Khorramirouz R, Akbarzadeh A, Seyedian SL, Pasalar P, Orangian S, Beigi RSH, Aryan Z (2015) Whole-organ tissue engineering: decellularization and recellularization of three-dimensional matrix liver scaffolds. J Biomed Mater Res Part A 103(4):1498–1508
2. Kajbafzadeh A-M, Javan-Farazmand N, Monajemzadeh M, Baghayee A (2013) Determining the optimal decellularization and sterilization protocol for preparing a tissue scaffold of a human-sized liver tissue. Tissue Eng Part C Methods 19(8):642–651
3. Sharifi E, Azami M, Kajbafzadeh A-M, Moztarzadeh F, Faridi-Majidi R, Shamousi A, Karimi R, Ai J (2016) Preparation of a biomimetic composite scaffold from gelatin/collagen and bioactive glass fibers for bone tissue engineering. Mater Sci Eng C 59:533–541
4. Kajbafzadeh A-M, Payabvash S, Salmasi AH, Sadeghi Z, Elmi A, Vejdani K, Tavangar SM, Tajik P, Mahjoub F (2007) Time-dependent neovasculogenesis and regeneration of different bladder wall components in the bladder acellular matrix graft in rats. J Surg Res 139(2):189–202
5. Kajbafzadeh A-M, Elmi A, Talab SS, Esfahani SA, Tourchi A (2010) Functional external anal sphincter reconstruction for treatment of anal incontinence using muscle progenitor cell auto grafting. Dis Colon Rectum 53(10):1415–1421
6. Semyari H, Rajipour M, Sabetkish S, Sabetkish N, Abbas FM, Kajbafzadeh A-M (2016) Evaluating the bone regeneration in calvarial defect using osteoblasts differentiated from adipose-derived mesenchymal stem cells on three different scaffolds: an animal study. Cell Tissue Bank 17(1):69–83
7. Kajbafzadeh A-M, Tourchi A, Mousavian A-A, Rouhi L, Tavangar SM, Sabetkish N (2014) Bladder muscular wall regeneration with autologous adipose mesenchymal stem cells on three-dimensional collagen-based tissue-engineered prepuce and biocompatible nanofibrillar scaffold. J Pediatr Urol 10(6):1051–1058
8. Talab SS, Kajbafzadeh AM, Elmi A, Tourchi A, Sabetkish S, Sabetkish N, Monajemzadeh M (2014) Bladder reconstruction using scaffold-less autologous smooth muscle cell sheet engineering: early histological outcomes for autoaugmentation cystoplasty. BJU Int 114 (6):937–945
9. Tabatabaei FS, Ai J, Kashi TSJ, Khazaei M, Kajbafzadeh A-M, Ghanbari Z (2013) Effect of dentine matrix proteins on human endometrial adult stem-like cells: in vitro regeneration of odontoblasts cells. Archives of oral biology 58 (7):871–879
10. Kajbafzadeh A-M, Sabetkish S, Heidari R, Ebadi M (2014) Tissue-engineered cholecyst-derived extracellular matrix: a biomaterial for in vivo autologous bladder muscular wall regeneration. Pediatr Surg Int 30(4):371–380
11. Amidi F, Hoseini MA, Nia KN, Habibi M, Kajbafzadeh AM, Mazaheri Z, Yamini N (2015) Male germ-like cell differentiation potential of human umbilical cord Wharton's jelly-derived mesenchymal stem cells in co-culture with human placenta cells in presence of BMP4 and retinoic acid. Iran J Basic Med Sci 18(4):325

12. Elmi A, Kajbafzadeh A-M, Tourchi A, Talab SS, Esfahani SA (2011) Safety, efficacy and health related quality of life of autologous myoblast transplantation for treatment of urinary incontinence in children with bladder exstrophy-epispadias complex. J Urol 186 (5):2021–2026
13. Kajbafzadeh A-M, Esfahani SA, Talab SS, Elmi A, Monajemzadeh M (2011) In-vivo autologous bladder muscular wall regeneration: application of tissue-engineered pericardium in a model of bladder as a bioreactor. J Pediatr Urol 7(3):317–323
14. Sharifi E, Ebrahimi-Barough S, Panahi M, Azami M, Ai A, Barabadi Z, Kajbafzadeh AM, Ai J (2016) In vitro evaluation of human endometrial stem cell-derived osteoblast-like cells' behavior on gelatin/collagen/bioglass nanofibers' scaffolds. J Biomed Mater Res Part A 104(9):2210–2219
15. Dehkordi MB, Madjd Z, Chaleshtori MH, Meshkani R, Nikfarjam L, Kajbafzadeh A-M (2016) A simple, rapid, and efficient method for isolating mesenchymal stem cells from the entire umbilical cord. Cell Transplant 25(7):1287–1297
16. Kajbafzadeh A-M, Esfahani SA, Sadeghi Z, Elmi A, Monajemzadeh M (2012) Application of different scaffolds for bladder wall regeneration: the bladder as a natural bioreactor. Tissue Eng Part A 18 (7–8):882–887
17. Kajbafzadeh A-M, Khorramirouz R, Kameli SM, Hashemi J, Bagheri A (2017) Decellularization of human internal mammary artery: biomechanical properties and histopathological evaluation. BioRes Open Access 6(1):74–84
18. Kajbafzadeh A-M, Sabetkish S, Sabetkish N, Muhammadnejad S, Akbarzadeh A, Tavangar SM, Mohseni MJ, Amanpour S (2015) In-vivo trachea regeneration: fabrication of a tissue-engineered trachea in nude mice using the body as a natural bioreactor. Surg Today 45(8):1040–1048
19. Hashemi J, Pasalar P, Soleimani M, Khorramirouz R, Fendereski K, Enderami SE, Kajbafzadeh AM (2018) Application of a novel bioreactor for in vivo engineering of pancreas tissue. J Cell Physiol 233 (5):3805–3816
20. Kajbafzadeh A-M, Tafti SHA, Mokhber-Dezfooli M-R, Khorramirouz R, Sabetkish S, Sabetkish N, Rabbani S, Tavana H, Mohseni MJ (2016) Aortic valve conduit implantation in the descending thoracic aorta in a sheep model: The outcomes of pre-seeded scaffold. Int J Surg 28:97–105
21. Kajbafzadeh A, Sabetkish N, Sabetkish S, Tavangar S, Beigi RH, Talebi M, Akbarzadeh A, Nikfarjam L (2015) Lung tissue engineering and preservation of alveolar microstructure using a novel casting method. Biotech Histochem 90(2):111–123
22. Khorramirouz R, Sabetkish S, Akbarzadeh A, Muhammadnejad A, Heidari R, Kajbafzadeh A-M (2014) Effect of three decellularisation protocols on the mechanical behaviour and structural properties of sheep aortic valve conduits. Adv Med Sci 59 (2):299–307
23. Kajbafzadeh A-M, Sabetkish S, Tourchi A, Amirizadeh N, Afshar K, Abolghasemi H, Elmi A, Talab SS, Eshghi P, Mohseni MJ (2014) The application of tissue-engineered preputial matrix and fibrin sealant for urethral reconstruction in rabbit model. Int Urol Nephrol 46(8):1573–1580
24. Kajbafzadeh A-M, Abbasioun R, Sabetkish S, Sabetkish N, Rahmani P, Tavakkolitabassi K, Arshadi H (2017) Future prospects for human tissue engineered urethra transplantation: decellularization and recellularization-based urethra regeneration. Ann Biomed Eng 45(7):1795–1806
25. Kajbafzadeh A-M, Kajbafzadeh M, Sabetkish S, Sabetkish N, Tavangar SM (2016) Tissue-engineered external anal sphincter using autologous myogenic satellite cells and extracellular matrix: functional and histological studies. Ann Biomed Eng 44 (5):1773–1784
26. Sabetkish N, Kajbafzadeh A-M, Sabetkish S, Tavangar SM (2014) Augmentation cystoplasty using decellularized vermiform appendix in rabbit model. J Pediatr Surg 49(3):477–483
27. Kajbafzadeh A-M, Khorramirouz R, Kameli SM, Fendereski K, Daryabari SS, Tavangar SM, Garajegayeh BA (2019) Three-year efficacy and patency follow-up of decellularized human internal mammary artery as a novel vascular graft in animal models. J Thorac Cardiovasc Surg 157(4):1494–1502

28. Kameli SM, Khorramirouz R, Eftekharzadeh S, Fendereski K, Daryabari SS, Tavangar SM, Kajbafzadeh AM (2018) Application of tissue-engineered pericardial patch in rat models of myocardial infarction. J Biomed Mater Res Part A 106 (10):2670–2678
29. Hashemi J, Pasalar P, Soleimani M, Arefian E, Khorramirouz R, Akbarzadeh A, Ghorbani F, Enderami SE, Kajbafzadeh AM (2018) Decellularized pancreas matrix scaffolds for tissue engineering using ductal or arterial catheterization. Cell Tissue Org 205:72–84
30. Nafisi N, Akbari ME, Mahjoub F, Mohseni MJ, Sabetkish S, Khorramirouz R, Tehrani M, Kajbafzadeh A-M (2017) Application of human acellular breast dermal matrix (ABDM) in implant-based breast reconstruction: an experimental study. Aesthetic Plast Surg 41(6):1435–1444
31. Elmi A, Kajbafzadeh A-M, Oghabian MA, Talab SS, Tourchi A, Khoei S, Rafie B, Esfahani SA (2014) Anal sphincter repair with muscle progenitor cell transplantation: serial assessment with iron oxide–enhanced MRI. Am J Roentgenol 202(3):619–625
32. Khorramirouz R, Kameli SM, Fendereski K, Daryabari SS, Kajbafzadeh A-M (2019) Evaluating the efficacy of tissue-engineered human amniotic membrane in the treatment of myocardial infarction. Regenerat Med 14(2):113–126
33. Kajbafzadeh A-M, Tafti SHA, Khorramirouz R, Sabetkish S, Kameli SM, Orangian S, Rabbani S, Oveisi N, Golmohammadi M, Kashani Z (2017) Evaluating the role of autologous mesenchymal stem cell seeded on decellularized pericardium in the treatment of myocardial infarction: an animal study. Cell Tissue Bank 18(4):527–538
34. Kajbafzadeh A-M, Abbasioun R, Sabetkish N, Sabetkish S, Habibi AA, Tavakkolitabassi K (2017) In vivo human corpus cavernosum regeneration: Fabrication of tissue-engineered corpus cavernosum in rat using the body as a natural bioreactor. Int Urol Nephrol 49(7):1193–1199
35. Kajbafzadeh A-M, Khorramirouz R, Sabetkish S, Talebi MA, Akbarzadeh A, Keihani S (2016) In vivo regeneration of bladder muscular wall using decellularized colon matrix: an experimental study. Pediatr Surg Int 32(6):615–622
36. Farahani F, Yazdi AK, Ghasemi M, Shariatpanahi E, Kajbafzadeh A-M, Amanpour S (2015) Results of acellular dermis matrix graft used for tympanoplasty in guinea pig model. Iran J Otorhinolaryngol 27(79):95
37. Kajbafzadeh A-M, Mozafarpour S, Seyedian SSL, Khorramirouz R, Hojjati HN (2015) Decellularized dermal strip as a suburethral sling in a rat model of stress urinary incontinence. Int Urol Nephrol 47(8):1303–1310
38. Kajbafzadeh AM, Khorramirouz R, Akbarzadeh A, Sabetkish S, Sabetkish N, Saadat P, Tehrani M (2015) A novel technique for simultaneous whole-body and multi-organ decellularization: umbilical artery catheterization as a perfusion-based method in a sheep foetus model. Int J Exp Pathol 96(2):116–132
39. Sabetkish N, Sabetkish S, Mohseni MJ, Kajbafzadeh A-M (2019) Prevention of renal scarring in acute pyelonephritis by probiotic therapy: an experimental study. Probiot Antimicrob Proteins 11(1):158–164
40. Khorramirouz R, Kameli SM, Eftekharzadeh S, Kajbafzadeh AM (2017) Application of omentum as an in vivo bioreactor for regeneration of decellularized human internal mammary artery. J Biomed Mat Res Part A 105(10):2685–2693
41. Kajbafzadeh A-M, Khorramirouz R, Masoumi A, Keihani S, Nabavizadeh B (2018) Decellularized human fetal intestine as a bioscaffold for regeneration of the rabbit bladder submucosa. J Pediatr Surg 53(9):1781–1788
42. Sabetkish S, Sabetkish N, Talebi MA, Halimi S, Kajbafzadeh A-M (2018) The role of nonautologous and autologous adipose-derived mesenchymal stem cell in acute pyelonephritis. Cell Tissue Bank 19(3):301–309
43. Salari Sedigh H, Saffarpour A, Jamshidi S, Ashouri M, Nassiri SM, Sharifi D, Torkzaban P, Kajbafzadeh A (2010) Canine periodontal stem cells: Isolation, differentiation potential and electronic microscopic characterization. Iran J Vet Surg 5(1–2):19–28
44. Sabetkish S, Sabetkish N, Kajbafzadeh A-M (2019) In-vivo regeneration of bladder muscular wall with whole decellularized bladder matrix: A novel hourglass technique for duplication of bladder volume in rabbit model. J Pediatr Surg

45. Akbarzadeh A, Kianmanesh M, Fendereski K, Ebadi M, Daryabari SS, Masoomi A, Ghazisaeedi F, Beigi RSH, Sheikh R, Kajbafzadeh A-M (2019) Decellularised whole ovine testis as a potential bio-scaffold for tissue engineering. Reprod Fertil Develop 31 (11):1665–1673
46. Fatourechi M, Fattahian H, Kajbafzadeh A (2015) The experimental study of bio-engineered free-cell ostrich cornea as xenograft.
47. Kajbafzadeh A-M, Elmi A, Talab SS, Sadeghi Z, Emami H, Sotoudeh M (2010) Autografting of renal progenitor cells ameliorates kidney damage in experimental model of pyelonephritis. Cell Med 1(3):115–122
48. Elmi A, Kajbafzadeh A, Talab SS, Esfahani S, Tourchi A, Oghabian M, Khooi S (2010) In vivo magnetic resonance tracking of transplanted myoblasts labeled with magnetic iron oxide nanoparticle in a rabbit model of fecal incontinence. J Pediatr Urol 6:S28
49. Saman ST, Kajbafzadeh A, Elmi A, Shadi AE (2010) Autoaugmentation Cystoplasty Using Autologous Smooth Muscle Cell Sheet Engineering. J Pediatr Urol 6:S26–S27
50. Kajbafzadeh A-M, Azadeh E, Emami H, Shafaattalab S, Esfahani SA, Tourchi A (2009) Restoration of functional anal sphincter by muscle progenitor cells transplantation in a rabbit model of fecal incontinence. J Pediatr Urol 5:S85–S86
51. Vejdani K, Kajbafzadeh A, Amanpour S, Golestani M (2005) Human bladder acellular matrix for augmentation cystoplasty in the rabbit.
52. Daryabari SS, Kajbafzadeh A-M, Fendereski K, Ghorbani F, Dehnavi M, Rostami M, Garajegayeh BA, Tavangar SM (2019) Development of an efficient perfusion-based protocol for whole-organ decellularization of the ovine uterus as a human-sized model and in vivo application of the bioscaffolds. J Assist Reprod Genet 36(6):1211–1223
53. Ebrahim Soltani AR, Kajbafzadeh A, Ezzati M, Ebrahim Soltani Z, Hosseinifar N, Maleki A, Nezhad Sistani M (2019) Novel evaluation of sevoflurane anesthetic exposure on the testicular germ cells of neonatal male mice. Toxicol Res 8(6):988–993
54. Akbarzadeh A, Khorramirouz R, Ghorbani F, Beigi RSH, Hashemi J, Kajbafzadeh A-M (2019) Preparation and characterization of human size whole heart for organ engineering: scaffold microangiographic imaging. Regenerat Med 14(10):939–954
55. Kajbafzadeh A-M, Khorramirouz R, Nabavizadeh B, Seyedian S-SL, Akbarzadeh A, Heidari R, Masoumi A, Azizi B, Beigi RSH (2019) Whole organ sheep kidney tissue engineering and in vivo transplantation: Effects of perfusion-based decellularization on vascular integrity. Mat Sci Eng C 98:392–400
56. Kajbafzadeh A-M, Elmi A, Payabvash S, Salmasi AH, Saeedi P, Mohamadkhani A, Sadeghi Z, Nikfarjam L (2008) Transurethral autologous myoblast injection for treatment of urinary incontinence in children with classic bladder exstrophy. J Urol 180(3):1098–1105

Characterization Methods of Acellularized Tissue and Organs

1

Layla Shojaie, Yekta Rahimi, Masoumeh Majidi Zolbin, Faezeh Daghigh, and Abdol-Mohammad Kajbafzadeh

Abstract

The extracellular matrix (ECM) of mammalian organs and tissues has been applied as a substitute scaffold to simplify the restoration and reconstruction of several tissues. Such scaffolds are prepared in various arrangements including sheets, powders, and hydrogels. One of the more applicable processes is using natural scaffolds, for this purpose discarded tissues or organs are naturally derived by processes that comprised decellularization of following tissues or organs. Protection of the complex structure and 3D (three dimensional) ultrastructure of the ECM is extremely necessary but it is predictable that all protocols of decellularization end in disruption of the architecture and potential loss of surface organization and configuration. Tissue decellularization with conservation of ECM bioactivity and integrity can be improved by providing well-designed protocols regarding the agents and decellularization techniques operated during processing. An overview of the characterization of decellularized scaffolds and the role of reagnets can validate the applied methods' efficacy.

Keywords

Acellular · Tissue engineering · Scaffold · Characterize

1.1 History

Decellularized extracellular matrices (dECM) provide a possible supply of substances to generate different scaffolds. To date, there are no absolute criteria precisely to confirm decellularized ECM. Indeed effective decellularization methodology is ordained by various factors such as tissue density and structure, geometric and

L. Shojaie · Y. Rahimi · M. M. Zolbin · A.-M. Kajbafzadeh (✉)
Pediatric Urology and Regenerative Medicine Research Center, Section of Tissue Engineering and Stem Cells Therapy, Children's Hospital Medical Center, Pediatric Center of Excellence, Tehran University of Medical Sciences, Tehran, Iran
e-mail: kajbafzd@sina.tums.ac.ir

L. Shojaie
Department of Medicine Division of GI/Liver Keck School of Medicine, University of Southern California, Los Angeles, USA

F. Daghigh
Department of Physiology, Tabriz Branch, Islamic Azad University, Tabriz, Iran

F. Daghigh
Tuberculosis and Lung Disease Research Center, Tabriz University of Medical Sciences, Tabriz, Iran

Present Address:
A.-M. Kajbafzadeh
No. 62, Dr. Gharib's Street, Keshavarz Boulevard, (PANNEK, #6), 1419433151 Tehran, Iran

A. Kajbafzadeh (ed.), *Decellularization Methods of Tissue and Whole Organ in Tissue Engineering*, Advances in Experimental Medicine and Biology 1345,
https://doi.org/10.1007/978-3-030-82735-9_1

biologic characteristics needed for the targeted clinical purpose, as well as the specific characteristics of the tissue of the origin. Each tissue demands its specific characterization methods. Indeed, efficient cell and genetic elements elimination are crucial in preventing immune rejection of the construct to seeded cells. Quantitative metrics have not been described for the term decellularization yet. To evaluate the quality of decellularized tissue and its extracellular matrix (ECM), multiple aspects should be examined. Based on current literature and experiments in which in vivo constructive response was established, and immune rejection of the host did not occur, several criteria have been suggested. These minimal criteria are suggested to be exercised to assess the decellularization process and its efficacy. The basis of the suggested criteria lies in the amount and quality of genetic material that is remained in the ECM. First, the amount of dsDNA should be less than 50 nanograms per milligram of the dry weight of the ECM. Second, DNA fragments detected in the ECM should not be more than 200 base pairs. And lastly, histological evaluation should not be able to identify genetic material with hematoxylin-eosin staining (H & E) or 4′,6-diamidino-2-phenylindole (DAPI) (Medberry 2014).

The first and second principle is considered a quantifiable approach to assess the acellularized ECM, whereas qualitatively verified by histological stainings such as H&E or DAPI. Quantitive assessment is readily accomplished by available dsDNA intercalators. For the second criterion, endonucleases such as DNase and RNase are applied to break down nucleic acid base pair fragments. These enzymes, fortunately, decrease the length of fragments, but they do little to part the fragments of the ECM. In pursuit of decreasing immune response, the Intracellular membrane compartment (e.g., phospholipids) must be noticed via enzyme-based measurement.

ECM is a vital component during development, influencing cell differentiation, proliferation, and migration, and its prominent role in providing structural stability and support for cells and tissue is indispensable (Medberry 2014). ECM, a non-cellular element of the tissue microenvironment, comprises proteins such as collagens, laminins, fibronectin, and polysaccharide glycosaminoglycans (GAGs) (Cirulli et al. 2000). Collagens are the most abundant component in ECM; so far, they are the main aim to modify. Several methods have been applied to determine the amount and quality of ECM components. For example, to maintain the content of collagen, histological collagen stains are used, whereas Scanning electron microscopy (SEM) provides more information about the structure and architecture. Additionally, Second harmonic generation (SHG) detects structural changes in collagen fibers by loss of signals.

Proteins that help with structural abilities, as well as mechanical properties of the tissue, should match the original tissue to a reasonable extent so that the process would have more chance to be successful. It is assumed Some decellularization agents and protocols destroy basic ECM elements; for example, detergents may disrupt collagen structure; therefore mechanical strength of ECM undergoes changes or most detergents eliminate the amount of GAG, thus decrease the viscoelasticity feature of ECM (Kezwoń et al. 2016; Conconi et al. 2005).

Consequently, mechanical properties such as elastic modulus, viscous modulus, tensile strength, and yield strength should be assessed. However, all in all, the characteristic is mainly provisory on the type of the tissue or organ's sought function (Wang and Guan 2010).

The process of decellularization via a cell removal agent will alter the ECM composition and structure. The goal is to try and minimize these alterations to have a robust ECM. There are several types of agents and techniques available for decellularization. The agent can be chemical, biological, physical, or miscellaneous. Among the most commonly used chemical agents are Triton X-100, Triton X-200, and sodium dodecyl sulfate (SDS). Each reagent has its own characteristics. For example, SDS and Triton X-100 can remove more significant than 90% of nuclear remnants. Triton X-100 and Triton X-200 have mixed results; nevertheless, they both can remove cells more efficaciously from thin tissue; however, Triron X-200 tends to disrupt the ECM more considerably. SDS is more effective than both and can remove cells from dense tissues, but disrupts the ECM, damages the collagen, and removes GAGs (Merritt et al. 2010; Guo et al. 2010; Hudson et al. 2004). It is worth highlighting that all these outcomes depend on the target tissue. Biologic agents can be enzymes or chelating compounds. Enzymes like trypsin, nucleases, and dyspases are commonly used (Yang et al. 2010; Wainwright et al. 2010). Chelating agents are not effective when used alone and should be used with either enzymes or detergents to be effective. Examples of physical and miscellaneous agents are freezing and thawing, direct force and pressure, and electroporation. The common side effect of all the physical methods is ECM disruption which can happen directly and indirectly (Lehr et al. 2011; Sellaro et al. 2010; Hashimoto et al. 2010). The structure of the remaining ECM and the clearance of cell debris and genetic material can influence the host response's efficacy. Since the best reagents are still not 100% effective, wash buffers and extensive wash procedures have been suggested (Hashimoto et al. 2010; Xing et al. 2015).

1.1.1 Hematoxylin–Eosin Staining (H & E)

Hematoxylin and eosin (H&E), is well-known staining used to assess the overall histologic appearance of samples, showing cells, cytoplasm, nuclei, and ECM constitution. Hematoxylin and eosin have distinct functions. Hematoxylin stains nucleic acids and has a deep purple color. Eosin stains proteins are pink. In healthy tissue, nuclei are stained blue, whereas the cytoplasm and extracellular matrix have diverging degrees of pink staining. Nucleoli stain with eosin. If abundant polyribosomes are present, the cytoplasm will have a distinct blue cast. This stain reveals sufficient structural data with specific functional implications (Fischer et al. 2008).

For preparation, the samples are fixed in formalin solution before being embedded in paraffin. Transversal sections are cut, dewaxed, and stained. H&E staining is performed according to the manufacturer's protocols. Decellularized tissue can show a decrease or lack of hematoxylin-stained nuclei indicating adequate cell elimination. Such as seen, in many studies that by use of H & E stain, the first and second criteria for decellularized tissue can be established. Complete cell removal was also observed in a study by Pashos et al. within areas of dense collagen (Pashos et al. 2017). The H & E stain of the decellularized tissue is altering the color of pink with considerably less blue-purple portions, indicating the presence of the nucleus in the native tissue stain.

1.1.2 4′,6-Diamidino-2-Phenylindole (DAPI)

Another confirmation test for cell nucleus removal is staining that is broadly utilized for assessing decellularized tissue is 4′,6-diamidino-2-phenylindole (DAPI).

For visualization of intact and undamaged nuclei, DAPI staining is used. The tissue sections are prepared, as mentioned for H & E staining. Immunofluorescence staining will be performed, a fluorescent microscope with a Slide book software is utilized to get a photo of the slide. If intact nuclei are present, the fluorescence dye will manifest in the camera. In the majority of cases with successful decellularization, the DAPI image would show a much more limited fluorescent dye than in the native tissue, and it will approve the (Pashos et al. 2017; Crapo et al. 2011).

1.1.3 The MTT Cell Proliferation Assay

MTT (3 4 5 dimethylthiazol 2 y 2 5 diphenyl tetrazolium) is a yellow water-soluble tetrazolium that is reduced by mitochondria of lived cells. After reduction, it turns to water-insoluble purple/blue formazan product. Classically 10,000 cells suspended in 100 μl of media are incubated with 10 micro L of MTT reagent. After 3 h, detergents should be added to the media, to lyse the cells and dissolve the colored crystal. Ethanol or propanol, acid-isopropanol, acid-isopropanol plus 10% Triton X 100, mineral oil, dimethyl sulfoxide, all are suggested as solubilized agents. The amount of purple/blue formazan production then is detected by spectrophotometry at 570 nm and is directly proportional to a number of viable cells. MTT is a sensitive and quantitive assay for cell proliferation and determining the absence of viable cells (Purpura et al. 2018) the strength of using MTT assay to confirm decellularized ECM is even small changes in metabolic activity can provoke alteration in MTT.

1.1.4 Electron Microscopy

There are 2 main categories of electron microscopy: scanning electron microscope (SEM) and transmission electron microscope (TEM). The critical difference between these is in the optics. TEM conveys the beam of electrons through a thin sample onto a detector, then condensed by lenses and hit the sample. TEM is appropriate for imaging microscopic particles such as viruses and organelles inside the cells.

On the other hand, SEM is suitable for imaging the surface, such as tissues, bacteria, cells, and organisms. In order to image the surface of the sample, SEM utilizes a beam of electrons in a raster pattern through the sample and provides information about topography, composition, and directionality (Godwin et al. 2017).

In order to get accurate images, at first, the sample must be fixed in a serial solution such as glutaraldehyde or paraformaldehyde in either phosphate or cacodylate buffer. For biological samples, distilled water is the right choice. Afterward, the sample must be entirely dried, prior to placing in a high vacuum environment. These two tests confirm the preservance of extra cellular matrix and lack of distortion in the scaffold and more specifically the removal of organelle remanent in the de-cellular scaffold which is vital for implantation purpose (Keene and Tufa 2018).

1.1.5 Second Harmonic Generation

In 1962, Second harmonic imaging (SHG) introduced by Kleinman in crystalline quartz. SHG is nonlinear optical microscopy in which two high energy photons hit a medium and directly convert to a single photon with the same total energy at double frequency. SHG imaging is used for non-centrosymmetric spatial arrangements such as collagen-based structure or birefringent crystals, membranes, proteins. The strength of SHG is that stationing procedures are not required and also preserve the molecular architecture and polarization dependence. SHG intensity depends on both the size and organization of the collagen fibers. This imaging tool represents the integrity and uniformity of collagen fibers in the context of scaffold and any disorganization in the stroma of the tissue can be demonstrated with no need for any specific pretreatment and protocols, so it can be conducted in the fresh tissue structure (Keikhosravi et al. 2014; Leonard et al. 2018).

1.1.6 Mechanical Properties

ECM preserves the three-dimensional structure of the cells and mainly consists of collagen, elastin, proteoglycans. Cellular function determines the relative amount of each component.

It has been demonstrated besides these components, other molecules such as laminin, also play roles to maintain the structure, stiffness, and cell to cell adhesion. To assess the mechanical properties of ECM, there are 2 main categories of measurement:

1. Micro-scale measurement: atomic force microscopy, micro-stretching. For example, Combining SHG with a micro-stretching technique presents a practical evaluation of the collagen residue in decellularized ECM.
2. Macro-scale measurement: measurement of the force-length curve of a tissue. This curve is resulted from change in dimensions by a given material such as collagen sheets or elastin bands during applying a force or stress.

1.1.7 Zymography

Zymography is a substrate gel electrophoresis to assess the amount of matrix metalloproteinase Metalloproteinases are a group of proteolytic enzymes that contribute a key role in tissue remodeling. The zymography technique established the splitting up of proteins by nonreducing sodium dodecyl sulfate–polyacrylamide gel electrophoresis (SDS–PAGE). Most commonly, the entire gel is composed of gelatin or casein (Leonard et al. 2018) During electrophoresis, SDS non-proteolyticly activates MMPs. After being separated by electrophoresis and a renaturation step; the gel gets incubated in a buffer of ionized calcium and zinc at 37 °C that is optimized for measuring MMPs activity towards distinct substrates. In vascular remodeling and vascular disorders, zymography has also been employed to evaluate the alterations in MMP activity (Ren et al. 2017).

1.2 Conclusion

Nowadays these scaffolds are applied in clinical tissue engineering and the most important issue with them is the quality characterization of scaffolds in order to make them safe and compatible with body composition. Since the source tissues for biologic scaffolds are typically allogeneic or xenogeneic in derivation, highest decellularization is necessary. The biologic scaffold materials preparation of mammalian ECM entails decellularization of source tissues. Such decellularization typically comprises exposure to selected biologic chemical and non-physiologic agents such as detergents and enzymes and mechanical forces that unavoidably cause disruption of the associated ECM. The optimal choice of decellularization protocols can be reasonably selected if thorough information of the mechanism of disorderly action is considered and assumed.

References

Cirulli V, Beattie GM, Klier G, Ellisman M, Ricordi C, Quaranta V et al (2000) Expression and function of αvβ3 and αvβ5 integrins in the developing pancreas: roles in the adhesion and migration of putative endocrine progenitor cells. J Cell Biol 150(6):1445–1460

Conconi MT, De Coppi P, Di Liddo R, Vigolo S, Zanon GF, Parnigotto PP et al (2005) Tracheal matrices, obtained by a detergent-enzymatic method, support in vitro the adhesion of chondrocytes and tracheal epithelial cells. Transplant Int Official J Eur Soc Organ Transplant 18(6):727–734

Crapo PM, Gilbert TW, Badylak SF (2011) An overview of tissue and whole organ decellularization processes. Biomaterials 32(12):3233–3243

Fischer AH, Jacobson KA, Rose J, Zeller R (2008) Hematoxylin and eosin staining of tissue and cell sections. CSH protocols. 2008:pdb.prot4986

Godwin ARF, Starborg T, Sherratt MJ, Roseman AM, Baldock C (2017) Defining the hierarchical organisation of collagen VI microfibrils at nanometre to micrometre length scales. Acta Biomater 52:21–32

Guo S, Ren X, Wu B, Jiang T (2010) Preparation of the acellular scaffold of the spinal cord and the study of biocompatibility. Spinal Cord 48(7):576–581

Hashimoto Y, Funamoto S, Sasaki S, Honda T, Hattori S, Nam K et al (2010) Preparation and characterization of decellularized cornea using high-hydrostatic pressurization for corneal tissue engineering. Biomaterials 31 (14):3941–3948

Hudson TW, Liu SY, Schmidt CE (2004) Engineering an improved acellular nerve graft via optimized chemical processing. Tissue Eng 10(9–10):1346–1358

Keene DR, Tufa SF (2018) Ultrastructural analysis of the extracellular matrix. Methods Cell Biol 143:1–39

Keikhosravi A, Bredfeldt JS, Sagar AK, Eliceiri KW (2014) Second-harmonic generation imaging of cancer. Methods Cell Biol 123:531–546

Kezwoń A, Chromińska I, Frączyk T, Wojciechowski K (2016) Effect of enzymatic hydrolysis on surface activity and surface rheology of type I collagen. Colloids Surf B Biointerfaces 137:60–69

Lehr EJ, Rayat GR, Chiu B, Churchill T, McGann LE, Coe JY et al (2011) Decellularization reduces immunogenicity of sheep pulmonary artery vascular patches. J Thorac Cardiovasc Surg 141(4):1056–1062

Leonard AK, Loughran EA, Klymenko Y, Liu Y, Kim O, Asem M et al (2018) Methods for the visualization and analysis of extracellular matrix protein structure and degradation. Methods Cell Biol 143:79–95

Medberry CJ (2014) Central nervous system extracellular matrix as a therapeutic bioscaffold for central nervous system injury. University of Pittsburgh

Merritt EK, Hammers DW, Tierney M, Suggs LJ, Walters TJ, Farrar RP (2010) Functional assessment of skeletal muscle regeneration utilizing homologous extracellular matrix as scaffolding. Tissue Eng Part A 16(4):1395–1405

Pashos NC, Scarritt ME, Eagle ZR, Gimble JM, Chaffin AE, Bunnell BA (2017) Characterization of an acellular scaffold for a tissue engineering approach to the nipple-areolar complex reconstruction. Cells Tissues Organs 203(3):183–193

Purpura V, Bondioli E, Cunningham EJ, De Luca G, Capirossi D, Nigrisoli E et al (2018) The development of a decellularized extracellular matrix-based biomaterial scaffold derived from human foreskin for the purpose of foreskin reconstruction in circumcised males. J Tissue Eng 9:2041731418812613

Ren Z, Chen J, Khalil RA (2017) Zymography as a Research tool in the study of matrix metalloproteinase inhibitors. Methods Mol Biol (clifton, NJ). 1626:79–102

Sellaro TL, Ranade A, Faulk DM, McCabe GP, Dorko K, Badylak SF et al (2010) Maintenance of human hepatocyte function in vitro by liver-derived extracellular matrix gels. Tissue Eng Part A 16(3):1075–1082

Wainwright JM, Czajka CA, Patel UB, Freytes DO, Tobita K, Gilbert TW et al (2010) Preparation of cardiac extracellular matrix from an intact porcine heart. Tissue Eng Part C Methods 16(3):525–532

Wang F, Guan J (2010) Cellular cardiomyoplasty and cardiac tissue engineering for myocardial therapy. Adv Drug Deliv Rev 62(7–8):784–797

Xing Q, Yates K, Tahtinen M, Shearier E, Qian Z, Zhao F (2015) Decellularization of fibroblast cell sheets for natural extracellular matrix scaffold preparation. Tissue Eng Part C Methods 21(1):77–87

Yang B, Zhang Y, Zhou L, Sun Z, Zheng J, Chen Y et al (2010) Development of a porcine bladder acellular matrix with well-preserved extracellular bioactive factors for tissue engineering. Tissue Eng Part C Methods 16(5):1201–1211

Esophagus Decellularization

2

Lousineh Arakelian, William Godefroy, Lionel Faivre, and Pierre Cattan

Abstract

In pathologies of the esophagus such as esophageal atresia, cancers and caustic injuries, methods for full thickness esophageal replacement require the sacrifice of healthy intra-abdominal organs such as the stomach and the colon. These methods are associated with high morbidity, mortality and poor functional results. The reconstruction of an esophageal segment by tissue engineering (TE) could answer this problem. For esophageal TE, this approach has been explored mainly by a combination of matrices and cells. In this chapter, we will discuss the studies on full organ esophageal decellularization, including the animal models, the methods of decellularization and recellularization.

Keywords

Esophagus · Tissue engineering · Decellularization · Scaffold

2.1 History

The esophagus is a tubular hollow organ composed of four layers (innermost mucosa, submucosa, muscularis propria and adventitia) and different cell types including epithelial, glandular and muscle cells (Poghosyan et al. 2011; Kuo and Urma 2006). It has been shown previously that extracellular matrix (ECM) can induce the recruitment and differentiation of cells in their relative compartments through its biochemical and biomechanical properties (Reing et al. 2009). Decellularized organs have the advantage of preserving these complex properties (Crapo et al. 2011), and very early after the development of decellularization methods, research groups started working on decellularized tissues for esophageal TE.

Decellularized ECM of other organs such as skin (Bozuk et al. 2006), urinary bladder (Badylak et al. 2005) or small intestinal submucosa (SIS) (Badylak et al. 2011) have been tried in animal models and humans for several types of esophageal repair. These matrices were shown to

L. Arakelian (✉) · L. Faivre
Unité de Thérapie Cellulaire, Hôpital Saint-Louis, AP-HP, Paris, France
e-mail: lousineh.arakelian@aphp.fr

L. Faivre
e-mail: lionel.faivre@aphp.fr

L. Arakelian · L. Faivre · P. Cattan
Institut de Recherche Saint Louis, INSERM, CIC-BT1427 and UMR-U976, Hôpital St-Louis, Paris, France
e-mail: pierre.cattan@aphp.fr

W. Godefroy
Paris, France

P. Cattan
Department of Digestive Surgery, St-Louis, Hospital—Paris 7 University, Paris, France

A. Kajbafzadeh (ed.), *Decellularization Methods of Tissue and Whole Organ in Tissue Engineering*, Advances in Experimental Medicine and Biology 1345,
https://doi.org/10.1007/978-3-030-82735-9_2

be efficient for superficial lesions or partial defects (Badylak et al. 2011). However, no success has been reported for full thickness circumferential esophageal replacement with non-esophageal matrices, with or without cells. Therefore, researchers turned to organ specific ECM for esophageal TE.

The first report on decellularized esophagus dates back to 2005 in a rat model (Ozeki et al. 2006). Porcine esophagi have also been decellularized in several studies with success (Koch et al. 2012; Totonelli et al. 2013; Luc et al. 2018; Arakelian et al. 2019).

We will discuss the challenges faced for a clinical use of these ECM in humans (Fig. 2.1).

2.2 Material and Methods

2.2.1 Animal Models

Full esophageal decellularization has been mainly carried out on rat (Ozeki et al. 2006; Urbani et al. 2018) and porcine esophagi (Luc et al. 2018; Arakelian et al. 2019). Even though rat esophagi have served as an important proof of concept, protocols developed on this small animal model cannot be directly applied to esophagi corresponding to human size. Porcine esophagus has the advantage of a highly similar structure and size compared to the human one (Ziegler et al. 2016). For this reason, porcine esophagus seems to be a relevant model to develop decellularization methods that can be used for a human esophageal decellularization.

2.2.2 Decellularization

The decellularization of the esophagus has been mainly carried out using detergents including SDS, DEOX, triton X-100 or Chaps (Mallis et al. 2019). The detergent is used for rupturing the cell membranes and eliminating cell content. Calcium chelator EDTA has also been added to facilitate cell detachment and improve the

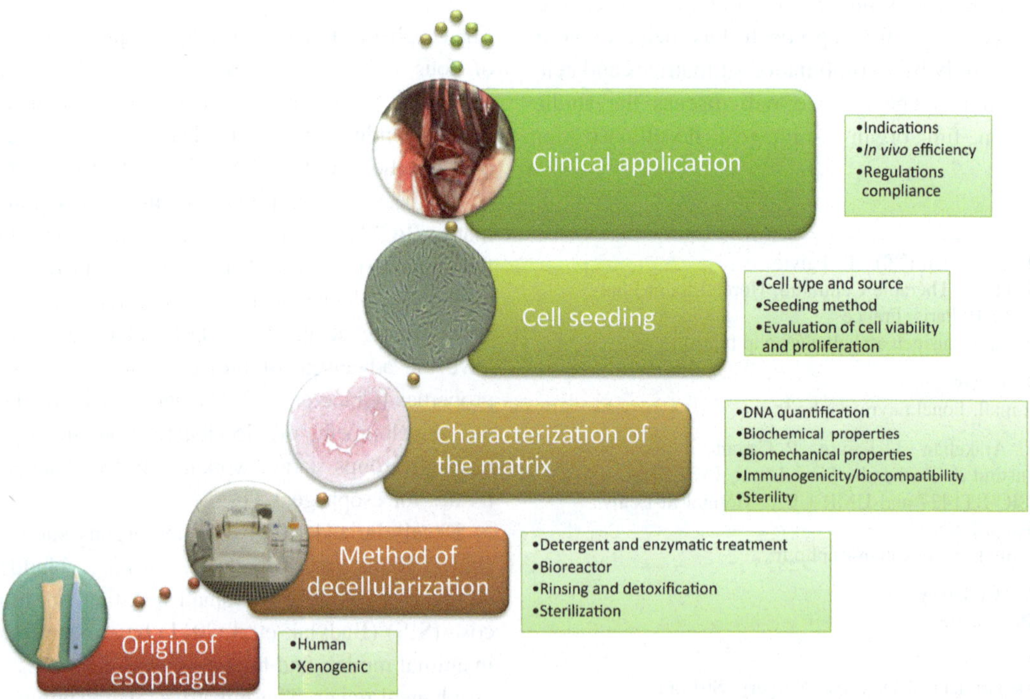

Fig. 2.1 Steps to produce a clinical grade esophageal decellularization: from bench to bedside

decellularization (Arakelian et al. 2019). The nature, concentration and treatment period can highly affect the quality of the final product. For smaller esophageal models, lower concentrations or mild detergents can be used for decellularization. However, for larger models, higher concentrations of stronger detergents such as SDS and DEOX were needed and the treatment period was extended to several days.

Decellularization protocols showed that even though in some experiments in the rat esophagi, DNA can be eliminated by cycles of detergent treatment (Mallis et al. 2019), in larger animal esophagi, the detergent alone does not remove DNA and the cell nuclei (Arakelian et al. 2019). Therefore, esophageal decellularization protocols include a DNase treatment. In the two recent decellularization studies in porcine model, one treated the decellularized matrix 12 h with 2000 Kunitz units of DNase-I (Sigma-Aldrich) (Luc et al. 2018), whereas the other team privileged a shorter 3 h treatment with 100 u/ml clinical grade DNase (Pulmozyme) (Arakelian et al. 2019).

At the end of decellularization, an efficient rinsing method should be developed in order to fully remove these detergents to avoid cytotoxicity. In small animal models, abundant rinsing with water or PBS was reported to be sufficient to remove these detergents. In bigger models, the rinsing cycles were much longer or it could be necessary to use an absorbing resin which significantly improved detergent removal and reduced cytotoxicity (Arakelian et al. 2019).

In the first attempts of decellularization, mechanical treatment, along with enzymatic and detergent treatment, was achieved by placing the esophagi under constant agitation (Ozeki et al. 2006) or by perfusing the organ using a speed roller pump (Totonelli et al. 2013). These methods increased detergent and enzyme infiltration within the esophagus and improved decellularization compared to static conditions. However, these technics worked better for smaller rat esophagi compared to larger and thicker porcine ones. Furthermore, these are open systems which require a high level of manual manipulation and an increased risk of contamination. The recent decellularization protocols included the use of bioreactors for liquid perfusion (Luc et al. 2018) or perfusion and axial rotation (Arakelian et al. 2019). These closed systems increased the efficiency of decellularization and reduced manual handling which may be an advantage for future clinical applications.

2.2.3 Sterilization

The esophagus is an organ which is in constant exchange with extracorporeal, non-sterile environment. It is therefore important to use a sterilization method to prevent bacterial and fungal growth throughout the decellularization or at the end of the process. For decontamination, a team used sodium azide (Luc et al. 2018), a molecule which can be highly toxic (Chang and Lamm 2003) and not recommended for clinical use. Others privileged the use of antibiotics (ATB) for an initial decontamination. For this purpose, a mix of ATB (gentamycin, clindamycin vancomycin and amphotericin B), previously used for vascular graft applications, was validated for esophageal decontamination (Arakelian et al. 2019). Due to the high concentration of the ATB, it is important to efficiently remove them at the end of the decellularization to avoid toxicity, while preserving the sterility of the decellularized matrix.

Another option is a final sterilization with chemical or physical treatments. Chemical treatments can include ethylene oxide or peracetic acid (PAA). The difficulty with these treatments is that these products may remain in the decellularized tissue and induce cytotoxicity (Lucas et al. 2017). Furthermore, it has been shown that PAA can prevent vascularization of soft tissues after implantation in vivo (Scheffler et al. 2008). Physical sterilization includes treatment with gamma rays. Even though this treatment efficiently removes bacterial, fungal and viral contaminations, it can compromise the biomechanical properties of the decellularized matrices (Witt et al. 2016).

2.2.4 Characterization of the Decellularized Matrix

As for other decellularized organs, the recommended criteria to define a complete decellularization are to validate the absence of residual cells, the elimination of DNA (less than 50 µg/mg of dry mass) and to make sure that no residual DNA fragments exceeding 200 bp remains in the tissue (Crapo et al. 2011). However, these recommendations can vary slightly according to the nature and the origin of the tissue. Furthermore, the general structure, the bioactive molecules and the biomechanical properties should also be maintained after decellularization.

2.2.5 DNA Quantification

In decellularized esophagi, DNA was extracted from the matrix and was then quantified. In all these studies, an efficient elimination of DNA was demonstrated (Luc et al. 2018; Arakelian et al. 2019). For DNA fragment size, an electrophoresis of the extracted DNA on agarose gel was carried out which showed that no large DNA fragments (more than 200 bp) was visible. The elimination of nuclei was also shown by DAPI staining (Luc et al. 2018; Arakelian et al. 2019; Mallis et al. 2019).

2.2.6 General Structure and Composition

For the demonstration of cell removal, histology (HES staining) remains the standard method of validation (Luc et al. 2018; Arakelian et al. 2019). Furthermore, it is important to show that the components of the ECM such as collagens, elastin fibers, glycosaminoglycans (GAGs) and other molecules are preserved after decellularization. In two studies of rat and porcine esophageal decellularization, collagen has been quantified using a hydroxyproline assay kit (Mallis et al. 2019) or stained with picrosirius red and analyzed by histochemistry (Arakelian et al. 2019). These studies showed that most of the collagen was preserved, despite some loss of structure. Elastin fibers have been stained with orcein after esophageal decellularization, and it was shown that they were highly preserved after decellularization. Finally, GAG quantification with dimethylmethylene blue assay (DMMB) or staining with toluidine blue (Arakelian et al. 2019) showed that there was a major loss of these molecules after decellularization. However, immunostaining with specific antibodies showed that the loss of GAGs was mainly related to chondroitin sulfates, whereas the heparan sulfates and dermatan sulfates were preserved (Arakelian et al. 2019). These last two categories of GAGs are the main ones involved in the biomechanical properties of the matrix, as well as the binding and the delivery of hormones and growth factors (Kjellén and Lindahl 2018). It is important to mention that the extent of loss of these molecules highly depends on the nature and concentration of the detergent, as well as the duration of the treatment (Mallis et al. 2019). It is therefore important to develop a protocol which allows an efficient decellularization without a major loss of structural molecules.

2.2.7 Biomechanical Properties

The biomechanical properties of the decellularized esophagi have been evaluated and compared to the native esophagi. The two methods that have been used to evaluate the biomechanical properties are burst pressure test and tensile strength. In the decellularized esophagi, porosity was detected which prevented the decellularized esophagi from reaching a burst point (Luc et al. 2018). Tensile tests showed that in the transversal orientation, the decellularized and native esophagi had similar properties. On the other hand, in the longitudinal orientation, the decellularized esophagi were stiffer than the native one (Luc et al. 2018; Arakelian et al. 2019). As for in vivo implantation, decellularized esophagi were easily handled for surgical procedures and were resistant to sutures.

2.2.8 Immunogenicity and Biocompatibility

One of the main purposes of decellularization is to reduce the immune reaction of the host to avoid graft rejection and fibrosis. To study these properties in decellularized esophagi, an *in vitro* assay was developed based on the proliferation of lymphocytes stained with fluorescent molecule and analyzed by flow cytometry (Arakelian et al. 2019).This assay showed that the decellularized esophagi did not induce lymphocyte proliferation and indicated the absence of an acute immunogenicity.

However, the immune reaction is a complex mechanism and true immunogenicity should be evaluated in vivo. In another study, this reaction was evaluated by a subcutaneous implantation of the matrix in non-immunosuppressed Winstar rats (Luc et al. 2018). After 14 days, an induction of inflammatory response with infiltration of mononuclear cells was shown.

2.2.9 Cytotoxicity

As the products used for decellularization such as detergents and a high dose of ATB are toxic for cells, it is important to make sure that they are efficiently removed after decellularization. It is therefore necessary to develop assays to answer these questions efficiently. In decellularized esophagi, the main assays used so far were based on the evaluation of cell viability, by direct or indirect methods (Iso 10993-5-2009). In the direct method, mesenchymal stromal cells (MSCs) were seeded on the decellularized esophagi and the viability and metabolic activity were evaluated by neutral red assay and MTT assay, respectively (Luc et al. 2018). In the indirect method, the decellularized esophagi were incubated with cell culture medium and the supernatant was then used for Balb/3T3 cell culture. The viability of these cells was evaluated by flow cytometry after annexin V and 7AAD staining (Arakelian et al. 2019).The difficulty with a direct MTT assay is that the resulting dye is absorbed by the matrix, and it is difficult to have accurate and reproducible results. The indirect method allows to overcome this difficulty and to evaluate the release of toxic substances by the matrix. Both methods can be used for short term cytotoxicity evaluation. However, the presence of detergents and toxic substances should be further evaluated by mass spectrometry and long-term cytotoxicity should also be evaluated in vivo.

2.3 Cell Seeding

2.3.1 Cell Types and Origin

Cell seeding on decellularized esophagi has been explored in order to functionalize these matrices and to evaluate the potential of cells to accelerate tissue regeneration. For in vivo applications, it is essential to question the cell types and their origin (autologous or allogeneic), as this choice conditions the desired mechanism of action. The first choice is to use differentiated cells, organ-specific or not, such as epithelial cells (Ozeki et al. 2006; Urbani et al. 2018; Asnaghi et al. 2009; Barron et al. 2016; Jensen et al. 2018; Poghosyan et al. 2015; Nakase et al. 2008), smooth muscle cells (Barron et al. 2016; Poghosyan et al. 2015; Takeoka et al. 2019) and endothelial cells (Takeoka et al. 2019). The functionalization of the decellularized esophagus by these cells can be induced either by a direct colonization of the ECM by the seeded cells or by paracrine effects. It has been shown that some cells can indeed secrete factors that can attract the host cells and accelerate tissue regeneration (Marzaro et al. 2006; Xiuunl et al. 2009).

The other option is to use non-differentiated cells. To date, no clear stem cell niche, able to give rise to all the cell types, has been identified in the adult esophagus (Seery 2002). Regarding stem cells, another possibility is to seed the matrix with MSCs either originating from adipose tissue or bone marrow (Hass et al. 2011). These cells promote the recruitment of patient cells in situ through paracrine effects, accelerate re-vascularization and reduce inflammatory and scarring processes (Luc et al. 2018; Arakelian et al. 2019; Asnaghi et al. 2009; Jensen et al. 2018; Poghosyan et al. 2015; Takeoka et al. 2019; Tan et al. 2013; Francesca et al. 2018; Catry et al. 2017). It was shown that bone marrow MSC seeded on a non-esophageal extracellular matrix accelerated muscle regeneration and re-epithelialisation in a patch esophagoplasty and a full thickness esophageal replacement models (Tan et al. 2013; Catry et al. 2017).

Beyond these mechanistic aspects, the origin of cells can lead to significant constraints. Indeed, autologous cells will require a sample of the patient, isolation, amplification and then the constitution of the substitute; while the use of allogeneic cells will reduce the production time, but raises the question of immunological rejection. Thanks to their immunomodulatory properties, MSCs are an interesting source for the recellularization of decellularized esophagi.

2.3.2 Seeding Methods

The decellularized esophagus is a cylindrical hollow tube with an inner and outer surface. The challenge is therefore to decide which layer should be seeded and how to distribute the cells evenly on the matrix.

Cell density, as well as the duration of cell culture *in-vitro* are further important parameters to ensure the colonization of the matrix by the cells and their infiltration. Five teams showed very variable culture times, ranging from 7 to 21 days (Ozeki et al. 2006; Luc et al. 2018; Arakelian et al. 2019; Urbani et al. 2018). The number of seeded cells varies from one study to another from 1.10^5 to several millions per cm^2 (Ozeki et al. 2006; Luc et al. 2018; Arakelian et al. 2019; Urbani et al. 2018). These parameters could be different according to cell types and their capacity to adhere and proliferate.

Some tubular esophageal substitutes were seeded under static conditions. Cells were deposited on the outer surface or were injected inside the lumen using a pipette (Catry et al. 2017; Poghosyan et al. 2013). However, in most studies, axial rotation was applied to homogenize cell distribution on the matrix. This rotation was achieved either manually at regular time intervals (Urbani et al. 2018; Barron et al. 2016; Jensen et al. 2018) or using a continuous rotation system (Ozeki et al. 2006; Arakelian et al. 2019; Urbani et al. 2018; Asnaghi et al. 2009; Francesca et al. 2018). These systems include: (1) an axial rotary bioreactor with partial liquid immersion of the substitute (Asnaghi et al. 2009), (2) an axial rotating stirrer with a filter plug tube (Arakelian et al. 2019), (3) a rotating bioreactor with a full liquid immersion of the matrix (Francesca et al. 2018) or (4) a Waverotor bioreactor (Thermonics, Tokyo, Japan) (Ozeki et al. 2006). The advantage of using a bioreactor for cell seeding is that it allows a homogeneous cell distribution, as well as reducing manual intervention and a better control of oxygenation, pH and cellular metabolism. These parameters are important for the reproducibility of cell seeding and for a future clinical application under GMP conditions. Urbani et al. clearly demonstrated the benefits of a dynamic culture (Urbani et al. 2018). However, the animal model used being the rat, the transposition to a human-sized esophagus remains to be demonstrated. Cell sheet technology is another option of cell seeding on the decellularized esophagi. This method has been explored using MSCs. To summarize, MSCs were cultured in a dish at a very high confluence and the cell sheet was rolled around a decelluarized esophagus (Luc et al. 2018). Cell sheet seeding can be improved using thermoresponsive polymers such as pNIPAM which allow a full cell sheet detachment upon changing the temperature. This method has already been validated in a clinical trial for superficial lesions using epithelial cells (Yamaguchi et al. 2017) and could be used for seeding of decellularized esophagi.

2.4 Clinical Applications

Commercialized non-esophageal decellularized natural ECM have previously been tested in clinical trials for treating esophageal leaks with decellularized skin or superficial esophageal lesions with SIS patches to prevent stenosis (Bozuk et al. 2006; Badylak et al. 2011).

However, these methods have never been successfully applied for full thickness circumferential replacement humans.

For the clinical application of decellularized esophagi, it is important to consider the regulatory aspect which will be applied. In Europe, for example, if the matrix is to be used alone, without cell seeding, it could be considered as an implantable medical device "IMD" or as "human cells, tissues, and cellular and tissue-based product (HCT/P)". One of the main determining criteria for choosing between these two categories is the origin of the decellularized matrix and the nature of the protocol. A final decontamination is mandatory for IMDs. A human matrix can be treated both as an IMD and a HCT/P, whereas a porcine decellularized esophagus can only be treated as an IMD. In both categories, it is necessary to show the sterility of the matrix and both can involve an initial decontamination with antibiotics and a final sterilization using gamma rays or chemicals such as ethylene oxide. For IMD, the quality controls should be carried out to obtain a CE marking and the matrix can be produced by pharmaceutical companies. A HCT/P, however, should be produced in special accredited facilities such as human tissue banks. In both categories, a long-term conservation method should be validated which could include the preparation of a frozen matrix bank.

If the decellularized esophageal matrix is to be seeded with cells before implantation, the final product is considered as an advanced therapy medicinal product (ATMP), corresponding to a new category of regulations (https://www.ema.europa.eu/en/human-regulatory/overview/advanced-therapy-medicinal-products-overview). This means that on top of evaluating the biological properties and the sterility of the matrix, the nature of the cells and the culture conditions on the matrix before and after in vivo implantation should be evaluated. The cells should be isolated and cultured in a clinical grade cell culture media, and the optimal cell density as well as in vitro maturation time should be clearly defined. Once implanted in the animal, the possible migration of the cells within different organs, as well as their tumorigenic potential, should be carefully evaluated. Unlike MD and tissue products, ATMPs need to be produced in authorized special facilities such as platforms or pharmaceutical industries.

For all the three categories, a pre-clinical trial in a big animal model is necessary to show the efficiency of the matrix in esophageal replacement.

One of the challenges for in vivo esophageal replacement is the method of vascularization. As the esophagus is composed of microvessels coming from the aorta and the surrounding organs, it is necessary to find a vascularization method to prevent organ necrosis. The option that has been tested in previous esophageal tissue engineering studies has been a maturation step in the omentum (Luc et al. 2018; Poghosyan et al. 2015). These studies showed that a tubular substitute composed of SIS for esophageal replacement was successfully vascularized by this option. This method has also been used successfully for the vascularization of a rat decellularized esophagus and a porcine one. However, long-term efficiency after organ replacement should be evaluated in vivo.

References

Arakelian L, Caille C, Faivre L, Corté L, Bruneval P, Shamdani S et al (2019) A clinical-grade acellular matrix for esophageal replacement. J Tissue Eng Regen Med. https://doi.org/10.1002/term.2983

Asnaghi MA, Jungebluth P, Raimondi MT, Dickinson SC, Rees LEN, Go T et al (2009) A double-chamber rotating bioreactor for the development of tissue-engineered hollow organs: From concept to clinical trial. Biomaterials 30(29):5260–5269

Badylak SF, Vorp DA, Spievack AR, Simmons-Byrd A, Hanke J, Freytes DO et al (2005) Esophageal reconstruction with ECM and muscle tissue in a dog model. J Surg Res

Badylak SF, Hoppo T, Nieponice A, Gilbert TW, Davison JM, Jobe BA (2011) Esophageal preservation in five male patients after endoscopic inner-layer circumferential resection in the setting of superficial cancer: a regenerative medicine approach with a biologic scaffold. Tissue Eng Part A [Internet] 17 (11–12):1643–50. Available from: http://www.liebertonline.com/doi/abs/https://doi.org/10.1089/ten.tea.2010.0739

Barron MR, Blanco EW, Aho JM, Chakroff J, Johnson J, Cassivi SD et al (2016) Full-thickness oesophageal regeneration in pig using a polyurethane mucosal cell seeded graft. J Tissue Eng Regen Med 12(1):175–185

Bozuk MI, Fearing NM, Leggett PL (2006) Use of decellularized human skin to repair esophageal anastomotic leak in humans. JSLS 10:83–85

Catry J, Luong-Nguyen M, Arakelian L, Poghosyan T, Bruneval P, Domet T et al (2017) Circumferential esophageal replacement by a tissue-engineered substitute using mesenchymal stem cells. Cell Transplant [Internet] 26(12):1831–1839. Available from: http://journals.sagepub.com/doi/https://doi.org/10.1177/0963689717741498

Chang S, Lamm SH (2003) Human health effects of sodium azide exposure: a literature review and analysis. Int J Toxicol

Crapo PM, Gilbert TW, Badylak SF (2011) An overview of tissue and whole organ decellularization processes. Biomaterials 32:3233–3243

Hass R, Kasper C, Böhm S, Jacobs R (2011) Different populations and sources of human mesenchymal stem cells (MSC): a comparison of adult and neonatal tissue-derived MSC. Cell Commun Signal 9(1):12

Jensen T, Wanczyk H, Sharma I, Mitchell A, Sayej WN, Finck C (2018) Polyurethane scaffolds seeded with autologous cells can regenerate long esophageal gaps: an esophageal atresia treatment model. J Pediatr Surg

Kjellén L, Lindahl U (2018) Specificity of glycosaminoglycan–protein interactions. Curr Opin Struct Biol 50:101–108

Koch H, Graneist C, Emmrich F, Till H, Metzger R, Aupperle H et al (2012) Xenogenic esophagus scaffolds fixed with several agents: Comparative in vivo study of rejection and inflammation. J Biomed Biotechnol

Kuo B, Urma D (2006) Esophagus—anatomy and development. GI Motil online [Internet], pp 1–20. Available from: http://www.nature.com/gimo/contents/pt1/full/gimo6.html

La Francesca S, Aho JM, Barron MR, Blanco EW, Soliman S, Kalenjian L et al (2018) Long-term regeneration and remodeling of the pig esophagus after circumferential resection using a retrievable synthetic scaffold carrying autologous cells. Sci Rep [Internet] 8(1):4123. Available from: http://www.nature.com/articles/s41598-018-22401-x

Luc G, Charles G, Gronnier C, Cabau M, Kalisky C, Meulle M et al (2018) Decellularized and matured esophageal scaffold for circumferential esophagus replacement: proof of concept in a pig model. Biomaterials 175:1–18

Lucas AD, Merritt K, Hitchins VM, Woods TO, Mcnamee SG, Lyle DB et al (2017) Residual ethylene oxide in medical devices and device material residual ethylene oxide in medical devices and device material

Mallis P, Chachlaki P, Katsimpoulas M, Stavropoulos-Giokas C, Michalopoulos E (2019) Optimization of decellularization procedure in rat esophagus for possible development of a tissue engineered construct. Bioengineering

Marzaro M, Vigolo S, Oselladore B, Conconi MT, Ribatti D, Giuliani S et al (2006) In vitro and in vivo proposal of an artificial esophagus. J Biomed Mater Res A 77(4):795–801

Nakase Y, Nakamura T, Kin S, Nakashima S, Yoshikawa T, Kuriu Y et al (2008) Intrathoracic esophageal replacement by in situ tissue-engineered esophagus. J Thorac Cardiovasc Surg 136(4):850–859

Ozeki M, Narita Y, Kagami H, Ohmiya N, Itoh A, Hirooka Y et al (2006) Evaluation of decellularized esophagus as a scaffold for cultured esophageal epithelial cells. J Biomed Mater Res Part A

Poghosyan T, Gaujoux S, Chirica M, Munoz-Bongrand N, Sarfati E, Cattan P (2011) Functional disorders and quality of life after esophagectomy and gastric tube reconstruction for cancer. J Visc Surg [Internet] 148: e327—e335. Available from: http://ac.els-cdn.com.gate2.inist.fr/S1878788611001093/1-s2.0-S1878788611001093-main.pdf?_tid=102588d6-8322-11e7-8a8b-00000aacb35f&acdnat=1502956978_4c9f951fb448fd427801b6ae238e4cf9

Poghosyan T, Gaujoux S, Vanneaux V, Bruneval P, Domet T, Lecourt S et al (2013) In vitro development and characterization of a tissue-engineered Conduit \nResembling esophageal wall using human and pig skeletal myoblast, Oral\nEpithelial Cells, and biologic scaffolds. Tissue Eng Part a 19(19–20):2242–2252

Poghosyan T, Sfeir R, Michaud L, Bruneval P, Domet T, Vanneaux V et al (2015) Circumferential esophageal replacement using a tube-shaped tissue-engineered substitute: an experimental study in minipigs. Surg (United States) [Internet] 158(1):266–277. https://doi.org/10.1016/j.surg.2015.01.020

Reing JE, Zhang L, Myers-irvin J, Ph D, Cordero KE, Freytes DO et al (2009) Degradation products of extracellular matrix affect cell migration and proliferation. 15(3)

Scheffler SU, Cmd JG, Cmd JK, Przybilla D (2008) Remodeling of ACL allografts is inhibited by peracetic acid sterilization. Clin Orthop Relat Res 466:1810–1818

Seery JP (2002) Stem cells of the oesophageal epithelium. J Cell Sci 115:783–1789

Takeoka Y, Matsumoto K, Taniguchi D, Tsuchiya T, Machino R, Moriyama M et al (2019) Regeneration of esophagus using a scaffoldfree biomimetic structure created with biothree-dimensional printing. PLoS ONE 14(3):1–12

Tan B, Wei RQ, Tan MY, Luo JC, Deng L, Chen XH et al (2013) Tissue engineered esophagus by mesenchymal stem cell seeding for esophageal repair in a canine model. J Surg Res 182(1):40–48

Totonelli G, Maghsoudlou P, Georgiades F, Garriboli M, Koshy K, Turmaine M et al (2013) Detergent enzymatic treatment for the development of a natural acellular matrix for oesophageal regeneration. Pediatr Surg Int

Urbani L, Camilli C, Phylactopoulos D, Crowley C, Natarajan D, Scottoni F et al (2018) Multi-stage bioengineering of a layered oesophagus with in vitro expanded muscle and epithelial adult progenitors. Nat Commun

Witt TA, Marler RJ, Pislaru SV, Robert D (2016) Low-dose gamma irradiation of decellularized heart valves results in tissue injury in vitro and in vivo. Ann Thorac Surg 101(2):667–674

Xiuunl I, Xiaoz UO, Weiz HI, Pengy A, Huiix IE, Zhiingy A (2009) Grafts of porcine small intestinal submucosa with cultured autologous oral mucosal epithelial cells for esophageal repair in a canine model. Exp Biol Med (Maywood) 234(4):453–461

Yamaguchi N, Isomoto H, Kobayashi S, Kanai N, Kanetaka K, Sakai Y et al (2017) Oral epithelial cell sheets engraftment for esophageal strictures after endoscopic submucosal dissection of squamous cell carcinoma and airplane transportation. Sci Rep 7(1)

Ziegler A, Gonzalez L, Blikslager A (2016) Large animal models: the key to translational discovery in digestive disease research. Cmgh [Internet] 2(6):716–24. https://doi.org/10.1016/j.jcmgh.2016.09.003

Fetal Lung Tissue Engineering

3

Masoumeh Majidi Zolbin, Faezeh Daghigh,
Layla Shojaie, Masoumeh Ekhtiyari,
and Abdol-Mohammad Kajbafzadeh

Abstract

Lung transplantation may be considered as a final treatment option for diseases such as chronic lung disease, pulmonary hypertension, bronchopulmonary dysplasia, pulmonary fibrosis, and end-stage lung disease. The five-year survival rate of lung transplants is nearly 50%. Unfortunately, many patients will die before a suitable lung donor can be found. Importantly, the shortage of donor organs has been a significant problem in lung transplantation. The tissue engineering approach uses de- and recellularization of lung tissue to create functional lung substitutes to overcome donor lung limitations. Decellularization is hope for generating an intact ECM in the development of the engineered lung. The goal of decellularization is to prepare a suitable scaffold of lung tissue that contains an appropriate framework for the functionality of regenerated lung tissue. In this chapter, we aim to describe the decellularization protocols for lung tissue regenerative purposes.

Keywords

Tissue engineering · Decellularization · Extracellular matrix · Regenerated lung

M. M. Zolbin · L. Shojaie · M. Ekhtiyari ·
A.-M. Kajbafzadeh (✉)
Pediatric Urology and Regenerative Medicine Research Center, Section of Tissue Engineering and Stem Cells, Therapy, Children's Hospital Medical Center, Tehran University of Medical Sciences, Tehran, Iran
e-mail: kajbafzd@sina.tums.ac.ir

F. Daghigh
Department of Physiology, Tabriz Branch, Islamic Azad University, Tabriz, Iran

F. Daghigh
Tuberculosis and Lung Disease Research Center, Tabriz University of Medical Sciences, Tabriz, Iran

L. Shojaie
Department of Medicine Division of GI/Liver Keck School of Medicine, University of Southern California, Los Angeles, USA

Present Address:
A.-M. Kajbafzadeh
No. 62, Dr. Gharib's Street, Keshavarz Boulevard, (PANNEK, #6), 1419433151 Tehran, Iran

3.1 History

The respiratory system is a biological system consisting of specific organs and structures used for gas exchange. It includes tracheal airways, lungs, and blood vessels. Respiratory disorders, including interstitial lung disease (ILD), pulmonary arterial hypertension, cystic fibrosis, or genetic condition such as α_1-antitrypsin disorder, are potential causes of destructive lung tissue. Nearly 1000–1500 lung transplantation are performed every year in the United States (Orens

A. Kajbafzadeh (ed.), *Decellularization Methods of Tissue and Whole Organ in Tissue Engineering*, Advances in Experimental Medicine and Biology 1345,
https://doi.org/10.1007/978-3-030-82735-9_3

and Garrity 2009). Tracheal defects or stenosis may result from trauma, congenital abnormalities, and pathologies such as infection or cancer. Surgical approaches, including slide tracheoplasty and tracheal mobilization in pediatric (Kocyildirim 2004) and adult patients have been associated with some success. However, regardless of the approach, high rates of complications and long term morbidities are common problems in tracheal surgeries (Birchall and Macchiarini 2008). Lung transplantation is an ultimate therapeutic approach for over 25 million patients with end-stage lung diseases such as COPD, CF, and IPF (Smith et al. 2012). Despite recent advances in the treatment of lung disease, irreversible structural alterations in lung tissues remain a significant problem for patients with end-stage lung disease. To be eligible for a lung transplant, patients must meet the following requirements: In general, they must be physiologically 60 years of age or less for bilateral lung transplantation and 65 years of age or less for single lung transplantation. Although organ transplantation is an effective treatment for destructive lung diseases, there are disadvantages to it, occurs months to years are:

Immunosuppression is a major concern of organ transplantation (Prakash et al. 2015). Moreover, significant short survival rates and limited number of lung donors, demonstrate a critical demand to explore new approaches (Orens and Garrity 2009).

3.2 Lung Tissue Engineering

Tissue engineering can be considered as a bridge among different medicine fields that relates the essentials of engineering and basic science towards the progress of biological substitutes that restore, maintain, or improve function. Lung is an organ with complex functional and structural compositions (Badylak et al. 2011) and ECM proteins such as laminin, fibronectin, collagen, elastin, and so on (Prakash et al. 2015). Lung regenerative engineering research appears to be a potential method for addressing the limitation of lung transplantation for organ failure ex vivo (Gilpin 2016). Ex vivo lung engineering is considered as a potential candidate for increasing the availability of lung tissue for transplantation (Gilpin 2016). Using patient's own cells with reducing immunosuppression problems, including infection and neurologic disorders is known as the advantage of engineered lung tissue (Nichols et al. 2009; Reichenspurner 2005). Importantly, using other acellular tissues such as skin in lung tissue engineering follows a successful approach in clinic (Crapo et al. 2011; Wainwright 1995).

Another approach in engineering is creating an appropriate biological scaffold for lung tissue (Vertrees 2008). Currently, decellularization methods have been used for preparing scaffold for tissue engineering in many organs as well as the lung. Decellularization is known as a new technique for removing cellular remanents and prepares intact whole organs, which is an important characteristic of the three-dimensional matrix for cellular migration (Jensen 2012; Petersen et al. 2010). During lung decellularization, the ECM cells will be removed from the matrix, and all ECM proteins and growth factors will be preserved in the lung in order to protect the native scaffold for the migration, proliferation, and differentiations of stem cells (Prakash et al. 2015).

Furthermore, researchers focus on engineered airway by utilizing a tracheal scaffold made from a synthetic polymer (Omori 2005) or decellularized tracheal allotransplant (Macchiarini 2008). Available studies have revealed that acellular scaffolds can be prepared from species, including non-human primates, pigs, rodents, and humans (Bombelli 2018). There are various types of decellularization methods, including enzymatic, physical, and chemical methods (Petersen 2009, 2012). Also, several protocols are existed for obtaining scaffolds of acellular human lung (Gilpin 2014; Nichols 2013,2017; Booth 2012; Petersen 2010; Balestrini 2016; Skolasinski and Panoskaltsis-Mortari 2017; Uhl et al. 2017; Wagner 2014; Tebyanian 2019). The aim of this chapter is an overview of the techniques of decellularized lung tissue engineering purposes.

3.3 Material and Methods

The study population can be evaluated in several species including rats, rabbits, monkeys, and human embryos which have been successfully decellularized in the laboratory.

3.4 Scaffold Preparation

Decellularized lung tissue can be prepared by harvesting the tissues from rat, rabbit, monkey, porcine, and animal models. Different types of detergents were successfully used for this goal, including Triton X-100 (nonionic), sodium deoxycholate (anionic), sodium dodecyl sulfate (SDS, anionic detergent), 3-[(3-cholamidopropyl) dimethyl ammonio]-1-propanesulfonate (CHAPS, zwitterionic detergent), and Tween (nonionic) in our center. Here we discussed some of species and in Table 3.1 different protocols are demonstrated in detail.

3.4.1 In Rat Lung Model

a pulsatile dynamic flow was utilized for the lung decellularization approach. Upon removal, the lung was cannulated via trachea and was connected to a solution containing 0.01% heparin in 0.2 l distilled water (at 38 °C for 1 h (0.05 ml/min). Then, phosphate buffered saline (PBS) (1 h) and 0.05% sodium dodecyl sulfate (SDS) (2 h) have been applied for washing the specimens (Kajbafzadeh 2015).

3.4.2 In Human and Porcine Model

after washing the lung section with 2X phosphate buffered saline (PBS) (15 min), using one of the following protocols can be done for lung tissue decellularization; (1) washing with 1.8 mM SDS (four—2 h), each followed by 2X PBS and dH2O (15 min), (2) washing with 1.8 mM CHAPS (four—2 h), each followed by dH2O and 2X PBS (15 min), (3) washing with 3% Tween-20 (2 h), then washing with 4% sodium deoxycholate (2 h), and finally washing with 0.1% peracetic acid (1 h). All slices were subjected to dH2O washes and alternating 1X PBS (2 of each). Finally, suitable sections of tissue can be considered as a good candidate for tissue evaluation.

3.4.3 In Rabbit Model

Sodium dodecyl sulfate (SDS) in phosphate buffered serum (PBS) has been added to lung specimens (24 h), After applying 1% Triton X-100 solution in PBS (12 h), and placing the specimens in PBS (2 h). The specimen was placed in PBS for 2 h. Finally, for sterilization, the specimens were kept in 70% ethanol (30 min, at 37 °C) (Gilbert et al. 2006). In the monkey model, the lung transverse section has been washed with phosphate buffered saline (PBS) and followed by the decellularization solution of 0.1% (v/v) diluted in sterile distilled water at 4 ° C (Nakayama 2011). Lung decellularization was continued by detergents such as SDS and Triton X-100 and three detergent concentrations of v/v (0.01%, 0.1%, and 1%) carried out at 4 °C. The solution has been changed 8 h after initial harvest and every 48 h until tissue transparency (approximately 20–24 days for lung tissues).

3.5 Casting and SEM of Casts

In this section, foam wound dressing (Cavi-Care; Smith & Nephew, Victoria, Australia) can be used for molding of native and decellularized lungs in order to compare airways structure before and after decellularization. The liquid components have been mixed and injected via the trachea into the lung in order to form a white foam which conformed to the lung microstructure. Both native and decellularized lungs were kept at room temperature for 24 h to allow all the injected materials to be dried. After the injected foam was hardened, lungs were kept in 90% acetic acid for 10 min in order to digest the tissue.

Table 3.1 The protocols, markers, and characterization method for the efficiency of decellularization

Spices	Decellularization Media	Duration	Histology Markers	Recellularization evaluation	Source of cell for recellularization	Excessive assessment tests	Year of study
Mice C57BL	Hypotonic buffer (10 mM Trizma base), 0.1% EDTA 0.2% SDS 10 mM Trizma base DNase	18 h 24 h 20 min 6 h	Col IV Col I – Alexa Fluor 555-conjugated donkey-anti-rabbit secondary - Hoechst 33,342 – Fibronectin Alexa Fluor 555-conjugated donkey-anti-rabbit – Alexa Fluor 568-conjugated donkey-anti-goat – ProLong™ Gold antifade mounting medium – Sudan Black B	Immunostaining of repopulated ECM slices - Staining of repopulated ECM slices with EdU - Homogenization of lung ECM	LFs 50,000 cells/cm^2	Extraction and quantification of DNA Quantification of tissue density SEM TEM Mass spectrometry analysis Isolation of primary mouse lung fibroblasts	2017
Rats	8 mM CHAPS benzonase endonuclease	1 h	Antibody for RTII-70 – pro-SPC – aquaporin-5 – Col IV – fibronectin – MMP-2 – a-actin – imentin Col IV		Human type II epithelial cells		2012
Rat	0.1% SDS 0.5% SDS		TP63 KRT5 E-cadherin Surfactant Protein-B pro-Surfactant Protein-C Aquaporin-5 Acetylated α-Tubulin α2β1 integrin α3β1 integrin KI67 CD31 Col IV		Primary pulmonary endothelial cells KRT5 + TP63 + basal epithelial stem cell	Quantitative PCR	2016
Porcine	0.5% sodium dodecyl sulfate 1% Triton X-100	5 days 24 h	Hematoxylin and eosin Verhoff van Gieson				2018

(continued)

Table 3.1 (continued)

Spices	Decellularization Media	Duration	Histology Markers	Recellularization evaluation	Source of cell for recellularization	Excessive assessment tests	Year of study
			Masson's trichrome Alcian blue Wright Giemsa CD45 CD3 CD4 CD20 CD68 CD86 CD206 CD31 Alpha smooth muscle actin biotinylated anti-rabbit IgG Biotinylated anti-mouse IgG, or biotinylated anti-goat IgG Vecta-Shield ABC Reagent				
Human	PBS containing sodium nitroprusside at 1 μg/ml 8 mM CHAPS, 1 M NaCl, and 25 mM EDTA in PBS benzonase nuclease	1 h Over night 1 h	Verhoeff Van and Gieson (VVG) Pro-Collα1 CD45 CD14lo or CD14hi monocytes Collagen		PBMCs (peripheral blood mononuclear cells)	Antibody-mediated Netrin blockade SEM PCR Flow Cytometry	2017
Rat	1 M NaCl sodium deoxycholate (0.01%, 0.05%, 0.1% benzonase	30 min 1 h	Laminin-_1, fibronectin, proliferating cell nuclear antigen collagen IV alpha		Rat lung microvascular endothelial cells	*Western Blot* *Liquid Chromatography* *Tandem Mass Spectrometry*	2016
C57BL/6 mice	1% sodium dodecyl sulfate (SDS) 1% SDS without agitation	24 h overnight	DAPI				2015
Porcine	0.1% Triton X-100 2% sodium deoxycholate DNase solution sodium chloride	24 h 24 h 60 min 60 min	– Elastin/collagen quantification – Mouse a-Galactosidase primary antibody stained with DAB chromogen solution – Glycaosamnioglycan (GAG) – a-galactosidase staining		Mesenchymal stem cells (hMSCs)	Picogreen double stranded DNA quantification SEM	2016

(continued)

Table 3.1 (continued)

Spices	Decellularization Media	Duration	Histology Markers	Recellularization evaluation	Source of cell for recellularization	Excessive assessment tests	Year of study
Rat	hypertonic detergent solution decellularization solution benzonase endonuclease	2–3 h				- Protein extraction - *Liquid Chromatography* *Tandem Mass Spectrometry*	2016
Human	1 M NaCl, and 25 mM EDTA in PBS 20 mM Tris–HCl, 2 mM Mg2 + , and 20 mM NaCl	4 h 30 min	Verhoeff van Gieson –ApopTag® Red Collagen type IV Ki-67 TUNEL IHC Elastic Van Gieson (EVG) Collagen type IV Galectin-3 Versican Decorin		Primary human lung fibroblasts	SEM Protein extraction from tissue Mass spectrometry DNA quantification	2017
Rat	0.1% Triton X-100 2% sodium deoxycholate 1 M NaCl 30 mg/mL DNase, 1.3 mM MgSO4, and 2 mM CaCl2	overnight overnight 1 h 1 h	Movat's pentachrome Alcian Blue staining of glycosaminoglycans (GAGs, blue), Immunohistochemistry (IHC) Col I Col IV laminin Elastin SMA Vitronectin	Ki-67	rat adipose-derived stem cells (rASCs)	Genomic DNA extraction Western blot Proteomics	2014
C57BL/6 mice	0.1% Triton X-100 2% sodium deoxycholate 1 M NaCl DNase	24 h 24 h 1 h 1 h	Collagen Elastin Laminin GAGs Chicken anti-laminin (Abcam) Anti-chicken secondary			SEM	2010

(continued)

Table 3.1 (continued)

Spices	Decellularization Media	Duration	Histology Markers	Recellularization evaluation	Source of cell for recellularization	Excessive assessment tests	Year of study
			Biotinylated anti-CK18 Goat-anti-mouse Pro-SPC FITC-labeled anti-CD11b Biotinylated anti-CD45 Rat-anti-mouse CD31 Goat-anti-rodent CC10 Donkey-anti-goat Ab Rabbit anti-mouse aquaporin-5 Goat-anti-rabbit secondary Rabbit anti-mouse vimentin				
Rat	Twice 0.05% sodium dodecyl sulfate (SDS)	2 h	DAPI Laminin TTF1 antibody Anti-GFP antibody Pentachrome Picro Sirius Col I			*SEM* *RQ-TRAP assay*	2014
Mice	0.1% Triton X-100 2% sodium deoxycholate 1 M NaCl DNAse	8 h 14 h 1 h 1 h	Sirius red stain or elastic van Gieson (EvG) Col IV			SEM	2016
Rat	0.1% Triton X-100 2% sodium deoxycholate		Verhoeff van Gieson DAPI		Rat lung microvessel ECs (RLMVEC)	SEM FITC-dextran barrier function assay	2016
Human	0.0035% Triton X-100 sodium deoxycholate from 0.01% to 0.1% Benzonase Nuclease in 1-M Tris-buffer with 1-mM MgCl2 0.5% Triton X-100 solution		SOX17 (endoderm marker DAPI-P63 NKX2.1 P63 DAPI- CK14 DAPI-CK5		iPSC-lung epithelial cells	Real-time quantitative reverse transcription-PCR	2019
Rat	0.1% sodium dodecyl sulfate 1% Triton X-100	2 h 15 min	Rabbit anti-Ki67 Rabbit anti-active/pro-caspase 3 Anti-rabbit IgG		Human adenocarcinoma cell lines H358, PC9, and SW1573		2015

(continued)

Table 3.1 (continued)

Spices	Decellularization Media	Duration	Histology Markers	Recellularization evaluation	Source of cell for recellularization	Excessive assessment tests	Year of study
			Rabbit anti-phospho- EGFR Biotinylated goat-anti-rabbit IgG Avidin–biotin peroxidase complex Diaminobenzidine				
Rat	8 mM CHAPS, 25 mM EDTA, and 1 M NaCl in PBS Benzonase	1 h	Fibronectin col I col IV laminin Tenascin, elastin GAGs	CD90 CD105 CD73 CD45	hBM-MSCs and hAT-MSCs	Real-time quantitative RT-PCR TEM	2014
Mice	8 mM CHAPS,1 M NaCl, 25 mM EDTA in PBS benzonase nuclease	Overnight 1 h	1-integrin-blocking antibody IgG isotype Masson's Trichrome VVG/Elastin Alsace Blue GAGs procollagen I_1 Ki67 Fibronectin col I col IV laminin TUNEL		Mouse A9 cells	*Fluorometric DNA assay* *Western blot*	2014
Human and porcine	0.1% Triton X-100 2% sodium deoxycholate (SDC) 1 M sodium chloride (NaCl) DNAse	24 h 24 h 1 h 1 h	EVG Fibronectin col I col IV laminin Elastin		Human lung fibroblasts (HLFs) human bone marrow-derived mesenchymal stromal cells (hMSCs)	Assessment of residual DNA Bronchoscopy and angioscopy Thermal imaging Mass spectrometry TEM	2014
Porcine	0.5% SDS	72 h	Collagen GAGs		Human lung epithelial cells (A549)	DNA content NADH dehydrogenase Nuclear factor NF-kappa-B Chromodomain-helicase-DNA-binding protein	2017

(continued)

Table 3.1 (continued)

Spices	Decellularization Media	Duration	Histology Markers	Recellularization evaluation	Source of cell for recellularization	Excessive assessment tests	Year of study
						ATP-binding cassette sub-family G member 4-like Multiple epidermal growth factor-like domains protein NFKB1 protein Copper-transporting ATPase Cytochrome c oxidase subunit I Neutralized-like protein 4 Sodium/glucose cotransporter 4 Acetylcholine receptor Retrotransposon gag domain-containing protein 1 Type III secretion translocator protein Glutamate synthase Fatty acid synthase Carbohydrate binding family protein PE-PGRS family protein GA27210, isoform A	
Rat	1 M NaCl 0.5% Triton X-100	20 min 15 min	H&E DAPI		Calu-3 cells	TEM	2019
	0.1% Triton X-100 2% sodium deoxycholate NaCl DNase	Overnight 2-day 3 h 3 h	Sudan Laminin col I col IV elastin GAGs fibronectin vitronectin Cy3 anti-mouse (Col I) Fluorescein isothiocyanate (FITC) Anti-rabbit (Col IV and laminin) FITC anti-mouse (elastin and vitronectin)			SEM	2015

(continued)

Table 3.1 (continued)

Spices	Decellularization Media	Duration	Histology Markers	Recellularization evaluation	Source of cell for recellularization	Excessive assessment tests	Year of study
			FITC anti-sheep (fibronectin) IgM a-GAL primary antibody Anti-mouse IgM secondary				
Rat	0.1% SDS 1% Triton X-100	2 h 10 min	T1α Ttf1 pro-SPC Vimentin		Human umbilical cord venous endothelial cells (HUVECs)	TEM	2010
B10.R and BALBc mice	Triton X-100, sodium deoxycholate NaCl DNase		Elastin GAGs Laminin Fibronectin Vitronectin		Human bronchial epithelial cells (hBECs) and lung fibroblasts (hLFs)	qRT-PCR Mass Spectrometry	2016
Porcine	Triton X-100 sodium dodecyl sulfate (SDS) sodium deoxycholate (SDC), CHAPS		Col I Elastin laminins Fibronectin GAGs DAB EVG peroxidase		A549 cells	DNA Quantification Western blotting	2016
Rat	Detergent 1% Triton X-100	12 h 30 min	C5b-9 CD68 Col I Col IV fibronectin laminin Cytokines GAGs Alcian blue			DNA content	
Rat	0.1% SDS Triton X-100	2 h 10 min	Clara cell-10 pro-surfactant Protein-C TTF1 CD31 ED1 Vimentin Mouse IgG		Rat fetal pneumocytes	*Morphometry and Stereology*	2011

(continued)

Table 3.1 (continued)

Spices	Decellularization Media	Duration	Histology Markers	Recellularization evaluation	Source of cell for recellularization	Excessive assessment tests	Year of study
Goat	0.25% trypsin–EDTA 0.1% SDS solution containing RNase and DNase	24 h 6–8 h 24 h	DAPI			SEM Chemical characterization by Fourier Transform Infra-Red (FTIR) Spectroscopy ALP activity and Alizarin test	2017
Goat	0.25% trypsin–EDTA 0.1% SDS solution containing RNase and DNase	24 h 6–8 h 24 h	CD105 CD73 CD44 Keratin18 β-Actin		Human Skin-derived Mesenchymal Stem cells (S-MSCs)	SEM Chemical characterization by Fourier Transform Infra-Red (FTIR) Spectroscopy ALP activity and Alizarin Test RNA harvest Reverse transcript (RT) PCR	2013
C57BL/6 mice	0.2% SDS DNase	24 h 3 h	Masson's Trichrome				2019

3.6 SEM

SEM can be used for detecting of the internal structure of terminal alveoli in the casts of both native and decellularized lung tissue. Samples of native and decellularized lung keeping in 2.5% glutaraldehyde (at 4 °C 1.5 h), were kept in PBS at 4 °C for 1.5 h. After washing, graded ethanol series have been used for 9 h in order to have dehydrated samples. In order to achieve electrical conductivity, a layer of gold (approximately 2 nm) was used for coating the tissue samples.

3.7 Mechanical Tests

Decellularized and native lung tissues can also be analyzed by Instron 5848 equipped with a 10-N load cell. Lung tissue sections (5–8 mm in length and 4–16 mm^2 in cross-sectional area) were used to cyclically prestretched (10 cycles to 20% strain). Tissue stretching was continued until failure to assess ultimate tensile strength (UTS) (Petersen 2012). A tensile testing device is used to identify the stability of decellularized lung scaffold and in order to compare it to the natural lung. Samples of natural and decellularized (2. 2 cm thickness) were subjected to uniaxial force with an acceleration rate of 4 mm/min until the disappearance of the load was identified by the device (Kajbafzadeh 2015). Using tissue dimensions, engineering strain and engineering stress were calculated from distance and force. The curve of stress vs strength was drawn; the maximal point in the curve demonstrated maximum pressure tolerance of natural and decellularized lung tissues.

3.8 DNA Staining and Quantification

The DNA content of decellularized lung tissues in lung specimens (O'Neill 2013), was suggested to be quantified by applying the Quant-iT Pico-Green dsDNA assay kit. The samples were weighed and kept in 125 mg/mL papain (Sigma) overnight (at 60 °C). Measurment of fluorescence was done at 520 nm with excitation at 480 nm. DNA was qualified using a standard curve prepared with λ-phage DNA and normalized to lung wet weight. In rat lung, blue-florescence 4', 6-diamidino-2-phenylindole (DAPI) were used to visualize double stranded DNA (dsDNA). In rat lung, according to Laird et al. (1991), lung samples were homogenized in a solution containing trypsin (0.25%) and EDTA (1 mM) in deionized water. After incubation of the homogenate with stirring (at 37 °C), Cell lysis was done using a solution containing 200 mM NaCl, 2% SDS, 5 mM EDTA, and 100 mM Tris–HCl (pH 8.5, for 24 h at 55 °C). DNA extraction was continued with isopropanol and dissolving in a solution of 0.1 mM EDTA and 10 mM Tris HCl. DNA amount was measured by spectrophotometry at 260 nm.

3.9 Collagen

In order to evaluate the efficiency of the method and tissue structure presence, lung tissues were weighed and kept in pepcin (0.1 mg/mL pepsin overnight at 4 °C). Then, according to this procedure, Sircol collagen assay kit can be used for collagen quantification. In another model, according to Bergman and Loxley (1970) and Stegemann (1967) (Bergman and Loxley 1970), the amount of hydroxiprolin was suggested as an indicator of collagen contents. Following hydrolyzing the samples in 5 ml 6 N HCl for 14–16 h (at 110 °C), Chloramine—T (0.5 ml) solution was added to the samples. Then, samples were incubated at room temperature (20 min). Finally, Ehrlich's (Sigma-Aldrich Co. Radeberg, Germany) (1 ml) solution was added to the sample. The resulting solution was incubated for 20 min at 65 °C. OD absorbance at 543 nm was assessed with a spectrophotometer. Concentration of hydroxyproline was expressed as mean ± SD μ g/mg of each tissue.

3.10 Sulfated Glycosaminoglycans

Dimethylene blue dye assay is recommended for the quantification of sulfated glycosaminoglycans (sGAG) in the lung. After weighing, the samples were digested in 125 μg/mL papain at 60 °C (overnight) and then were mixed with dimethylene blue dye (1:8 ratio). The absorbance was measured at 595 nm immediately and was normalized to the tissue sample wet weight.

3.11 Elastin

Elastin in the sample was quantified by fastin elastin assay Kit. The sample was weighed and a-elastin extraction was done with the combination of three hot 0.25-M oxalic acid extractions in each sample (35 mg tissue per 1 mL solution).

3.12 Cell Culture and Seeding

Lung tissue preparing from new born rats were chopped into nearly 1 mm ^{2} cubes using crossed scalpels. Tissue was incubated in type I collagenase (0.1%) for 30 min at 37 °C with shaking. 10% fetal bovine serum was used for neutralization of the solution and then the solution was transferred into a centrifuge tube and centrifuge at 5 min at 172 × g. After removing the supernatant, the pellet was re-suspended in approximately 5 ml of Dulbecco's modified Eagle's medium (DMEM) high glucose +15% FBS and then seeded in a 25 cm 2 flask. After 48 h, the medium was replaced. Subcultured cells were done when the cell culture reached 80–90% confluence (Kajbafzadeh 2015). In fibroblast culture, Human lung fibroblasts (hMRC-5s) were cultured in Dulbecco's modified Eagle medium (DMEM) with 1% pen/strep and 10% fetal bovine serum (FBS). The hMSCs were cultured in DMEM/F12 with 1% pen/strep and 10% FBS (1:1). Following the 70–80% cell confluency enreachment, the cells were seeded onto decellularized tissue punches with the density of 2.5 × 104 cells/mL, and were cultured for 7 days (O'Neill 2013). In another model, after preparing the blastemal cell, the scaffold was kept in the blastemal ring. The scaffold with the blastema ring was together transferred to a 12 well-plate in Dulbecco's Modified Eagle's Medium and was incubated at 37 °C in 5% CO_2 (Shahri 2013). In monkey lung seeding, scaffolds were cut into 8 mm diameter cylenderical biopsy and were seeded with 5 × 10^5 undifferentiated hESCs. Seeded scaffold was suspended on PET track-etched membrane cell culture inserts. Then, scaffolds were cultured in Dulbecco's Modified Eagle's Medium high glucose with 10% fetal bovine serum (FBS), 1% L-glutamine and 1% pen-strep at 37 °C and 5% CO_2. Finally, 48 h after seeding, it has been transferred to new culture media in order to quantify non-adherent cells remaining in the original inserts with viability assessed with trypan blue and a hemacytometer.

3.13 Bioreactor

The cells located in the ECM scaffold were needed to maintain in the organ-specific biomimetic environment after the cell attachment. For this purpose, the decellularized lung scaffolds were kept in petri dishes containing DMEM with either GFP negative or GFP positive lung cells. The cells were seeded at the density of 2 × 10^4 cells/cm^2.

Cells and scaffolds were transferred to the close system bioreactor, which was located in an incubator for seeding of lung cells. A glass bottle is connected to a pump with two ports on the cap, one port for circulation of the medium and the other for trachea cannula, was used for scaffold suspending.

The medium was circulated 2 min per hour with 58 min of resting time to facilitate cell adherence to the scaffold. On the first day, 70 ml medium containing 2 × 10^6 cells were transferred via trachea cannula into the scaffold with a flow rate of 5 ml/min at 37 °C in a 5% CO_2 atmosphere. On the next day, the medium was centrifuged and remained cells were added to the other 20×10^6 lung stem cells. Cell seeding on the matrix was conducted for 15 consecutive days and the medium was every three days. A shaker can be used in all steps in order to prevent cells adhesion to the container walls.

3.14 Histological Evaluation

To evaluate the efficacy of decellularization protocol, tissue sections were stained with Masson's trichrome, IHC, and Hematoxylin & Eosin (H & E), stainings. The pieces of native and decellularized lung (4 × 4 mm) were fixed in 10% neutral formalin (24 h at room temperature) were washed in distilled water, dehydrated via graded ethanol, and embedded in paraffin. After dehydration, Hematoxylin eosin and IHC staining will be ready to be performed for tissue evaluation.

Briefly, after fixing in 4% paraformaldehyde, samples were immersed in Triton X-100 diluted 1:100 in PBS to reduced penetration of the antibody. After blocking with 1% BSA/PBS, samples were incubated with anti-laminin in order to tissue architecture preservation. H & E staining and IHC staining with thyroid transcription factor-1 (anti-TTF1) and anti-GFP antibodies are proposed for assessing the efficacy of the mentioned protocol for regenerating recellularized scaffold (Kajbafzadeh 2015).

Briefly, samples were stained in 0.1% Sirius red solution in a saturated picric acid, and then samples were rapidly washed in 0.01 N HCl. Under polarized light, type I collagen and type III collagen fibers turned into strongly birefringent yellow–red and weakly birefringent and greenish, respectively. Picro Sirius red staining can be applied for collagen typing.

In the human and porcine model, the distribution of matrix proteins was also suggested to be evaluated by the use of Alcian blue (proteoglycans), Masson's trichrome (collagen), and Van Gieson's (elastic fibers) stains. Of note, the distribution of collagen IV, laminin, fibronectin, and elastin in native and decellularized can be assessed by IHC (O'Neill 2013) (Figs. 3.1 and 3.2).

3.14.1 Histomorphological Evaluation

The main staining method for the evaluation of the cells' nuclei morphology was suggested to be evaluated using a fluorescence microscope at an excitation wavelength of 350 nm. 4′-6-Diamidino-2-phenylindole (DAPI) was known to form fluorescent complexes and showing fluorescence specificity with natural double stranded DNA. The cells were washed with PBS

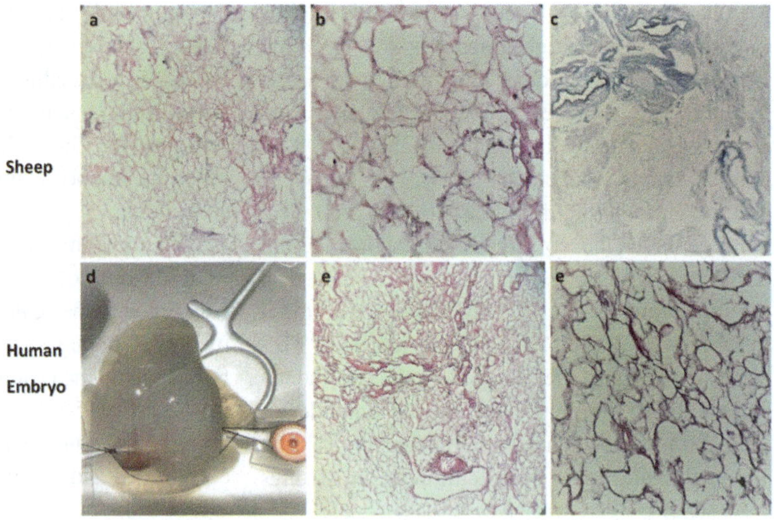

Fig. 3.1 Sheep and human embryo decellularization approval tests H&E and Trichrome staining

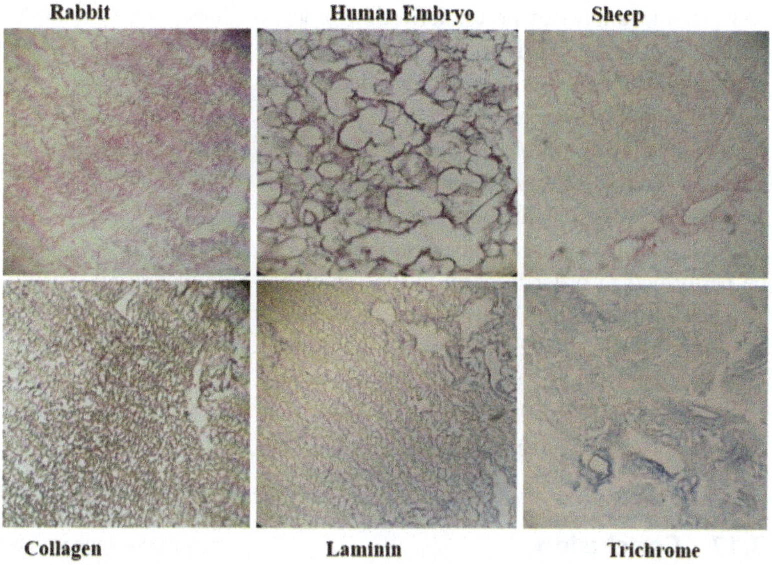

Fig. 3.2 Different species lung Acellularization: upper row represent the Rabbit, Human Embryo and sheep decellular tissue and second row represents the immunohistochemistry staining of the lung tissue after decellularization, trichrome stating to demonstrate the collagen fibers integrity in decellularization process

and re-suspended in PBS containing 0.1% Triton and incubated on ice (10 min). Then, cells were spun down and re-suspended at 5000 cells/μl in 4% PBS paraformaldehyde solution containing 10 μg/ml DAPI. The suspension (10 μl) was placed on the slide and then covered with a coverslip. The cell' morphology is ready to be assessed by fluorescence microscope. Another main staining is immunohistochemistry staining. Lung tissue can be stained with H&E and trichrome and further confirmation can be conducted with specific ECM markers such as laminin (Fig. 3.1).

3.15 Radiography, MRI, and CT Scan

In order to verify that the structure of bronchial trees was preserved in the scaffold of lung tissue after decellularization, dimeglumine gadopentetate was injected into the trachea to obtain terminal alveoli visualization on the radiograph. Radiography (antero-posterior) can be performed after 2 min injection of gadopentetate dimeglumine solution.

Superconductive MRI and CT scan can be performed for comparing the structure of natural and decellularized lungs. Briefly, specimens evaluated by MRI were soaked in Magnevist (Gadoteric acid) 0.5 mmol/ml for 2 h before analysis. Also, samples subjected to CT were kept in PBS. MRI scanner was used for three-dimensional spoiled gradient recalled images (GRASS) to create ultrathin serial sections with a special high signal-to-noise ratio. The images were prepared using pulse sequences (TR = 30–50 ms, TE = 2–8 ms), flip angle (30° d), and echo train length (8 moderate). The data were downloaded to an independent workstation which was equipped with volume rendering software. Multiplanar re-formation and 3D reconstruction can be performed to reformat the achieved images.

3.16 Limitation

The lung is one of the best sites to test the implantation of engineered tissues in the body. It is well known that oxygen diffusion through synthetic and natural lung matrices depends on the matrix ultrastructure and can be varied between materials (Androjna 2008). Thus, the decellularized lung is the best matrix to support lung tissue engineering. However, there are some limitations that we have to consider before using decellularized extracellular matrix. We have to consider the impact of decellularization process on the lung tissue matrix. It has been suggested that the matrix structure may be degraded during decellularization process (Gilbert et al. 2006). On the other hand, the biological efficacy of recellularized scaffolds has not been examined in vivo studies. Moreover, evaluation of surfactant protein and measuring of gas exchange is needed to be performed in order to confirm the functional property. Importantly, many unresolved issues have been remained such as the ideal duration of recellularization process and decellularization method, necessary cell types, and minimum numbers of cells for recellularization in vivo.

3.17 Conclusion

To address the shortage of lungs for transplantation, many approaches have been evaluated. Decellularized lung with the preservation of lung ECM has been considered as an ideal scaffold for cell seeding. Lung cells seeding into decellularized lung scaffolds with the functional properties of the cells may confirm the engineered lung tissue process. Thus, the recellularization of lung regeneration may be an effective way for the treatment of lung diseases in the future.

References

Androjna C et al (2008) Oxygen diffusion through natural extracellular matrices: implications for estimating "critical thickness" values in tendon tissue engineering. Tissue Eng Part A 14(4):559–569

Badylak SF, Taylor D, Uygun K (2011) Whole-organ tissue engineering: decellularization and recellularization of three-dimensional matrix scaffolds. Annu Rev Biomed Eng 13:27–53

Balestrini JL et al (2016) Comparative biology of decellularized lung matrix: implications of species mismatch in regenerative medicine. Biomaterials 102:220–230

Bergman I, Loxley R (1970) The determination of hydroxyproline in urine hydrolysates. Clin Chim Acta 27(2):347–349

Birchall M, Macchiarini P (2008) Airway transplantation: a debate worth having? Transplantation 85(8):1075–1080

Bombelli S et al (2018) Nephrosphere-derived cells are induced to multilineage differentiation when cultured on human decellularized kidney scaffolds. Am J Pathol 188(1):184–195

Booth AJ et al (2012) Acellular normal and fibrotic human lung matrices as a culture system for in vitro investigation. Am J Respir Crit Care Med 186(9):866–876

Crapo PM, Gilbert TW, Badylak SF (2011) An overview of tissue and whole organ decellularization processes. Biomaterials 32(12):3233–3243

Gilbert TW, Sellaro TL, Badylak SF (2006) Decellularization of tissues and organs. Biomaterials 27(19):3675–3683

Gilpin SE et al (2014) Perfusion decellularization of human and porcine lungs: bringing the matrix to clinical scale. J Heart Lung Transplant 33(3):298–308

Gilpin SE et al (2016) Bioengineering lungs for transplantation. Thorac Cardiovasc Surg 26(2):163–171

Jensen T et al (2012) A rapid lung de-cellularization protocol supports embryonic stem cell differentiation in vitro and following implantation. Tissue Eng Part C Methods 18(8):632–646

Kajbafzadeh A et al (2015) Lung tissue engineering and preservation of alveolar microstructure using a novel casting method. Biotech Histochem 90(2):111–123

Kocyildirim E et al (2004) Long-segment tracheal stenosis: slide tracheoplasty and a multidisciplinary approach improve outcomes and reduce costs. J Thorac Cardiovasc Surg 128(6):876–882

Laird PW et al (1991) Simplified mammalian DNA isolation procedure. Nucleic Acids Res 19(15):4293

Macchiarini P et al (2008) Clinical transplantation of a tissue-engineered airway. The Lancet 372(9655):2023–2030

Nakayama KH et al (2011) Renal tissue engineering with decellularized rhesus monkey kidneys: age-related differences. Tissue Eng Part A 17(23–24):2891–2901

Nichols JE, Niles JA, Cortiella J (2009) Design and development of tissue engineered lung: progress and challenges. Organogenesis 5(2):57–61

Nichols JE et al (2013) Production and assessment of decellularized pig and human lung scaffolds. Tissue Eng Part A 19(17–18):2045–2062

Nichols JE et al (2017) Giving new life to old lungs: methods to produce and assess whole human paediatric bioengineered lungs. J Tissue Eng Regen Med 11(7):2136–2152

Omori K et al (2005) Regenerative medicine of the trachea: the first human case. Ann Otol Rhinol Laryngol 114(6):429–433

O'Neill JD et al (2013) Decellularization of human and porcine lung tissues for pulmonary tissue engineering. Ann Thorac Surg 96(3):1046–1056

Orens JB, Garrity ER Jr (2009) General overview of lung transplantation and review of organ allocation. Proc Am Thorac Soc 6(1):13–19

Petersen T (2009) In vitro development of engineered lung tissue

Petersen TH et al (2010) Tissue-engineered lungs for in vivo implantation. Science, p 1189345

Petersen TH et al (2010) Tissue-engineered lungs for in vivo implantation. Science 329(5991):538–541

Petersen TH et al (2012) Matrix composition and mechanics of decellularized lung scaffolds. Cells Tissues Organs 195(3):222–231

Prakash Y, Tschumperlin DJ, Stenmark KR (2015) Coming to terms with tissue engineering and regenerative medicine in the lung. Am J Physiol Lung Cell Mol Physiol 309(7):L625–L638

Reichenspurner H (2005) Overview of tacrolimus-based immunosuppression after heart or lung transplantation. J Heart Lung Transplant 24(2):119–130

Shahri NM et al (2013) In vitro decellularization of rabbit lung tissue. Cell J (yakhteh) 15(1):83

Skolasinski S, Panoskaltsis-Mortari A (2017) Decellularization of intact lung tissue through vasculature and airways using negative and positive pressure. Decellularized scaffolds and organogenesis. Springer, Berlin, pp 307–315

Smith J et al (2014) Organ procurement and transplantation network (OPTN) and scientific registry of transplant recipients (SRTR). OPTN/SRTR 2012 annual data report. Am J Transplant 14(suppl 1):97–111

Tebyanian H et al (2019) Rat lung decellularization using chemical detergents for lung tissue engineering. Biotech Histochem 94(3):214–222

Uhl FE, Wagner DE, Weiss DJ (2017) Preparation of decellularized lung matrices for cell culture and protein analysis. Fibrosis. Springer, Berlin, pp 253–283

Vertrees RA et al (2008) Cellular differentiation in three-dimensional lung cell cultures. Cancer Biol Ther 7(3): 404–411

Wagner DE et al (2014) Three-dimensional scaffolds of acellular human and porcine lungs for high throughput studies of lung disease and regeneration. Biomaterials 35(9):2664–2679

Wainwright D (1995) Use of an acellular allograft dermal matrix (AlloDerm) in the management of full-thickness burns. Burns 21(4):243–248

Lung Extracellular Matrix as a Platform for Lung Organ Bioengineering: Design and Development of Tissue Engineered Lung

4

Shabnam Sabetkish
and Abdol-Mohammad Kajbafzadeh

Abstract

Since lung tissue is not able to be reconstructed after substantial injury, lung transplantation often is the only alternative for treatment. Antibiotic-resistant organisms that remain in donor lungs causing infection in the immunosuppressed recipient are among the complications following transplantation. Development of strategies for whole lung regeneration is a pleasing choice particularly in patients with end-stage lung diseases. Reconstruction of lung tissue in vitro for transplantation received increased attention which could deal with the shortage of donor organs. Recent advancements in the field of tissue engineering and regenerative medicine have paved the road for beneficial alternative therapies. Our group has extensive experience with regard to the structure of the lung tissue, which makes us to our decision to continue with the preparation of lung, with the aim of developing a new ECM scaffold. Herein, we aim to review the state-of-art and the tissue engineering and regenerative medicine technology highlighting the major achievements toward the production of a bioengineered lung obtained decellularization and recellularization techniques. We have strong hopes that recent developments in the engineering of lung will lead to similar breakthroughs in the engineering of distal lung components in the future.

Keywords

Decellularization · Recellularization · Lung matrices · Scaffold · Lung stem cells

4.1 History

The lung is an extremely complex and dynamic organ with number of diverse cell types with distinct functions. Respiratory diseases have been mentioned as the third leading cause of death globally which is predicted to augment in the near future. In addition, more than €380 billion annually is dedicated to respiratory disease in the EU (Kreuter et al. 2018). In the field of tissue engineering, both synthetic scaffolds such as polyhedral oligomeric silsesquioxane poly (carbonate-urea) urethane (POSS-PCU), polyglycolic acid (PGA), pluronic F-127 and biologically derived scaffolds have distinct pros

S. Sabetkish · A.-M. Kajbafzadeh (✉)
Pediatric Urology and Regenerative Medicine Research Center, Section of Tissue Engineering and Stem Cells Therapy, Children's Medical Center, Pediatric Center of Excellence, Tehran University of Medical Sciences, Tehran, Iran,
No. 62, Dr. Gharib's Street, Keshavarz Boulevard, 1419433151 Tehran, Iran
e-mail: kajbafzd@sina.tums.ac.ir

A. Kajbafzadeh (ed.), *Decellularization Methods of Tissue and Whole Organ in Tissue Engineering*, Advances in Experimental Medicine and Biology 1345,
https://doi.org/10.1007/978-3-030-82735-9_4

and cons. These scaffolds are considered to have improved storage stability and can be constructed with excellent accuracy; however, most are deficient in providing biological possessions such as native integrin binding sites and bioactive cues for the purpose of cellular attachment, proliferation and differentiation (Santis et al. 2018). In addition, synthetic polymers are not successful in driving the differentiation of cultured cells and will not direct seeded cells into required fate (Mondrinos et al. 2006). Due to the complex 3D structure of the parenchymal lung tissue and interconnected pores need for proficient gas exchange (Weibel 2017), manufacturing a scaffold with geometrical parameters suitable for parenchymal lung tissue is difficult.

A variety of diverse skills such as decellularization for biological scaffolds, casting, electrospinning, cryogelation, and microfabrication techniques has been recently introduced. The purpose of current tissue engineering and, indeed, regenerative medicine as a translational approach is to reconstruct nonfunctional tissues and organs after disease or traumatic injury (Mandrycky et al. 2017). Decellularized lung scaffolds are obtained by eliminating the cellular material from natural tissue as well as maintaining the macroscopic structure and extracellular matrix (ECM) composition via the method of decellularization (Wagner et al. 2013; Gilpin et al. 2016; Calle et al. 2017). The ECM is one of the chief components of the microenvironment which is able to direct migration, proliferation and differentiation of cells after recellularization process (Crapo et al. 2011; Wagner et al. 2014a; Fischer et al. 2011). Remained microscopic ECM proteins and associated carbohydrates after decellularization process are biologically active and serve as a surface for further cells attachment and also as a signalling platform for the related cells. These cells within an ECM contains combinations of ECM proteins and glycosaminoglycans (GAGs) which provide structure support and help direct repair and regeneration after damage (Burgstaller et al. 2017). Lung biofabrication is based on seeding cells into a decellularized scaffold and on culturing the cells in an especially designed bioreactor. The most advanced method uses whole lung organ as the scaffold. Whole organs obtained from donor or cadavers can be decellularized and then re-seeded with a patient's own cells (Badylak et al. 2011; Scarritt et al. 2015). Furthermore, there are encouraging reports of recellularization and generation of functional lung organ. The first successful tissue-engineered airway replacement was performed in a patient in 2008, by implantation of a tissue-engineered trachea to maintain lung function of a patient with end-stage left-main bronchus malacia (Macchiarini et al. 2008). Douglas et al. primarily described the first 3D culture of lung cells. In their study, they demonstrated rat fetal lung cells were cultured on a collagen matrix in order to study lung epithelial cell biology (Douglas et al. 1976). Decellularization of a biocompatible whole lung scaffold with terminal preserved alveoli is a novel approach to functional lung regeneration which was performed in our center. The results of the previous study done in this center by the senior author (AMK) demonstrated preservation of micro-architecture and terminal alveoli after decellularization process of rat three-dimensional lung scaffold which may be able to increase the probability of an effective cell seeding process (Kajbafzadeh et al. 2015a). Due to previously obtained successes in whole lung tissue engineering, this technique can be considered as a potential therapeutic approach for further organ transplantation.

In this chapter, we are about to converse the lung as a subject for tissue engineering paradigm, discuss about the recent developments made in the fields of lung tissue-engineering and stem cells, as well as emphasize the challenges which we may confront with when applying such constructs in a clinical setting.

4.2 Materials and Methods

4.2.1 Decellularization Technology

The choice of the reagents in decellularization method is critically important, due to the fact that they can harm the microstructure and composition of the obtained scaffold and may indirectly affect the mechanical properties of the final product. Frequent techniques for decellularization of lungs consist of different physical (osmotic shock, sonication, freezing–thawing), ionic, chemical and enzymatic methods (trypsin, endo- and exo-nucleases) (Wagner et al. 2013).

Chelating agents, including ethylenediaminetetraacetic acid (EDTA) and ethylene glycol tetraacetic acid (EGTA), are able to detach cells from ECM. EDTA and EGTA are regularly applied in combination with trypsin. The combination of these two agents is recommended in a multistep protocol as they are not particularly effective if used alone (Tudorache et al. 2007). Trypsin as the most frequently used proteolytic enzyme, in decellularization protocols, will remove ECM components, and still supports EC growth in vitro. However, the duration of exposure to trypsin–EDTA treatments should be limited in order to minimize the disruptive effects on the ultrastructure of the ECM.

Detergent-based perfusion technique has been prevalently used to produce decellularized lung scaffolds. Triton X-100, sodium deoxycholate (SDC), sodium dodecyl sulfate (SDS), and 3-[(3-cholamidopropyl) dimethylammonio]-1-propanesulfonate (CHAPS) are among the commonly used detergents, which are used with or without hypertonic sodium chloride and DNase and/or RNase solutions (Wagner et al. 2013). In our center working with more than 24 organs, we eliminate the use of DNase and RNase in the decellularization process, due to the fact that these enzymes remain in the scaffold and can raise toxicity, which may result in limited cell seeding. In addition, the concentration and volume of the mentioned detergents are dissimilar in different protocols and species. In our center, we were able to produce a biocompatible three-dimensional (3D) lung scaffold with comparative histological and biomechanical properties of native lung ECM, using a perfusion-based decellularization method (Kajbafzadeh et al. 2015a).

We should be able to use the most satisfactory decellularization protocol to obtain the best histological features and reach to satisfactory content of both ECM and other retained proteins. However, it still remains vague how different lung decellularization methods are able to affect recellularization of the implanted scaffold (Brown et al. 2009; Badylak et al. 2008). Table 1 summarizes the type of the scaffold and different decellularization and recellularization processes used for lung tissue (Kajbafzadeh et al. 2015a; Kuttan et al. 1981; Petersen et al. 2010; Ott et al. 2010; Wallis et al. 2012; Booth et al. 2012; Nichols et al. 2013; Sokocevic et al. 2013; Wagner et al. 2014b; Gilpin et al. 2014; Ren et al. 2015; Zhou et al. 2018).

4.3 Tests to Confirm the Efficacy of the Decellularization Process

Macroscopic appearance alone is insufficient to verify the extent of decellularization. Absence of visible cellular or nuclear components on histological examination (DAPI or H&E), less than 50 ng dsDNA per 1 mg of dry weight of the ECM scaffold, and remnant DNA shorter than 200 bp should be taken into consideration (Crapo et al. 2011). However, the criteria of cytocompatibility sterility, composition, and mechanical properties of the decellularized scaffolds should be also taken into account. Mass spectrometry proteomics is used to characterize the composition of decellularized scaffolds. But large amounts cytoskeletal elements and cell-associated proteins are still remained in the scaffold after decellularization (Wagner et al. 2014a). Scanning electron microscopy (SEM) is a valuable examination used to determine the detailed internal structure of terminal alveoli in lung tissue after the decellularization process. Tensile test is also used to measure the stability of the decellularized lung scaffold and to compare it to the natural one. To verify the maintenance of the structure of the bronchial branch in the scaffolds after decellularization, radiography, MRI and CT scan are also valuable examinations. As previously described (Neuman and Logan 1950), the collagen contents of the native and decellularized lungs can be assessed by determining the amount of hydroxyproline. Picrosirius red staining and Russell–Movat's

Table 1 Previous studies regarding advances within the field of in-vitro and in-vivo lung tissue engineering

Species	Method of decellularization/Recellularization	Results	Reference
Alveolar basement membrane (various origin)	Perfusion-based 4% Triton X-100 /SDC	Ultrastructurally heterogeneous consisting of amorphous basement membranes and interstitial collagen and microfibrillar components	**21**
Rat lung	Perfusion-based 0.05% SDS Seeded with GFP positive lung cells	Complete cell and DNA clearance, preservation of ECM components, preservation of micro-architecture and terminal alveoli	**17**
Rat lung	Perfusion-based CHAPS Seeded with pulmonary epithelium and vascular endothelium in- vitro and in-vivo	Repopulation of lung as viable strategy for lung regeneration showing the implanted lung into rats could participated in gas exchange	**22**
Rat lung	Perfusion-based 0.1% SDS Seeded with epithelial and endothelial cells in- vitro and in-vivo	Gas exchange in vivo for up to 6 h after extubation.	**23**
Mouse lung	Slices 0.1% Triton-X 100, 2% SDC, NaCl, DNAse 0.1% SDS CHAPS	Complete cells removal and no significant differences between detergent-based protocols for mouse lung de- and recellularization	**24**

Table 1 (continued)

	Seeded with bone marrow–derived MScs or with C10 mouse lung epithelial cells in-vitro		
Human lung	Slices detergents, salts, and DNase Seeded with fibroblasts in-vitro	Complete cell and DNA clearance, preservation of ECM components, while retaining native dimensionality and stiffness of lung tissue, ability to induce angiogenesis, conservation of the framework of the innate vasculature, and immunogenicity	**25**
Human and porcine lung	Slices freezing and 2% SDS	Cellular material removal+ preservation of ECM proteins and the native vascular tree	**26**
Mouse lung	Slices Triton/SDC, salts, DNase, MgSO4, CaCl2	Production of decellularized scaffold with retained ECM components	**27**
Human and porcine lung	3D lung segments Triton/SDC, salts, MgSO4, CaCl2, DNase, PAA	Small segments to retain 3D lung structure in acellular scaffolds from large animals and human origin for physiologic recellularisation	**28**
Rat and human lung	Slices and whole organ Constant-pressure perfusion 0.1% SDS Reseeded with human iPSC-derived lung progenitor cells in- vitro and in-vivo	Cellular material removal and preservation of ECM proteins/ cells cultured on decellularized lung slices demonstrated proliferation and lineage commitment	**29**

Table 1 (continued)

Rat and human lung	Perfusion-based 1% Triton X-100/ 0.1%SDS Reseeded with endothelial and perivascular cells derived from induced pluripotent stem cells in-vitro and in-vivo	Complete decellularization and regeneration of functional pulmonary vasculature	**30**
Porcine lung	Sequential single-pass perfusion Triton/ 0.5% SDS Reseeded with human airway epithelial progenitor cells	Generation of pulmonary vasculature with mature endothelial lining and sufficient anti-thrombotic function/. Creation of a living, functioning gas exchange graft.	**31**

pentachrome staining are another employed techniques for collagen typing, and demonstrating different constituents of connective tissue including collagen, elastin, muscle, mucin and finbrin, respectively (Khorramirouz et al. 2014; Kajbafzadeh et al. 2015b, 2006). Casting of decellularized lungs has been reported firstly in our previous study (Kajbafzadeh et al. 2015a). In our study, both native and decellularized lungs were molded with special foam in order to evaluate the airway structure of lung before and after decellularization by using foam wound dressing (Cavi-Care; Smith & Nephew, Victoria, Australia). This casting technique may be useful for developing useful engineered whole lung following the cell seeding process by autologous cells.

4.4 Cell Seeding

In addition to recognizing techniques to manufacture appropriate scaffolds for lung tissue engineering, a supplementary challenge is in finding cell sources and obtaining sufficient cell numbers. Selecting the right cell types responsible for fundamental functions of the lung is still the biggest uncertainty in the path toward regenerating lung organ. Therefore, the use of decellularized lungs is challenging due to the complex 3-D structure of the lung organ and its various functions. As a result, recellularization technique of lung tissue in a spatiotemporally and functionally appropriate manner is difficult due to the large number of cell types with various functions. The decellularized lung could recellularize with autologous cells or an allogeneic source. A careful selection of a cell source is a crucial step for the generation of precise biological functions. Embryonic stem cells (ESC) (Ali et al. 2002; Coraux et al. 2005; Samadikuchaksaraei et al. 2006; Rippon et al. 2006), endogenous pulmonary stem cells (Cortiella et al. 2006; Majka et al. 2005), extrapulmonary stem cells (Wong et al. 2009; Sueblinvong et al. 2008; Jiang et al. 2002), and mesenchymal stromal, or stem cells (MSCs) (Goolaerts et al. 2014; Gotts and Matthay 2012; McAuley et al. 2014), are among the probable cell sources for engineering of functional lung replacement tissues and for any stem cell-based therapy.

iPS cells are considered as an alternative approach to endogenous progenitor cells. Recent studies have demonstrated the potential of human iPS cells differentiation into cells expressing a distal pulmonary epithelial cell phenotype and will be re-seeded into decellularized human lung scaffolds (Huang et al. 2014; Ghaedi et al. 2013).

ESCs are flourishing to produce upper airway epithelial cells and AECII-like cells which are able to differentiate into alveolar epithelial type II cells (AECI) (Garcia et al. 2012). It has been verified that the microenvironment surrounding the ESCs has the ability to affect the cell fate. The capability of pulmonary mesenchyme on differentiation of murine ESCs has been examined in the study of Vranken et al. (2005). The type of mesenchyme can also play a crucial role in plasticity of pulmonary epithelium (Shannon and Hyatt 2004). In addition, growth factors known to be sequestered by the lung ECM are required for differentiation of pluripotent stem cells into proximal or distal lung epithelial cells. The capability of the re-seeded cells to survive, proliferate and differentiate is vital for evaluating short- and long-term cytocompatibility of the scaffolds (Shojaie et al. 2015). Several studies have depicted that single cell suspensions from fetal lung homogenates and re-endothelialized with immortalized endothelial cell lines or iPS-derived endothelium and epithelium can be used for recellularization or transplantation of the lung organ (Ghaedi et al. 2018). Directed differentiation protocols have introduced distal lung cells from human iPS cells (Ghaedi et al. 2013) and proximal airway cells from human ES cells (Wong et al. 2012).

MSCs have the ability to be isolated from bone marrow or the fat of patients as peripheral tissues, expanded in large scale, and then is differentiated into other cell types in culture. Hence, autologous MSCs for patient-specific organ engineering (McAuley et al. 2014). However, realization of this potential is dependent on capability of MSCs to recapitulate anatomical and functional properties of the diverse lung cell types. Several studies have also demonstrated that MSCs can attach and proliferate in lung scaffolds (Nichols et al. 2013; Daly et al. 2012; Mendez et al. 2014). It is of great importance to use one or more types of mesenchymal or stromal cells in order to support scaffold maintenance (Barkauskas et al. 2013). For better engraftment of transplanted embryonic epithelial lung progenitor cells in mice (Rosen et al. 2015) and also in order to enhance the barrier function of an endothelial-reseeded decellularized lung (Ren et al. 2015), adding stromal cells to lung cultures would be of great importance.

Although obtaining adult lung cells from surgical resections, and transplant recipients, for reseeding of scaffolds seems easy, proliferative capacities of these adult cells are restricted and are therefore improbable to be practical for reseeding large surface areas of a bioengineered human lung (Nichols et al. 2013; O'Neill et al. 2013; Sullivan et al. 2012).

Human amniotic fluid stem cells (hAFSCs) are applied for lung tissue engineering as a prosperous and capable resource for primitive cells. It has been verified that these cells play a crucial role in differentiating into pulmonary lineages following lung damage. As these cells can be obtained from umbilical cord blood cells, immediately after birth, they can be widely used for correction of congenital lung abnormalities (Azargoon and Negahdari 2018; Chang et al. 2009).

Autologous cells from the patient could be a potential source to diminish post-transplantation immune reactions and also minimize the need for immunosuppressive medications which decrease the risk of infection, cancer, and other possible toxicities.

For the recellularization procedure, seeded cells should be distributed in a spatial configuration as their innate conformation, in order to function properly in vivo. For this purpose, an appropriate type, mixture, and number of cells should be seeded following intramural injection or by infusion of cells into the vasculature (Badylak et al. 2011).

Growth factors also play significant roles for differentiation of cells to distal epithelium, control reprogramming of stem cells in lung engineering, and signalling molecules mediating the autocrine and paracrine signals and specifically drive the formation of lung tissue. Retinoic acid (RA), bone morphogenetic protein (BMP), hepatocyte growth factor (HGF), granulocyte colony-stimulating factor (G-CSF), the keratinocyte growth factor (KGF), adrenomedullin, and fibroblast growth factors (FGFs) are among the growth factors and signalling pathways that play fundamental roles in inducing the cells to specific lung cells (Tebyanian et al. 2019).

In our center, we recellularized rat lung satisfactory using a mixture of cells obtained from homogenized rat lungs. For this purpose, tracheal cannulation was preferred to vascular perfusion to create an efficient cell seeding process from the bronchial structures down to the terminal alveoli (Kajbafzadeh et al. 2015a). Additionally, we use a bioreactor in which the lung cells were suspended in the scaffolds for 2 min/h. Therefore, the lung cells have sufficient time to adhere to the scaffold. By using this novel technique, we improved the seeding process and minimized the seeding time. Large-scale production and culture of cells using suspension culture bioreactors and rotating wall bioreactors has been recently reported (Ghaedi et al. 2013; Raredon et al. 2015). In this regard, bioreactor technology ranges from small-scale apparatuses (Hoganson et al. 2011, 2008) to whole lung bioreactors as designed in our center (Kajbafzadeh et al. 2015a). In this technology, the lung organ will undergo decellularization process by perfusion to maintaining vascular structures and following recellularization in sterile, closed systems with consequent ventilation and perfusion (Tebyanian et al. 2019; Panoskaltsis-Mortari 2015; Doryab et al. 2017). However, the method of decellularization and recellularization using bioreactors needs to be optimized for a cost-effective and less labour-intensive large-scale production of cells. In order to diminish the disadvantages by the application of distinct cell seeding techniques in vitro, using the human body as a bioreactor has been also verified as a promising advance for bioengineering of the upper airways. It should be also taken into consideration that after the cell seeding process, the scaffold should be incubated for a proper length of time for better cell attachment, proliferation and differentiation. We are now only able to produce grafts with limited gas exchange and viability. Further improvements are required to improve the reseeding of decellularized lung tissue.

4.5 Clinical Applications

There has not been an advanced progress in the clinical applications of tissue-engineered distal lung tissue; however, endogenous and exogenous stem cell populations have been recently striking attention to be beneficial to be used as a source for development of engineered tissues. Decellularized cadaveric trachea seeded with mesenchymal stem cells has been recently applied to engineer upper respiratory tract, which has been considered as a foremost step toward the regeneration of lung tissue that may be able to improve the patient's quality of life which has always been the goal of regenerative medicine (Macchiarini et al. 2008). Additionally, some centers performed transplantation of single lobes which could be valuable for patient with COPD and A1AD (Sato et al. 2014). Although size matching is challenging with adult donors, transplantation of single lobe from an adult donor has been performed in pediatrics (Takahashi et al. 2013). This approach may be only useful for patients that a portion of the parenchyma or airway is affected. Moreover, the use of an appropriate large animal model with long-term outcomes would be helpful before translating these approaches to the clinic. It is obvious that for clinical achievements, the tissue-engineered lung is dependent on creating a stable and functional blood supply (Orlando et al. 2012). At the present time, in vitro lung perfusion systems are used by many investigators to have a better understanding of the lung tissue regeneration. Human-sized bioreactors are now presented for decellularization and recellularization processes of lung tissue which is also perfectly designed in our center as well. The bioreactor designed by the senior author (AMK) has the ability to completely mimic the environment of lung such as CO_2 and O_2 exchange, pressure, and so on. These in vitro lung perfusion systems are also used in clinical trials of lung transplantation as system for lung preservation. However, the current state of the art in whole-lung bioengineering is at a proof-of-concept stage, with considerable queries and roadblocks remaining. From a clinical translational viewpoint, we are going to pave the road to customize clinical-relevant size decellularized lung in a way slightly different from the one in experimental studies so far that can be regenerated with autologous patient-own cellular population and then transplanted in the same patient. This approach could address the limitations that today affect treatment of patients with severe respiratory diseases. We may also be able to create recellularized scaffolds with functional cells which can be transplanted in the patient whose cells were harvested, without any immunological reactions. This technique may limit the use of immunosuppressive drug regimes in related patients.

4.6 Limitations

The lung organ has numerous cell types and the bench to bedside application of engineered lung tissues is still challenging due to the complexity of lung tissue and various cell types that make up this organ. As highlighted above, the type and possessions of the lung scaffold and suitable cells for further recellularization process are predominantly important. However, after addressing these issues, maintaining and optimizing the tissue-engineered lung prior to transplantation still remain to be answered. Regarding immunogenicity, the balance between reducing rejection response and failure of the transplanted organ against preserving the innate immune functions of the lung tissue and its inhabitant cells is a mainly challenging barrier which should be lessened in the near future. Several fundamental obstacles are not yet overcome and require imperative improvements, above all in vivo short- and long-term functional testing.

4.7 Conclusion

As the lung is a complex organ, advances in development of standardized techniques for production of decellularized and recellularized matrices are of great importance. Additionally, remarkable scientific and technical development has been performed in the area of bioreactors for the purpose of studying lung regenerative strategies. Cell selection, precise and appropriate bioreactor design, and accurate control of the volume and frequency of the perfusion during the tissue culture have decisive impact on the development of regeneration of the lung tissue. We are still in need of to design satisfactory methods to produce engineered lung tissues into the injured lung environment before realizing our dream of applying tissue-engineered lung organs in humans. Further studies are still required to understand how the remaining ECM and residual protein composition may affect seeded cell types over time. This will be an interesting line of future research to produce normal scaffold which may help revert the phenotype of cells obtained from diseased patients. While we move forward, it is important to keep in mind the complex anatomy of the lung organ and continue to improve our tissue engineering models to better benefit healthcare.

References

Ali NN, Edgar AJ, Samadikuchaksaraei A, Timson CM, Romanska HM, Polak JM et al (2002) Derivation of type II alveolar epithelial cells from murine embryonic stem cells. Tissue Eng 8(4):541–550

Azargoon A, Negahdari B (2018) Lung regeneration using amniotic fluid mesenchymal stem cells. Artif Cells, Nanomed, Biotechnol 46(3):447–451

Badylak SF, Gilbert TW (eds) (2008) Immune response to biologic scaffold materials. Seminars in immunology. Elsevier

Badylak SF, Taylor D, Uygun K (2011) Whole-organ tissue engineering: decellularization and recellularization of three-dimensional matrix scaffolds. Annu Rev Biomed Eng 13:27–53

Barkauskas CE, Cronce MJ, Rackley CR, Bowie EJ, Keene DR, Stripp BR et al (2013) Type 2 alveolar cells are stem cells in adult lung. J Clin Investig 123 (7):3025–3036

Booth AJ, Hadley R, Cornett AM, Dreffs AA, Matthes SA, Tsui JL et al (2012) Acellular normal and fibrotic human lung matrices as a culture system for in vitro investigation. Am J Respir Crit Care Med 186(9):866–876

Brown BN, Valentin JE, Stewart-Akers AM, McCabe GP, Badylak SF (2009) Macrophage phenotype and remodeling outcomes in response to biologic scaffolds with and without a cellular component. Biomaterials 30(8):1482–1491

Burgstaller G, Oehrle B, Gerckens M, White ES, Schiller HB, Eickelberg O (2017) The instructive extracellular matrix of the lung: basic composition and alterations in chronic lung disease. Eur Respir J 50 (1):1601805

Calle EA, Leiby KL, Raredon MB, Niklason LE (2017) Lung regeneration: steps toward clinical implementation and use. Curr Opin Anaesthesiol 30(1):23

Chang YS, Oh W, Choi SJ, Sung DK, Kim SY, Choi EY et al (2009) Human umbilical cord blood-derived mesenchymal stem cells attenuate hyperoxia-induced lung injury in neonatal rats. Cell Transplant 18 (8):869–886

Coraux C, Nawrocki-Raby B, Hinnrasky J, Kileztky C, Gaillard D, Dani C et al (2005) Embryonic stem cells generate airway epithelial tissue. Am J Respir Cell Mol Biol 32(2):87–92

Cortiella J, Nichols JE, Kojima K, Bonassar LJ, Dargon P, Roy AK et al (2006) Tissue-engineered lung: an in vivo and in vitro comparison of polyglycolic acid and pluronic F-127 hydrogel/somatic lung progenitor cell constructs to support tissue growth. Tissue Eng 12 (5):1213–1225

Crapo PM, Gilbert TW, Badylak SF (2011) An overview of tissue and whole organ decellularization processes. Biomaterials 32(12):3233–3243

Daly AB, Wallis JM, Borg ZD, Bonvillain RW, Deng B, Ballif BA et al (2012) Initial binding and recellularization of decellularized mouse lung scaffolds with bone marrow-derived mesenchymal stromal cells. Tissue Eng Part A 18(1–2):1–16

De Santis MM, Bölükbas DA, Lindstedt S, Wagner DE (2018) How to build a lung: latest advances and emerging themes in lung bioengineering. Eur Respir J 52(1)

Doryab A, Heydarian M, Amoabediny G, Sadroddiny E, Mahfouzi S (2017) Recellularization on acellular lung tissue scaffold using perfusion-based bioreactor: an online monitoring strategy. J Med Biol Eng 37(1):53–62

Douglas WH, Moorman GW, Teel RW (1976) The formation of histotypic structures from monodisperse fetal rat lung cells cultured on a three-dimensional substrate. Vitro 12(5):373–381

Fischer SN, Johnson JK, Baran CP, Newland CA, Marsh CB, Lannutti JJ (2011) Organ-derived coatings on electrospun nanofibers as ex vivo microenvironments. Biomaterials 32(2):538–546

Garcia O, Carraro G, Navarro S, Bertoncello I, McQualter J, Driscoll B et al (2012) Cell-based therapies for lung disease. Br Med Bull 101(1):147–161

Ghaedi M, Calle EA, Mendez JJ, Gard AL, Balestrini J, Booth A et al (2013) Human iPS cell–derived alveolar epithelium repopulates lung extracellular matrix. J Clin Investig 123(11):4950–4962

Ghaedi M, Le AV, Hatachi G, Beloiartsev A, Rocco K, Sivarapatna A et al (2018) Bioengineered lungs generated from human iPSCs-derived epithelial cells on native extracellular matrix. J Tissue Eng Regen Med 12(3):e1623–e1635

Gilpin SE, Ren X, Okamoto T, Guyette JP, Mou H, Rajagopal J et al (2014) Enhanced lung epithelial specification of human induced pluripotent stem cells on decellularized lung matrix. Ann Thorac Surg 98 (5):1721–1729

Gilpin SE, Charest JM, Ren X, Ott HC (2016) Bioengineering lungs for transplantation. Thorac Cardiovasc Surg 26(2):163–171

Goolaerts A, Pellan-Randrianarison N, Larghero J, Vanneaux V, Uzunhan Y, Gille T et al (2014) Conditioned media from mesenchymal stromal cells restore sodium transport and preserve epithelial permeability in an in vitro model of acute alveolar injury. Am J Physiol-Lung Cell Mol Physiol 306(11):L975–L985

Gotts JE, Matthay MA (2012) Mesenchymal stem cells and the stem cell niche: a new chapter. Am J Physiol-Lung Cell Mol Physiol 302(11):L1147–L1149

Hoganson DM, Pryor HI, Vacanti JP (2008) Tissue engineering and organ structure: a vascularized approach to liver and lung. Pediatr Res 63(5):520–526

Hoganson DM, Pryor HI II, Bassett EK, Spool ID, Vacanti JP (2011) Lung assist device technology with physiologic blood flow developed on a tissue engineered scaffold platform. Lab Chip 11(4):700–707

Huang SX, Islam MN, O'neill J, Hu Z, Yang Y-G, Chen Y-W et al (2014) Efficient generation of lung and airway epithelial cells from human pluripotent stem cells. Nat Biotechnol. 32(1):84

Jiang Y, Jahagirdar BN, Reinhardt RL, Schwartz RE, Keene CD, Ortiz-Gonzalez XR et al (2002) Pluripotency of mesenchymal stem cells derived from adult marrow. Nature 418(6893):41–49

Kajbafzadeh A-M, Payabvash S, Salmasi AH, Monajemzadeh M, Tavangar SM (2006) Smooth muscle cell apoptosis and defective neural development in congenital ureteropelvic junction obstruction. J Urol 176(2):718–723

Kajbafzadeh A, Sabetkish N, Sabetkish S, Tavangar S, Beigi RH, Talebi M et al (2015a) Lung tissue engineering and preservation of alveolar microstructure using a novel casting method. Biotech Histochem 90(2):111–123

Kajbafzadeh AM, Khorramirouz R, Akbarzadeh A, Sabetkish S, Sabetkish N, Saadat P et al (2015b) A novel technique for simultaneous whole-body and multi-organ decellularization: umbilical artery catheterization as a perfusion-based method in a sheep foetus model. Int J Exp Pathol 96(2):116–132

Khorramirouz R, Sabetkish S, Akbarzadeh A, Muhammadnejad A, Heidari R, Kajbafzadeh A-M (2014) Effect of three decellularisation protocols on the mechanical behaviour and structural properties of sheep aortic valve conduits. Adv Med Sci 59(2):299–307

Kreuter M, Witt S, Polke M, Waelscher J, Schwarzkopf L (2018) Financial burden of interstitial lung disease-a claims data based study. C41. Health services research in pulmonary disease. American Thoracic Society. p A4927-A

Kuttan R, Spall R, Duhamel R, Sipes I, Meezan E, Brendel K (1981) Preparation and composition of alveolar extracellular matrix and incorporated basement membrane. Lung 159(1):333–345
Macchiarini P, Jungebluth P, Go T, Asnaghi MA, Rees LE, Cogan TA et al (2008) Clinical transplantation of a tissue-engineered airway. Lancet 372 (9655):2023–2030
Majka SM, Beutz MA, Hagen M, Izzo AA, Voelkel N, Helm KM (2005) Identification of novel resident pulmonary stem cells: form and function of the lung side population. Stem Cells 23(8):1073–1081
Mandrycky C, Phong K, Zheng Y (2017) Tissue engineering toward organ-specific regeneration and disease modeling. MRS Commun 7(3):332–347
McAuley DF, Curley GF, Hamid UI, Laffey JG, Abbott J, McKenna DH et al (2014) Clinical grade allogeneic human mesenchymal stem cells restore alveolar fluid clearance in human lungs rejected for transplantation. Am J Physiol-Lung Cell Mol Physiol 306(9):L809–L815
Mendez JJ, Ghaedi M, Steinbacher D, Niklason LE (2014) Epithelial cell differentiation of human mesenchymal stromal cells in decellularized lung scaffolds. Tissue Eng Part A 20(11–12):1735–1746
Mondrinos MJ, Koutzaki S, Jiwanmall E, Li M, Dechadarevian J-P, Lelkes PI et al (2006) Engineering three-dimensional pulmonary tissue constructs. Tissue Eng 12(4):717–728
Neuman RE, Logan MA (1950) The determination of hydroxyproline. J Biol Chem 184(1):299–306
Nichols JE, Niles J, Riddle M, Vargas G, Schilagard T, Ma L et al (2013) Production and assessment of decellularized pig and human lung scaffolds. Tissue Eng Part A 19(17–18):2045–2062
O'Neill JD, Anfang R, Anandappa A, Costa J, Javidfar J, Wobma HM et al (2013) Decellularization of human and porcine lung tissues for pulmonary tissue engineering. Ann Thorac Surg 96(3):1046–1056
Orlando G, Farney AC, Iskandar SS, Mirmalek-Sani S-H, Sullivan DC, Moran E et al (2012) Production and implantation of renal extracellular matrix scaffolds from porcine kidneys as a platform for renal bioengineering investigations. Ann Surg 256(2):363–370
Ott HC, Clippinger B, Conrad C, Schuetz C, Pomerantseva I, Ikonomou L et al (2010) Regeneration and orthotopic transplantation of a bioartificial lung. Nat Med 16(8):927
Panoskaltsis-Mortari A (2015) Bioreactor development for lung tissue engineering. Current Transplant Rep 2 (1):90–97
Petersen TH, Calle EA, Zhao L, Lee EJ, Gui L, Raredon MB et al (2010) Tissue-engineered lungs for in vivo implantation. Science 329(5991):538–541
Raredon MSB, Ghaedi M, Calle EA, Niklason LE (2015) A rotating bioreactor for scalable culture and differentiation of respiratory epithelium. Cell Med 7 (3):109–121
Ren X, Moser PT, Gilpin SE, Okamoto T, Wu T, Tapias LF et al (2015) Engineering pulmonary vasculature in decellularized rat and human lungs. Nat Biotechnol 33(10):1097
Rippon HJ, Polak JM, Qin M, Bishop AE (2006) Derivation of distal lung epithelial progenitors from murine embryonic stem cells using a novel three-step differentiation protocol. Stem Cells 24(5):1389–1398
Rosen C, Shezen E, Aronovich A, Klionsky YZ, Yaakov Y, Assayag M et al (2015) Preconditioning allows engraftment of mouse and human embryonic lung cells, enabling lung repair in mice. Nat Med 21 (8):869
Samadikuchaksaraei A, Cohen S, Isaac K, Rippon HJ, Polak JM, Bielby RC et al (2006) Derivation of distal airway epithelium from human embryonic stem cells. Tissue Eng 12(4):867–875
Sato M, Okada Y, Oto T, Minami M, Shiraishi T, Nagayasu T et al (2014) Registry of the Japanese society of lung and heart-lung transplantation: official Japanese lung transplantation report, 2014. Gen Thorac Cardiovasc Surg 62(10):594–601
Scarritt ME, Pashos NC, Bunnell BA (2015) A review of cellularization strategies for tissue engineering of whole organs. Front Bioeng Biotechnol 3:43
Shannon JM, Hyatt BA (2004) Epithelial-mesenchymal interactions in the developing lung. Annu Rev Physiol 66:625–645
Shojaie S, Ermini L, Ackerley C, Wang J, Chin S, Yeganeh B et al (2015) Acellular lung scaffolds direct differentiation of endoderm to functional airway epithelial cells: requirement of matrix-bound HS proteoglycans. Stem Cell Rep 4(3):419–430
Sokocevic D, Bonenfant NR, Wagner DE, Borg ZD, Lathrop MJ, Lam YW et al (2013) The effect of age and emphysematous and fibrotic injury on the recellularization of de-cellularized lungs. Biomaterials 34(13):3256–3269
Sueblinvong V, Loi R, Eisenhauer PL, Bernstein IM, Suratt BT, Spees JL et al (2008) Derivation of lung epithelium from human cord blood–derived mesenchymal stem cells. Am J Respir Crit Care Med 177(7):701–711
Sullivan DC, Mirmalek-Sani S-H, Deegan DB, Baptista PM, Aboushwareb T, Atala A et al (2012) Decellularization methods of porcine kidneys for whole organ engineering using a high-throughput system. Biomaterials 33(31):7756–7764
Takahashi K, Chen F, Ikeda T, Doi H, Nakamura H, Yanagi S et al (2013) Single-lobe lung transplantation for rapidly deteriorating pulmonary venoocclusive disease. Ann Thorac Surg 95(2):689–691
Tebyanian H, Karami A, Nourani MR, Motavallian E, Barkhordari A, Yazdanian M et al (2019) Lung tissue engineering: an update. J Cell Physiol 234(11):19256–19270
Tudorache I, Cebotari S, Sturz G, Kirsch L, Hurschler C, Hilfiker A et al. (2007) Tissue engineering of heart valves: biomechanical and morphological properties of decellularized heart valves. J Heart Valve Dis 16 (5):567–73. discussion 74

Vranken BV, Romanska H, Polak J, Rippon H, Shannon J, Bishop A (2005) Coculture of embryonic stem cells with pulmonary mesenchyme: a microenvironment that promotes differentiation of pulmonary epithelium. Tissue Eng 11(7–8):1177–1187

Wagner DE, Bonvillain RW, Jensen T, Girard ED, Bunnell BA, Finck CM et al (2013) Can stem cells be used to generate new lungs? Ex vivo lung bioengineering with decellularized whole lung scaffolds. Respirology 18(6):895–911

Wagner DE, Bonenfant NR, Parsons CS, Sokocevic D, Brooks EM, Borg ZD et al (2014a) Comparative decellularization and recellularization of normal versus emphysematous human lungs. Biomaterials 35 (10):3281–3297

Wagner DE, Bonenfant NR, Sokocevic D, DeSarno MJ, Borg ZD, Parsons CS et al (2014b) Three-dimensional scaffolds of acellular human and porcine lungs for high throughput studies of lung disease and regeneration. Biomaterials 35(9):2664–2679

Wallis JM, Borg ZD, Daly AB, Deng B, Ballif BA, Allen GB et al (2012) Comparative assessment of detergent-based protocols for mouse lung decellularization and re-cellularization. Tissue Eng Part C Methods 18(6):420–432

Weibel ER (2017) Lung morphometry: the link between structure and function. Cell Tissue Res 367(3):413–426

Wong AP, Keating A, Lu W-Y, Duchesneau P, Wang X, Sacher A et al (2009) Identification of a bone marrow–derived epithelial-like population capable of repopulating injured mouse airway epithelium. J Clin Investig 119(2):336–348

Wong AP, Bear CE, Chin S, Pasceri P, Thompson TO, Huan L-J et al (2012) Directed differentiation of human pluripotent stem cells into mature airway epithelia expressing functional CFTR protein. Nat Biotechnol 30(9):876

Zhou H, Kitano K, Ren X, Rajab TK, Wu M, Gilpin SE et al (2018) Bioengineering human lung grafts on porcine matrix. Ann Surg 267(3):590–598

5 Cardiac Extracellular Matrix as a Platform for Heart Organ Bioengineering: Design and Development of Tissue-Engineered Heart

Aram Akbarzadeh, Shabnam Sabetkish, and Abdol-Mohammad Kajbafzadeh

Abstract

The field of tissue engineering and regenerative medicine is able to depict the mechanism of cardiac repair and development of cardiac function as well, in order to reveal findings to new therapeutic designs for clinical treatment. The foremost approach of this scientific field is the fabrication of scaffolds, which contain cells that can be used as cardiac grafts in the body, to have the preferred recovery. Cardiac tissue engineering has not been completely organized for routine clinical usages. Hence, engineering innovations hold promise to character research and treatment options in the years to come. Our group has extensive experience with regard to the structure of the heart, which makes us to our decision to continue with the preparation of heart, with the aim of developing a new ECM scaffold. Herein, we aim to assess the state-of-the-art fabrication methods, advances in decellularization and recellularization techniques. We also highlight the major achievements toward the production of a bioengineered heart obtained from decellularization and recellularization techniques.

Keywords

Decellularization · Recellularization · Heart matrices · Scaffold · Cardiac stem cells

5.1 History

Heart is among the organs with least regenerative capacity, and cardiomyocytes (CMs) are susceptible to damage by several factors, such as necrosis, apoptosis, and oncosis (or ischemic cell death), culminating in heart failure (Heallen and Martin 2018; Mohamed et al. 2018). Myocardial infarction causes scar tissue, regions where CMs are replaced with fibrillar collagen and/or fibroblast-like cells (Frangogiannis 2016). About 38 million people globally were affected by heart failure; as of 2017 (Tzahor and Poss 2017), about 6.5 million of those are in the USA (Benjamin et al. 2017). According to World Health Organization (WHO), cardiovascular diseases are still the leading cause of mortality with a rate of 23 million new cases diagnosed universal every year (Bui et al. 2011). Such diseases can result in irreversible damages to the heart tissue that usually leads to heart failure, with a decrease

A. Akbarzadeh · S. Sabetkish ·
A.-M. Kajbafzadeh (✉)
Pediatric Urology and Regenerative Medicine Research Center, Section of Tissue Engineering and Stem Cells Therapy, Children's Medical Center, Pediatric Center of Excellence, Tehran University of Medical Sciences, No. 62, Dr. Gharib's Street, Keshavarz Boulevard, 1419433151 Tehran, Iran
e-mail: kajbafzd@sina.tums.ac.ir

A. Kajbafzadeh (ed.), *Decellularization Methods of Tissue and Whole Organ in Tissue Engineering*, Advances in Experimental Medicine and Biology 1345,
https://doi.org/10.1007/978-3-030-82735-9_5

in contractile capacity below a critical threshold (Chaudhry 2019). Therefore, the heart is one of the most essential subjects for tissue engineering research. Currently, despite abundant efforts to progress options for cardiac damage treatment, there is no effective therapy for heart failure, except heart transplantation; however, due to the invasive nature of the surgery and the shortage of organ donors, it is appropriate for a limited cohort of patients. Besides, impediments of state-of-the-art immunotherapeutic drugs and high risk of rejection limit the option of healing.

The tissue engineering and regenerative medicine techniques show enormous prospective as alternative options that produce constructs for repairing or replacing cardiovascular tissues (Kharaziha et al. 2016; Cutts et al. 2015; Tijore et al. 2018).

In this technology, we will focus on four important issues of (1) scaffold material selection; (2) scaffold material production; (3) cell selection; and (4) cell culture. Fiber production methods, such as electrospinning (Gabriel et al. 2017; Rockwood et al. 2008) and rotary-jet spinning, (Cardoso et al. 2014) as well as cell sheet engineering (Shimizu et al. 2003), are among the techniques that have been investigated in order to create grafts to be implanted in the heart. Besides that, the most efficient and recent approach is decellularization, aiming to obtain three-dimensional structures that not only may regenerate the existing heart, but be used to create an entire bioartificial organ. Firstly, it is essential to identify the best scaffold for cardiac regeneration. Some desired properties are adjustable degradation rates, good porosity, biocompatibility, hemocompatibility, and good cell adhesion, mechanical and elastic properties compatible with the natural heart (McDevitt et al. 2003; Baheiraei et al. 2014). The second most important point is to select the most promising technique to construct the scaffold in which cells are going to be seeded before the implantation. It has been considered that ischemic or damaged heart can be repaired by using decellularized scaffolds as an appropriate modality to deliver cardiac stem cells to the tissue, create a functional tissue substitute, and restore cardiac function after MI. However, the application of an appropriate extracellular matrix (ECM) as an appropriate suitable microenvironment for cells should be explored in order to overcome probable complications after transplantation and also to increase cell survival.

This chapter will outline the progress to date recorded for approaches of converse the heart as a subject for tissue engineering paradigm, discuss about the recent developments made in the fields of cardiac tissue engineering and stem cells, as well as emphasize the challenges which we may confront with when applying such constructs in a clinical setting.

In the next sections of this chapter, several techniques of decellularization and recellularization approaches will be introduced and discussed followed by methods for scaffold fabrication. Updates of upcoming and ongoing heart tissue engineering applications will be then broadly covered.

5.2 Materials and Methods

Decellularization is a process that consists of removing all cells from tissues or organs while preserving the extra cellular matrix (ECM) structure via different physical, chemical and enzymatic methods. Triton X-100, as a nonionic detergent, may affect lipid–lipid and lipid–protein interactions. However, this detergent can keep the proteins within an organ in a functional conformation (Wang et al. 2017). Anionic detergents including sodium deoxycholate (SDC) and sodium dodecyl sulfate (SDS) can be also used for the complete removal of nuclear remnants and cytoplasmic proteins. This detergent preserves the structure of the natural tissue while reducing GAG concentration and collagen integrity (Sabetkish et al. 2015). Enzyme such nucleases (DNases or RNases) are also able to reduce nucleotides after cell lysis (Moore et al. 1997). Ethylenediaminetetraacetic acid (EDTA) and ethylene glycol tetraacetic acid (EGTA) are among the non-enzymatic agents, which are able to detach cells from ECM (O'Connor Mooney et al. 2016). It has been demonstrated that the application of trypsin results in damage to the ECM components after decellularization process (Grauss et al. 2003).

The ECM is composed of functional and structural proteins such as collagen, elastin, laminin, fibronectin, proteoglycans and many other glycoproteins which should be preserved during the decellularization process (Laurie et al. 1989; Young et al. 2019). There are two kinds of processes containing static- and perfusion-based decellularization and recellularization methods. However, perfusion-based technique has been shown to be more efficient in maintaining the three-dimensional structure of tissues/organs while removing the cells with a more even distribution of decellularization agents (Tapias and Ott 2014; Keane et al. 2015). This perfusion-based technique has been the most commonly applied for whole heart bioengineering, owing in part to the anatomical complexity of the macro- and microanatomy of the heart organ, through the decellularization approach.

The ECM plays a crucial role in normal cardiac functioning and homeostasis and cellular behavior. Ideally, the scaffolds should faultlessly mimic natural cardiac ECM structures and present a physiological microenvironment for cells. The cardiac ECM consists of a compound arrangement of proteins, of which three-dimensional scaffolds have been created from decellularized cardiac ECM. Natural scaffolds play a crucial role in anchoring cells to produce functional tissues (Bhutani et al. 2018; Shevach et al. 2014; Martinelli et al. 2018; Huang et al. 2019). These decellularized scaffolds serve as a framework material for proliferation and differentiation of the desired tissue. Carrier substances facilitate cells to fabricate the ECM that holds growth factors in cardiac remodeling and renovate (Dolan et al. 2019; Neto et al. 2019; Mewhort et al. 2017). In the same way, scaffolds as porous matrices form a biomimetic ECM which promotes cell adhesion and differentiation, as well as 3D organotypic cultures. These scaffolds also act as a substitute for missing tissues/organs in the body (Liu et al. 2019; Wade et al. 2015). Typically, biomaterials for tissue engineering are synthesized or modified from primary natural materials. These biomaterials include polyglycolic acid (PGA) (Bruder et al. 2018), poly(L)-lactic acid (PLA) (Muniyandi et al. 2020; Tomecka et al. 2017; Flaig et al. 2020), poly(DL) glycolate (PLGA) (Martins et al. 2018; Bertuoli et al. 2019). Collagens, alginate, chitosan, fibrin and hyaluronic acids are among the natural biomaterials.

In cell sheet engineering, temperature-responsive polymer surfaces are used to facilitate the controlled release of cell monolayers; free-floating sheet of cohesive cells to be placed onto the epicardium (Haraguchi et al. 2014). This scaffold-free technology can be applied to all cell types that are competent of shaping cardiomyocytes for contractile maintenance and non-myocytes for the delivery of secreted factors (Matsuura et al. 2007; Gao et al. 2019).

A suitable and applicable scaffold for cardiac regeneration is required to sustain tissue reconstruction by active support for cell-to-tissue procedures by supporting cell–cell adhesion, proliferation and differentiation. Foremost technical progression in the field of cardiac tissue engineering is the ability to fabricate a physical framework of biocompatible resources and the control of mechanical characteristics, which can be efficiently used clinically.

Several investigations such as transthoracic echocardiography, scanning electron microscopy (SEM) (Hilbert et al. 2004; Kasimir et al. 2005), histological (hematoxylin–eosin (H&E) and Masson's trichrome) and immunohistochemical examination, DAPI staining, DNA quantification, mechanical properties, hydroxyproline assay, and 2D electrophoresis are used to evaluate the efficacy of the decellularization process. Movat pentachrome staining can be used to demonstrate the ECM components such as collagen, elastin and GAGs. Cytotoxicity assay, metabolic activity and viability tests (MTS assay) are among other valuable tests that should be performed after heart valve decellularization. The aortic heart valve architecture has a naturally three-layered arrangement including the lamina ventricularis, lamina spongiosa and fibrosa. The above-mentioned investigations can afford critical data on the effective cellular removal as well as the biological and structural properties of the decellularized matrix intended to seed.

Prior to recellularization, it is essential to ensure that the decellularized scaffolds are effectively sterilized to avoid cross-contamination and eliminate the risk of infection. Ethylene oxide, gamma irradiation and electron beam irradiation are among the sterilization techniques used in conventional medical implants. Nevertheless, these sterilization techniques may change the mechanical properties of the scaffolds and may also cause adverse immune response (Bonenfant et al. 2013).

5.3 Cell Seeding

Stem cell transplantation strategy, which can enhance tissue perfusion, angiogenesis, and preserve or regenerate myocardial tissue, has been proved to enhance cardiac function in patients with sophisticated heart failure after MI (Suncion et al. 2014; Xu et al. 2014; Yau et al. 2019). This technology was first applied to treat MI in 2001 with promising and encouraging results.

To date, autologous and allogeneic adult stem cell transplants had promising results in cardiac treatments in some reported cases (Sanz-Ruiz and Fernández-Avilés 2018; Barker et al. 2018). In current techniques of stem cell transplantation, cells are seeded onto 3D polymer scaffolds after electrical, mechanical or chemical stimulation such as heparin and hyaluronic acid to promote the differentiation of stem cells and restore the function of injured heart tissues (Hirt et al. 2014; Aslani et al. 2020; Kenar et al. 2019; Shiekh et al. 2018). However, due to limitations in the usage of stem cell-based therapies for human heart failure, immune tolerance and growth of stem cells on novel biomaterials have recently been considered as a capable approach for cardiac repair (Shiekh et al. 2018; Li et al. 2016).

Captivatingly, it has been confirmed that new CMs are able to arise from presented CMs and progenitor or stem cells early on periods of embryo growth (Yoon et al. 2018; Sereti et al. 2018; Malandraki-Miller et al. 2018; Radisic et al. 2006; Allegue et al. 2011). Cardiac stem cells (CSCs) (Rikhtegar et al. 2019; Su et al. 2018; Tang et al. 2017), embryonic stem cells (Alagarsamy et al. 2019; Wang et al. 2011), bone marrow-derived mesenchymal stem cells such as mesenchymal, endothelial and hematopoietic stem/ progenitor cells (Blondiaux et al. 2017; Joshi et al. 2018), cord-derived mesenchymal stem cells (Lim et al. 2018; Wu et al. 2018; Pushp et al. 2020; Zhang et al. 2019; Mao et al. 2017), and adipose tissue (ASC)-derived mesenchymal cells (Tang et al. 2016) are indispensable cell sources used in cell transplantation for research associated with MI.

Differentiation of stem-cell-derived CMs into the preferred lineages needs numerous features of the scaffold assembles, and cell's fate and environment (Richards et al. 2016; Hansen et al. 2018; Birket et al. 2015; Hosoyama et al. 2018; Maiullari et al. 2018). Human iPSCs (hiPSCs) have been showed to differentiate successfully into mature CMs with optimal protocols, which can be a probable advance toward heart regeneration methods. Fetal hiPSCs can be differentiated into pure CMs as well. Cardiac fibroblasts, embryonic stem cells (ESCs), and muscle cells can potentially be replaced for CMs for cardiovascular diseases.

The route of cell delivery is another critical subject in optimizing cardiomyoplasty. Intramyocardial injection has been investigated via sternotomy (Mathiasen et al. 2012), the endomyocardial route (Hashemi et al. 2008), and the intracoronary route (Revilla et al. 2011). The in vitro cell culture of the selected cell types is performed in specialized cell culture facilities, to encourage increased cellular proliferation, differentiation and maturation. The use of cell bioreactors, for the purpose of improving, refining and optimizing the quality and expansion of the cell itself has been recently taken into consideration. Bioreactors are considered as systems with controlled conditions and parameters that facilitate the stimulation of cell growth (Paez-Mayorga et al. 2019). The most competent technology to offer the proliferation and differentiation of these cells is the bioreactor.

In our center, we were able to produce a biocompatible heart scaffold with comparative histological and biomechanical properties of native cardiac ECM, using a perfusion-based decellularization method. In our recent study, we limited low transplanted cell retention and survival within the ischemic tissue by using decellularized pericardium patch in an animal model of MI. We also assessed the hypothesis that tissue-engineered pericardial patch containing autologous ADMSC would be beneficial for the treatment of MI with desirable properties in a rabbit model compared to the application of non-seeded decellularized pericardium (Kajbafzadeh et al. 2017). We also demonstrated that decellularized human internal mammary artery could be applied as a resourceful small-diameter vascular alternate with high patency. This decellularized internal mammary artery was considered as a novel vascular graft for small-diameter bypass surgeries (Kajbafzadeh et al. 2019). In another study, we demonstrated the efficacy of ADMSC-seeded human amniotic membrane cardiac patches as scaffolds for treatment of acute MI in rat models (Khorramirouz et al. 2019). Pre-seeded decellularized aortic valve conduit with bone marrow-derived MSCs depicted satisfactory outcomes in postoperative cell seeding capabilities with promising functional potentiality, which provides a new era of biological grafts in cardiovascular surgery (Kajbafzadeh et al. 2016). Advantages and disadvantages of different implanted cells are depicted in Table 1. An overview of the heart decellularization and recellularization literature is provided in Table 2 (Mirsadraee et al. 2006; Singelyn et al. 2012; Wainwright et al. 2010; Weymann et al. 2011; Akhyari et al. 2011; Oberwallner et al. 2014; Leyh et al. 2003; Grauss et al. 2005; Dainese et al. 2012; Malone et al. 1984; Akbarzadeh et al. 2019).Some of the most commonly used protocols of heart organ decellularization and recellularization processes

5.4 Clinical Applications

The first clinical implantation of a tissue-engineered heart valve was carried out in 2000. An allograft pulmonary heart valve was decellularized and underwent the cell seeding process in bioreactor. In the next step, the decellularized scaffold was implanted in a 43-year-old man. The neo-aortic heart valve demonstrated appropriate function in different follow-ups with no evidence of regurgitation (Hoerstrup et al. 2000). In the study of Cebotari et al., pulmonary heart valves were decellularized with trypsin/EDTA and reseeded with peripheral mononuclear cells that were isolated from human blood. The scaffolds were implanted into two pediatric patients affecting congenital pulmonary valve failure. They obtained promising postoperative results with no degenerative signs (Cebotari et al. 2006).

In clinical studies, the concerns of histocompatibility of regenerated cardiac cells and stem cell-derived pro-arrhythmic substrates (Chen et al. 2018) have restricted the application of stem cell-based therapies for human heart failure. Recent clinical studies showed that cell sheet technology improved the ejection fraction, regenerated the dysfunctional cardiac wall, increased vasculargenesis, and diminished fibrosis in heart disease models (Sawa et al. 2012; Sawa and Miyagawa 2013; Miyagawa et al. 2017; Yoshikawa et al. 2018; Yamamoto et al. 2019). From 2001, some clinical studies have indicated that stem cells are safe and demonstrate few treatment-related complications compared to control groups (Jackson et al. 2001; Segers and Lee 2008). However, the clinical use of tissue-engineered constructs in myocardial regeneration is still at an early phase. Most of the clinical studies over decellularized xenograft heart valves suggested for investigating the presentation of decellularized xenograft heart valves in human to conquer the challenge that allograft and homograft heart valves are in short supply, especially for pediatric population.

5.5 Limitations

Despite valuable tissue engineering approaches which may improve cell or tissue preservation, the difficulties with sources of autologous cell and survival in the host tissue still remain challenging (Naderi et al. 2011). In addition, the quality and number of cells, comorbidities,

Table 1 Some of the most commonly used protocols of heart organ decellularization and recellularization processes

Species	Method of decellularization/Recellularization	Results	Reference
Human pericardium from cadaveric donors	Decellularization: Hypotonic buffer, SDS in hypotonic buffer, and nuclease solution Recellularization: In-vitro seeding of human dermal fibroblasts and A549 cells	Promising results in glycosaminoglycan content and mechanical properties	88
Porcine ventricular myocardial tissue	Decellularization: SDS and Triton X-100. Pepsin-solubilization of the myocardial matrix Recellularization: In-vitro seeding of neonatal rat cardiomyocytes and in-vivo injection in left ventricle of rat models	Preserved glycosaminoglycan content and satisfactory cell-conductivity	89
Whole adult porcine heart	Decellularization: Aortic perfusion. Serial perfusion of enzymatic, non-ionic and ionic detergent, hypotonic and hypertonic solutions Recellularization: In-vitro seeding of chicken cardiomyocyte	Preserved collagen, elastin, and glycosaminoglycans, and mechanical integrity	90
Porcine whole heart	Decellularization: Perfusion of Trypsin/EDTA and TritonX 100/deoxycholic acid (DCA) Recellularization: none	Retained collagen, proteoglycan and elastin	91
Adult rat heart	Decellularization: 1) SDS/TritonX100-based v/s 2) Trypsin plus Triton/DCA-based v/s 3) SDS/DCA/saponin-based Recellularization: Reseeding with C2C12 myoblasts in-vitro	Detection of Laminin in all groups. Collagen IV removed in group 2, No elastin detection in group 3	92
Human Left ventricular myocardium tissue	Decellularization: SDS-based, Triton X-100-based, DCA-based, hypo/hypertonic solution-based decellularization protocols Recellularization: In-vitro culture with mesenchymal stem cells, iPS-derived cardiomyocytes and native neonatal mouse cardiomyocytes	Cell viability and growth in both protocols. More satisfactory cell removal and ECM architecture maintenance with SDS-based protocol	93
Porcine and sheep pulmonary valve conduits	Decellularization: Trypsin/EDTA digestion Recellularization: Orthotopic implantation in sheep	Reconstitution of surface endothelial cell monolayer and interstitial myofibroblasts. Calcifications were also noted	94
Porcine aortic valves	Decellularization: Triton X-100 v/s Trypsin Recellularization: In-vitro EC seeding	Changes in the extracellular matrix constitution in all methods, EC-mediated ECM deposition.	95
Aortic homograft leaflets	Decellularization: Trypsin Recellularization: In-vitro seeding with cardiac mesenchymal stromal cell	Rescuing most of the original cell density and differentiation towards endothelial lineage	96
Dog arterial segment	1° detergent step with Triton X-100, 6 h at room temperature • Protease inhibitor step	The results of allogeneic implant depicted well incorporated tissue appearance with complete endothelial layer after 90 d post-implantation	97

(continued)

Table 1 (continued)

Species	Method of decellularization/Recellularization	Results	Reference
	• 2° detergent step with SDS, 72 h at room temperature • Washing step with ethanol • Fixation treatment with carbodiimide; Detergents concentration not mentioned		
Ovine heart	perfusion with a 1% SDS in distilled water for 72 h at room temperature/ 1% Triton X-100 in distilled water for 24 h Recellularization: In-vivo implantation of decellularized matrix scaffold into the omentum of rats	Preserved the structure and composition of cardiac ECM and vascular structures within the scaffold without residual cellular components Implantation led to proper vascularization	98

Table 2 Advantages and Disadvantages of Implanted Cells

Cell type	Advantages	Disadvantages
Skeletal myoblasts	Easily isolated/High rate of proliferation/Hypoxia-resistant/Autologous	High occurrence of arrhythmias
Bone marrow-derived stem cells	Autologous/Easily isolated/Multipotent/Low immune response	Restricted accessibility/bone or cartilage formation in the myocardium
Adipose tissue-derived stem cells	Easily isolated/High availability Multipotent/Low immune response	Low survival
Cardiac stem cells	Multipotent/Autologous	Inadequate accessibility
Embryonic stem cells	Pluripotent/straightforward to develop	Teratogenic/Limited availability/Host immune response/Ethical problems
iPSC	Pluripotent/Easy to expand/Superior availability/Autologous	Potentially teratogenic/Possible oncogenic potential
Fetal cardiomyocytes	Cardiomyocyte phenotype	Limited availability Low survival Host immune response Ethical problems

iPSC, induced pluripotent stem cells

genetic defects, and gender are among the factors that affect the cell/tissue survival by the host tissue environment (Perrino et al. 2020). Other drawback is the high costs of superior therapy medicinal products in general as well as the failure of some scaffolds to convene translationally appropriate requirements. Remarkable inflammation, foreign body reaction, and arrhythmogenic potential are other limitations that commonly occur in long-term follow-ups after scaffold transplantation, discouraging the therapeutic effects (Shimizu et al. 2001; Christman and Lee 2006). These drawbacks should be investigated and completely addressed before clinical applications.

Despite several progressions in the field of heart tissue engineering, the capability and significance of adult mammalian cardiomyocytes and CSCs regeneration remain controversial (Aquila et al. 2018; Kretzschmar et al. 2018; Lee 2018). In addition, although human ESC-derived CMs have been considered as principal supply of adult human cardiac myocyte for medical beneficials, being well-organized and distributed, and functional transverse tubules (T-tubules) are among the essential features that still lack (Parikh et al. 2017).

Issues regarding cell sheet engineering technology are the limited number of sheets which can be stacked on each other without cell death and the weakness of these sheets which may ground their folding or tearing during manipulations (Zurina et al. 2020).

5.6 Conclusion

In this chapter, we discussed many essential achievements associated with tissue engineering and regenerative medicine technology for cardiac repair. The heart is tremendously compound organ, and the scaffold material selection, scaffold material production, cellular selection and sell seeding process both in vitro and in vivo are among many variables that can influence its regeneration. These techniques generally focus on the scaffold material selection, scaffold material production, cellular selection and cellular cultivation in vitro. With the progress of tissue engineering technique for heart organ, increasing stem cell-derived methods have already been studied in basic research and clinical trials. The presence of CSC population in adult hearts is still contentious; however, differentiating other stem cells into mature cardiomyocytes is of great importance in cardiac therapies. Due to progressive improvements regarding cardiac tissue engineering, we believe that the promising applications of stem cell-derived cell therapy in MI will be increasingly attracted in the next decade. However, more studies remain to be performed to better understand and explain the challenges, improve existing techniques and develop new techniques, protocols and methods. The combination of three-dimensional scaffolds, bioreactors and excellent stem cells can pave the road for the development of the next-generation human organ.

References

Akbarzadeh A, Khorramirouz R, Ghorbani F, Beigi RSH, Hashemi J, Kajbafzadeh AM (2019) Preparation and characterization of human size whole heart for organ engineering: scaffold microangiographic imaging. Regen Med 14(10):939–954. https://doi.org/10.2217/rme-2018-0111

Akhyari P, Aubin H, Gwanmesia P, Barth M, Hoffmann S, Huelsmann J et al (2011) The quest for an optimized protocol for whole-heart decellularization: a comparison of three popular and a novel decellularization technique and their diverse effects on crucial extracellular matrix qualities. Tissue Eng Part C Methods 17(9):915–926. https://doi.org/10.1089/ten.TEC.2011.0210

Alagarsamy KN, Yan W, Srivastava A, Desiderio V, Dhingra S (2019) Application of injectable hydrogels for cardiac stem cell therapy and tissue engineering. Rev Cardiovasc Med 20(4):221–230. https://doi.org/10.31083/j.rcm.2019.04.534

Allegue C, Gil R, Blanco-Verea A, Santori M, Rodríguez-Calvo M, Concheiro L et al (2011) Prevalence of HCM and long QT syndrome mutations in young sudden cardiac death-related cases. Int J Legal Med 125(4):565–572. https://doi.org/10.1007/s00414-011-0572-7

Aquila I, Marino F, Cianflone E, Marotta P, Torella M, Mollace V et al (2018) The use and abuse of Cre/Lox recombination to identify adult cardiomyocyte renewal rate and origin. Pharmacol Res 127:116–128. https://doi.org/10.1016/j.phrs.2017.06.012

Aslani S, Kabiri M, HosseinZadeh S, Hanaee-Ahvaz H, Taherzadeh ES, Soleimani M (2020) The applications of heparin in vascular tissue engineering. Microvasc Res 104027

Baheiraei N, Yeganeh H, Ai J, Gharibi R, Azami M, Faghihi F (2014) Synthesis, characterization and antioxidant activity of a novel electroactive and biodegradable polyurethane for cardiac tissue engineering application. Mater Sci Eng, C Mater Biol Appl 44:24–37. https://doi.org/10.1016/j.msec.2014.07.061

Barker RA, Carpenter MK, Forbes S, Goldman SA, Jamieson C, Murry CE et al (2018) The challenges of first-in-human stem cell clinical trials: what does this mean for ethics and institutional review boards? Stem Cell Reports. 10(5):1429–1431

Benjamin EJ, Blaha MJ, Chiuve SE, Cushman M, Das SR et al (2017) Heart disease and stroke statistics—2017 update: a report from the American heart association. Circulation. 135(10):e146

Bertuoli PT, Ordoño J, Armelin E, Pérez-Amodio S, Baldissera AF, Ferreira CA et al (2019) Electrospun conducting and biocompatible uniaxial and Core-Shell fibers having poly (lactic acid), poly (ethylene glycol), and polyaniline for cardiac tissue engineering. ACS Omega 4(2):3660–3672

Bhutani S, Nachlas AL, Brown ME, Pete T, Johnson CT, García AJ et al (2018) Evaluation of hydrogels presenting extracellular matrix-derived adhesion peptides and encapsulating cardiac progenitor cells for cardiac repair. ACS Biomater Sci Eng 4(1):200–210

Birket MJ, Ribeiro MC, Kosmidis G, Ward D, Leitoguinho AR, van de Pol V et al (2015) Contractile defect caused by mutation in MYBPC3 revealed under conditions optimized for human PSC-cardiomyocyte function. Cell Rep 13(4):733–745. https://doi.org/10.1016/j.celrep.2015.09.025

Blondiaux E, Pidial L, Autret G, Rahmi G, Balvay D, Audureau E et al (2017) Bone marrow-derived mesenchymal stem cell-loaded fibrin patches act as a reservoir of paracrine factors in chronic myocardial infarction. J Tissue Eng Regen Med 11(12):3417–3427. https://doi.org/10.1002/term.2255

Bonenfant NR, Sokocevic D, Wagner DE, Borg ZD, Lathrop MJ, Lam YW et al (2013) The effects of storage and sterilization on de-cellularized and re-cellularized whole lung. Biomaterials 34(13):3231–3245. https://doi.org/10.1016/j.biomaterials.2013.01.031

Bruder L, Spriestersbach H, Brakmann K, Stegner V, Sigler M, Berger F et al (2018) Transcatheter decellularized tissue-engineered heart valve (dTEHV) grown on polyglycolic acid (PGA) scaffold coated with P4HB shows improved functionality over 52 weeks due to polyether-ether-ketone (PEEK) insert. J Funct Biomater 9(4):64

Bui AL, Horwich TB, Fonarow GC (2011) Epidemiology and risk profile of heart failure. Nat Rev Cardiol 8 (1):30–41. https://doi.org/10.1038/nrcardio.2010.165

Cardoso GB, Machado-Silva AB, Sabino M, Jr Santos AR, Zavaglia CA (2014) Novel hybrid membrane of chitosan/poly (ε-caprolactone) for tissue engineering. Biomatter. 4. https://doi.org/10.4161/biom.29508.

Cebotari S, Lichtenberg A, Tudorache I, Hilfiker A, Mertsching H, Leyh R et al (2006) Clinical application of tissue engineered human heart valves using autologous progenitor cells. Circulation 114(1 Suppl): I132–I137. https://doi.org/10.1161/circulationaha.105.001065

Chaudhry MA (2019) Heart failure. Curr Hypertens Rev 15(1):7. https://doi.org/10.2174/157340211501190129144451

Chen K, Vigliotti A, Bacca M, McMeeking RM, Deshpande VS, Holmes JW (2018) Role of boundary conditions in determining cell alignment in response to stretch. Proc Natl Acad Sci USA 115(5):986–991. https://doi.org/10.1073/pnas.1715059115

Christman KL, Lee RJ (2006) Biomaterials for the treatment of myocardial infarction. J Am Coll Cardiol 48(5):907–913. https://doi.org/10.1016/j.jacc.2006.06.005

Cutts J, Nikkhah M, Brafman DA (2015) Biomaterial approaches for stem cell-based myocardial tissue engineering. Biomarker Insights. 10(Suppl 1):77–90. https://doi.org/10.4137/bmi.s20313

Dainese L, Guarino A, Burba I, Esposito G, Pompilio G, Polvani G et al (2012) Heart valve engineering: decellularized aortic homograft seeded with human cardiac stromal cells. J Heart Valve Dis 21(1):125–134

Dolan EB, Hofmann B, de Vaal MH, Bellavia G, Straino S, Kovarova L et al (2019) A bioresorbable biomaterial carrier and passive stabilization device to improve heart function post-myocardial infarction. Mater Sci Eng: C 103:109751

Flaig F, Ragot Hln, Simon A, Revet Gl, Kitsara M, Kitasato L et al. (2020) Design of functional electrospun scaffolds based on poly (glycerol sebacate) elastomer and poly (lactic acid) for cardiac tissue engineering. ACS Biomater Sci Eng 6(4):2388–400

Frangogiannis NG (2016) The functional pluralism of fibroblasts in the infarcted myocardium. Am Heart Assoc

Gabriel LP, Rodrigues AA, Macedo M, Jardini AL, Maciel FR (2017) Electrospun polyurethane membranes for tissue engineering applications. Mater Sci Eng, C Mater Biol Appl 72:113–117. https://doi.org/10.1016/j.msec.2016.11.057

Gao B, Matsuura K, Shimizu T (2019) Recent progress in induced pluripotent stem cell-derived cardiac cell sheets for tissue engineering. Biosci Trends 13 (4):292–298. https://doi.org/10.5582/bst.2019.01227

Grauss RW, Hazekamp MG, van Vliet S, Gittenberger-de Groot AC, DeRuiter MC (2003) Decellularization of rat aortic valve allografts reduces leaflet destruction and extracellular matrix remodeling. J Thorac Cardiovasc Surg 126(6):2003–2010. https://doi.org/10.1016/s0022-5223(03)00956-5

Grauss RW, Hazekamp MG, Oppenhuizen F, van Munsteren CJ, Gittenberger-de Groot AC, DeRuiter MC (2005) Histological evaluation of decellularised porcine aortic valves: matrix changes due to different decellularisation methods. Eur J Cardiothorac Surg: Official J Eur Assoc Cardiothorac Surg 27(4):566–571. https://doi.org/10.1016/j.ejcts.2004.12.052

Hansen KJ, Laflamme MA, Gaudette GR (2018) Development of a contractile cardiac fiber from pluripotent stem cell derived cardiomyocytes. Front Cardiovasc Med. 5:52. https://doi.org/10.3389/fcvm.2018.00052

Haraguchi Y, Shimizu T, Matsuura K, Sekine H, Tanaka N, Tadakuma K et al (2014) Cell sheet technology for cardiac tissue engineering. Methods Mol Biol (clifton, NJ) 1181:139–155. https://doi.org/10.1007/978-1-4939-1047-2_13

Hashemi SM, Ghods S, Kolodgie FD, Parcham-Azad K, Keane M, Hamamdzic D et al (2008) A placebo controlled, dose-ranging, safety study of allogenic mesenchymal stem cells injected by endomyocardial delivery after an acute myocardial infarction. Eur

Heart J 29(2):251–259. https://doi.org/10.1093/eurheartj/ehm559
Heallen TR, Martin JF (2018) Heart repair via cardiomyocyte-secreted vesicles. Nat Biomed Eng. 2 (5):271–272
Hilbert SL, Yanagida R, Souza J, Wolfinbarger L, Jones AL, Krueger P et al (2004) Prototype anionic detergent technique used to decellularize allograft valve conduits evaluated in the right ventricular outflow tract in sheep. J Heart Valve Dis 13(5):831–840
Hirt MN, Hansen A, Eschenhagen T (2014) Cardiac tissue engineering: state of the art. Circ Res 114(2):354–367
Hoerstrup SP, Sodian R, Sperling JS, Vacanti JP, Mayer JE Jr (2000) New pulsatile bioreactor for in vitro formation of tissue engineered heart valves. Tissue Eng 6(1):75–79. https://doi.org/10.1089/107632700320919
Hosoyama K, Ahumada M, McTiernan CD, Davis DR, Variola F, Ruel M et al (2018) Nanoengineered electroconductive collagen-based cardiac patch for infarcted myocardium repair. ACS Appl Mater Interfaces 10(51):44668–44677. https://doi.org/10.1021/acsami.8b18844
Huang D, Huang Y, Xiao Y, Yang X, Lin H, Feng G et al (2019) Viscoelasticity in natural tissues and engineered scaffolds for tissue reconstruction. Acta Biomater 97:74–92
Jackson KA, Majka SM, Wang H, Pocius J, Hartley CJ, Majesky MW et al (2001) Regeneration of ischemic cardiac muscle and vascular endothelium by adult stem cells. J Clin Investig 107(11):1395–1402. https://doi.org/10.1172/jci12150
Joshi J, Brennan D, Beachley V, Kothapalli CR (2018) Cardiomyogenic differentiation of human bone marrow-derived mesenchymal stem cell spheroids within electrospun collagen nanofiber mats. J Biomed Mater Res, Part A 106(12):3303–3312. https://doi.org/10.1002/jbm.a.36530
Kajbafzadeh AM, Ahmadi Tafti SH, Mokhber-Dezfooli MR, Khorramirouz R, Sabetkish S, Sabetkish N et al (2016) Aortic valve conduit implantation in the descending thoracic aorta in a sheep model: the outcomes of pre-seeded scaffold. Int J Surg (London, Engl) 28:97–105. https://doi.org/10.1016/j.ijsu.2016.02.061
Kajbafzadeh AM, Tafti SHA, Khorramirouz R, Sabetkish S, Kameli SM, Orangian S et al (2017) Evaluating the role of autologous mesenchymal stem cell seeded on decellularized pericardium in the treatment of myocardial infarction: an animal study. Cell Tissue Banking 18(4):527–538. https://doi.org/10.1007/s10561-017-9629-2
Kajbafzadeh AM, Khorramirouz R, Kameli SM, Fendereski K, Daryabari SS, Tavangar SM et al (2019) Three-year efficacy and patency follow-up of decellularized human internal mammary artery as a novel vascular graft in animal models. J Thorac Cardiovasc Surg 157(4):1494–1502. https://doi.org/10.1016/j.jtcvs.2018.08.106
Kasimir MT, Weigel G, Sharma J, Rieder E, Seebacher G, Wolner E et al (2005) The decellularized porcine heart valve matrix in tissue engineering: platelet adhesion and activation. Thromb Haemost 94(3):562–567. https://doi.org/10.1160/th05-01-0025
Keane TJ, Swinehart IT, Badylak SF (2015) Methods of tissue decellularization used for preparation of biologic scaffolds and in vivo relevance. Methods (san Diego, Calif). 84:25–34. https://doi.org/10.1016/j.ymeth.2015.03.005
Kenar H, Ozdogan CY, Dumlu C, Doger E, Kose GT, Hasirci V (2019) Microfibrous scaffolds from poly (l-lactide-co-ε-caprolactone) blended with xeno-free collagen/hyaluronic acid for improvement of vascularization in tissue engineering applications. Mater Sci Eng, C 97:31–44
Kharaziha M, Memic A, Akbari M, Brafman DA, Nikkhah M (2016) Nano-enabled approaches for stem cell-based cardiac tissue engineering. Adv Healthc Mater 5(13):1533–1553. https://doi.org/10.1002/adhm.201600088
Khorramirouz R, Kameli SM, Fendereski K, Daryabari SS, Kajbafzadeh AM (2019) Evaluating the efficacy of tissue-engineered human amniotic membrane in the treatment of myocardial infarction. Regen Med 14(2):113–126. https://doi.org/10.2217/rme-2018-0024
Kretzschmar K, Post Y, Bannier-Hélaouët M, Mattiotti A, Drost J, Basak O et al (2018) Profiling proliferative cells and their progeny in damaged murine hearts. Proc Natl Acad Sci USA 115(52):E12245–E12254. https://doi.org/10.1073/pnas.1805829115
Laurie GW, Horikoshi S, Killen PD, Segui-Real B, Yamada Y (1989) In situ hybridization reveals temporal and spatial changes in cellular expression of mRNA for a laminin receptor, laminin, and basement membrane (type IV) collagen in the developing kidney. J Cell Biol 109(3):1351–1362. https://doi.org/10.1083/jcb.109.3.1351
Lee RT (2018) Adult cardiac stem cell concept and the process of science. Circulation 138(25):2940–2942. https://doi.org/10.1161/circulationaha.118.036407
Leyh RG, Wilhelmi M, Rebe P, Fischer S, Kofidis T, Haverich A et al (2003) In vivo repopulation of xenogeneic and allogeneic acellular valve matrix conduits in the pulmonary circulation. Ann Thorac Surg 75(5):1457–63; discussion 63. https://doi.org/10.1016/s0003-4975(02)04845-2.
Li X, Tamama K, Xie X, Guan J (2016) Improving cell engraftment in cardiac stem cell therapy. Stem Cells Int 2016
Lim M, Wang W, Liang L, Han ZB, Li Z, Geng J et al (2018) Intravenous injection of allogeneic umbilical cord-derived multipotent mesenchymal stromal cells reduces the infarct area and ameliorates cardiac function in a porcine model of acute myocardial infarction. Stem Cell Res Ther 9(1):129. https://doi.org/10.1186/s13287-018-0888-z

Liu H, Wang Y, Cui K, Guo Y, Zhang X, Qin J (2019) Advances in hydrogels in organoids and organs-on-a-chip. Adv Mater 31(50):1902042

Maiullari F, Costantini M, Milan M, Pace V, Chirivì M, Maiullari S et al (2018) A multi-cellular 3D bioprinting approach for vascularized heart tissue engineering based on HUVECs and iPSC-derived cardiomyocytes. Sci Rep 8(1):13532. https://doi.org/10.1038/s41598-018-31848-x

Malandraki-Miller S, Lopez CA, Al-Siddiqi H, Carr CA (2018) Changing metabolism in differentiating cardiac progenitor cells—can stem cells become metabolically flexible cardiomyocytes? Front Cardiovasc Med. 5:119

Malone JM, Brendel K, Duhamel RC, Reinert RL (1984) Detergent-extracted small-diameter vascular prostheses. J Vasc Surg 1(1):181–191. https://doi.org/10.1067/mva.1984.avs0010181

Mao C, Hou X, Wang B, Chi J, Jiang Y, Zhang C et al (2017) Intramuscular injection of human umbilical cord-derived mesenchymal stem cells improves cardiac function in dilated cardiomyopathy rats. Stem Cell Res Ther 8(1):18. https://doi.org/10.1186/s13287-017-0472-y

Martinelli V, Bosi S, Peña B, Baj G, Long CS, Sbaizero O et al (2018) 3D carbon-nanotube-based composites for cardiac tissue engineering. ACS Appl Bio Mater 1 (5):1530–1537

Martins C, Sousa F, Araújo F, Sarmento B (2018) Functionalizing PLGA and PLGA derivatives for drug delivery and tissue regeneration applications. Adv Healthcare Mater 7(1):1701035

Mathiasen AB, Jørgensen E, Qayyum AA, Haack-Sørensen M, Ekblond A, Kastrup J (2012) Rationale and design of the first randomized, double-blind, placebo-controlled trial of intramyocardial injection of autologous bone-marrow derived mesenchymal stromal cells in chronic ischemic heart failure (MSC-HF Trial). Am Heart J 164(3):285–291. https://doi.org/10.1016/j.ahj.2012.05.026

Matsuura K, Masuda S, Shimizu T (2014) Cell sheet-based cardiac tissue engineering. Anat Rec (Hoboken, NJ: 2007). 297(1):65–72. https://doi.org/10.1002/ar.22834

McDevitt TC, Woodhouse KA, Hauschka SD, Murry CE, Stayton PS (2003) Spatially organized layers of cardiomyocytes on biodegradable polyurethane films for myocardial repair. J Biomed Mater Res, Part A 66 (3):586–595. https://doi.org/10.1002/jbm.a.10504

Mewhort HE, Svystonyuk DA, Turnbull JD, Teng G, Belke DD, Guzzardi DG et al (2017) Bioactive extracellular matrix scaffold promotes adaptive cardiac remodeling and repair. JACC: Basic Transl Sci 2 (4):450–64

Mirsadraee S, Wilcox HE, Korossis SA, Kearney JN, Watterson KG, Fisher J et al (2006) Development and characterization of an acellular human pericardial matrix for tissue engineering. Tissue Eng 12(4):763–773. https://doi.org/10.1089/ten.2006.12.763

Miyagawa S, Domae K, Yoshikawa Y, Fukushima S, Nakamura T, Saito A et al (2017) Phase I clinical trial of autologous stem cell-sheet transplantation therapy for treating cardiomyopathy. J Am Heart Assoc 6(4). https://doi.org/10.1161/jaha.116.003918.

Mohamed TM, Ang Y-S, Radzinsky E, Zhou P, Huang Y, Elfenbein A et al (2018) Regulation of cell cycle to stimulate adult cardiomyocyte proliferation and cardiac regeneration. Cell 173(1):104–16, e12

Moore A, Mercer J, Dutina G, Donahue CJ, Bauer KD, Mather JP et al (1997) Effects of temperature shift on cell cycle, apoptosis and nucleotide pools in CHO cell batch cultues. Cytotechnology 23(1–3):47–54. https://doi.org/10.1023/a:1007919921991

Muniyandi P, Palaninathan V, Veeranarayanan S, Ukai T, Maekawa T, Hanajiri T et al (2020) ECM mimetic electrospun porous poly (L-lactic acid)(PLLA) scaffolds as potential substrates for cardiac tissue engineering. Polymers 12(2):451

Naderi H, Matin MM, Bahrami AR (2011) Review paper: critical issues in tissue engineering: biomaterials, cell sources, angiogenesis, and drug delivery systems. J Biomater Appl 26(4):383–417. https://doi.org/10.1177/0885328211408946

Neto MD, Oliveira MB, Mano JF (2019) Microparticles in contact with cells: from carriers to multifunctional tissue modulators. Trends Biotechnol 37(9):1011–1028

Oberwallner B, Brodarac A, Choi YH, Saric T, Anić P, Morawietz L et al (2014) Preparation of cardiac extracellular matrix scaffolds by decellularization of human myocardium. J Biomed Mater Res, Part A 102 (9):3263–3272. https://doi.org/10.1002/jbma.35000

O'Connor Mooney R, Davis NF, Hoey D, Hogan L, McGloughlin TM, Walsh MT (2016) On the automatic decellularisation of porcine aortae: a repeatability study using a non-enzymatic approach. Cells Tissues Organs 201(4):299–318. https://doi.org/10.1159/000445107

Paez-Mayorga J, Hernández-Vargas G, Ruiz-Esparza GU, Iqbal HMN, Wang X, Zhang YS et al (2019) Bioreactors for cardiac tissue engineering. Adv Healthc Mater 8(7):e1701504. https://doi.org/10.1002/adhm.201701504

Parikh SS, Blackwell DJ, Gomez-Hurtado N, Frisk M, Wang L, Kim K et al (2017) Thyroid and glucocorticoid hormones promote functional T-tubule development in human-induced pluripotent stem cell-derived cardiomyocytes. Circ Res 121(12):1323–1330. https://doi.org/10.1161/circresaha.117.311920

Perrino C, Ferdinandy P, Bøtker HE, Brundel B, Collins P, Davidson SM et al (2020) Improving translational research in sex-specific effects of comorbidities and risk factors in ischemic heart disease and cardioprotection: position paper and recommendations of the ESC working group on cellular biology of the heart. Cardiovasc Res. https://doi.org/10.1093/cvr/cvaa155

Pushp P, Sahoo B, Ferreira FC, Sampaio Cabral JM, Fernandes-Platzgummer A, Gupta MK (2020)

Functional comparison of beating cardiomyocytes differentiated from umbilical cord-derived mesenchymal/stromal stem cells and human foreskin-derived induced pluripotent stem cells. J Biomed Mater Res, Part A 108(3):496–514. https://doi.org/10.1002/jbm.a.36831
Radisic M, Park H, Chen F, Salazar-Lazzaro JE, Wang Y, Dennis R et al (2006) Biomimetic approach to cardiac tissue engineering: oxygen carriers and channeled scaffolds. Tissue Eng 12(8):2077–2091
Revilla A, López J, Arnold R, Sánchez PL, Villa A, Pinedo M et al (2011) Long-term changes in left ventricular function following intracoronary stem cell transplantation for acute myocardial infarction. Rev Esp Cardiol 64(4):334–337. https://doi.org/10.1016/j.recesp.2010.06.009
Richards DJ, Tan Y, Coyle R, Li Y, Xu R, Yeung N et al (2016) Nanowires and electrical stimulation synergistically improve functions of hiPSC cardiac spheroids. Nano Lett 16(7):4670–4678. https://doi.org/10.1021/acs.nanolett.6b02093
Rikhtegar R, Pezeshkian M, Dolati S, Safaie N, Afrasiabi Rad A, Mahdipour M et al. Stem cells as therapy for heart disease: iPSCs, ESCs, CSCs, and skeletal myoblasts. Biomed Pharmacothe = Biomed Pharmacotherapie. 2019;109:304–13. https://doi.org/10.1016/j.biopha.2018.10.065.
Rockwood DN, Akins RE Jr, Parrag IC, Woodhouse KA, Rabolt JF (2008) Culture on electrospun polyurethane scaffolds decreases atrial natriuretic peptide expression by cardiomyocytes in vitro. Biomaterials 29 (36):4783–4791. https://doi.org/10.1016/j.biomaterials.2008.08.034
Sabetkish S, Kajbafzadeh AM, Sabetkish N, Khorramirouz R, Akbarzadeh A, Seyedian SL et al (2015) Whole-organ tissue engineering: decellularization and recellularization of three-dimensional matrix liver scaffolds. J Biomed Mater Res, Part A 103 (4):1498–1508. https://doi.org/10.1002/jbm.a.35291
Sanz-Ruiz R, Fernández-Avilés F (2018) Autologous and allogeneic cardiac stem cell therapy for cardiovascular diseases. Pharmacol Res 127:92–100
Sawa Y, Miyagawa S (2013) Present and future perspectives on cell sheet-based myocardial regeneration therapy. Biomed Res Int 2013:583912. https://doi.org/10.1155/2013/583912
Sawa Y, Miyagawa S, Sakaguchi T, Fujita T, Matsuyama A, Saito A et al (2012) Tissue engineered myoblast sheets improved cardiac function sufficiently to discontinue LVAS in a patient with DCM: report of a case. Surg Today 42(2):181–184. https://doi.org/10.1007/s00595-011-0106-4
Segers VF, Lee RT (2008) Stem-cell therapy for cardiac disease. Nature 451(7181):937–942. https://doi.org/10.1038/nature06800
Sereti K-I, Nguyen NB, Kamran P, Zhao P, Ranjbarvaziri S, Park S et al (2018) Analysis of cardiomyocyte clonal expansion during mouse heart development and injury. Nat Commun 9(1):1–13
Shevach M, Fleischer S, Shapira A, Dvir T (2014) Gold nanoparticle-decellularized matrix hybrids for cardiac tissue engineering. Nano Lett 14(10):5792–5796
Shiekh PA, Singh A, Kumar A (2018) Engineering bioinspired antioxidant materials promoting cardiomyocyte functionality and maturation for tissue engineering application. ACS Appl Mater Interfaces 10 (4):3260–3273
Shimizu T, Yamato M, Kikuchi A, Okano T (2001) Two-dimensional manipulation of cardiac myocyte sheets utilizing temperature-responsive culture dishes augments the pulsatile amplitude. Tissue Eng 7(2):141–151. https://doi.org/10.1089/107632701300062732
Shimizu T, Yamato M, Kikuchi A, Okano T (2003) Cell sheet engineering for myocardial tissue reconstruction. Biomaterials 24(13):2309–2316. https://doi.org/10.1016/s0142-9612(03)00110-8
Singelyn JM, Sundaramurthy P, Johnson TD, Schup-Magoffin PJ, Hu DP, Faulk DM et al (2012) Catheter-deliverable hydrogel derived from decellularized ventricular extracellular matrix increases endogenous cardiomyocytes and preserves cardiac function post-myocardial infarction. J Am Coll Cardiol 59(8):751–763. https://doi.org/10.1016/j.jacc.2011.10.888
Su T, Huang K, Daniele MA, Hensley MT, Young AT, Tang J et al (2018) Cardiac stem cell patch integrated with microengineered blood vessels promotes cardiomyocyte proliferation and neovascularization after acute myocardial infarction. ACS Appl Mater Interfaces 10(39):33088–33096. https://doi.org/10.1021/acsami.8b13571
Suncion VY, Ghersin E, Fishman JE, Zambrano JP, Karantalis V, Mandel N et al (2014) Does transendocardial injection of mesenchymal stem cells improve myocardial function locally or globally?: An analysis from the percutaneous stem cell injection delivery effects on neomyogenesis (POSEIDON) randomized trial. Circ Res 114(8):1292–1301. https://doi.org/10.1161/circresaha.114.302854
Tang J, Sun GY, Chen T, Wang YD, Zhang J, Qi XQ (2016) Effect of intracoronary autologous bone marrow mononuclear cells transplantation on arrhythmia in canines. Zhonghua Xin Xue Guan Bing Za Zhi 44 (12):1030–1035. https://doi.org/10.3760/cma.j.issn.0253-3758.2016.12.008
Tang J, Cui X, Caranasos TG, Hensley MT, Vandergriff AC, Hartanto Y et al (2017) Heart repair using nanogel-encapsulated human cardiac stem cells in mice and pigs with myocardial infarction. ACS Nano 11(10):9738–9749. https://doi.org/10.1021/acsnano.7b01008
Tapias LF, Ott HC (2014) Decellularized scaffolds as a platform for bioengineered organs. Curr Opin Organ Transplant 19(2):145–152. https://doi.org/10.1097/mot.0000000000000051
Tijore A, Irvine SA, Sarig U, Mhaisalkar P, Baisane V, Venkatraman S (2018) Contact guidance for cardiac tissue engineering using 3D bioprinted gelatin patterned hydrogel. Biofabrication 10(2):025003. https://doi.org/10.1088/1758-5090/aaa15d

Tomecka E, Wojasinski M, Jastrzebska E, Chudy M, Ciach T, Brzozka Z (2017) Poly (l-lactic acid) and polyurethane nanofibers fabricated by solution blow spinning as potential substrates for cardiac cell culture. Mater Sci Eng, C 75:305–316

Tzahor E, Poss KD (2017) Cardiac regeneration strategies: staying young at heart. Science 356(6342):1035–1039

Wade RJ, Bassin EJ, Gramlich WM, Burdick JA (2015) Nanofibrous hydrogels with spatially patterned biochemical signals to control cell behavior. Adv Mater 27(8):1356–1362

Wainwright JM, Czajka CA, Patel UB, Freytes DO, Tobita K, Gilbert TW et al (2010) Preparation of cardiac extracellular matrix from an intact porcine heart. Tissue Eng Part C Methods 16(3):525–532. https://doi.org/10.1089/ten.TEC.2009.0392

Wang H, Hao J, Hong CC (2011) Cardiac induction of embryonic stem cells by a small molecule inhibitor of Wnt/β-catenin signaling. ACS Chem Biol 6(2):192–197. https://doi.org/10.1021/cb100323z

Wang F, Zhang J, Wang R, Gu Y, Li J, Wang C (2017) Triton X-100 combines with chymotrypsin: a more promising protocol to prepare decellularized porcine carotid arteries. Bio-Med Mater Eng 28(5):531–543. https://doi.org/10.3233/bme-171694

Weymann A, Loganathan S, Takahashi H, Schies C, Claus B, Hirschberg K et al (2011) Development and evaluation of a perfusion decellularization porcine heart model–generation of 3-dimensional myocardial neoscaffolds. Circ J 75(4):852–860. https://doi.org/10.1253/circj.cj-10-0717

Wu KH, Wang SY, Xiao QR, Yang Y, Huang NP, Mo XM et al (2018) Small-molecule-based generation of functional cardiomyocytes from human umbilical cord-derived induced pluripotent stem cells. J Cell Biochem. https://doi.org/10.1002/jcb.27094

Xu R, Ding S, Zhao Y, Pu J, He B (2014) Autologous transplantation of bone marrow/blood-derived cells for chronic ischemic heart disease: a systematic review and meta-analysis. Can J Cardiol 30(11):1370–1377. https://doi.org/10.1016/j.cjca.2014.01.013

Yamamoto R, Miyagawa S, Toda K, Kainuma S, Yoshioka D, Yoshikawa Y et al (2019) Long-term outcome of ischemic cardiomyopathy after autologous myoblast cell-sheet implantation. Ann Thorac Surg 108(5):e303–e306. https://doi.org/10.1016/j.athoracsur.2019.03.028

Yau TM, Pagani FD, Mancini DM, Chang HL, Lala A, Woo YJ et al (2019) Intramyocardial injection of mesenchymal precursor cells and successful temporary weaning from left ventricular assist device support in patients with advanced heart failure: a randomized clinical trial. JAMA 321(12):1176–1186. https://doi.org/10.1001/jama.2019.2341

Yoon C, Song H, Yin T, Bausch-Fluck D, Frei AP, Kattman S et al (2018) FZD4 marks lateral plate mesoderm and signals with NORRIN to increase cardiomyocyte induction from pluripotent stem cell-derived cardiac progenitors. Stem Cell Reports. 10 (1):87–100

Yoshikawa Y, Miyagawa S, Toda K, Saito A, Sakata Y, Sawa Y (2018) Myocardial regenerative therapy using a scaffold-free skeletal-muscle-derived cell sheet in patients with dilated cardiomyopathy even under a left ventricular assist device: a safety and feasibility study. Surg Today 48(2):200–210. https://doi.org/10.1007/s00595-017-1571-1

Young BM, Shankar K, Tho CK, Pellegrino AR, Heise RL (2019) Laminin-driven Epac/Rap1 regulation of epithelial barriers on decellularized matrix. Acta Biomater 100:223–234. https://doi.org/10.1016/j.actbio.2019.10.009

Zhang C, Zhou Y, Lai X, Zhou G, Wang H, Feng X et al (2019) Human umbilical cord mesenchymal stem cells alleviate myocardial endothelial-mesenchymal transition in a rat dilated cardiomyopathy model. Transpl Proc 51(3):936–941. https://doi.org/10.1016/j.transproceed.2019.01.080

Zurina IM, Presniakova VS, Butnaru DV, Svistunov AA, Timashev PS, Rochev YA (2020) Tissue engineering using a combined cell sheet technology and scaffolding approach. Acta Biomater 113:63–83. https://doi.org/10.1016/j.actbio.2020.06.016

6 The Most Ideal Pancreas Extracellular Matrix as a Platform for Pancreas Bioengineering: Decellularization/ Recellularization Protocols

Shabnam Sabetkish
and Abdol-Mohammad Kajbafzadeh

Abstract

Natural scaffold appears to have extensive functions in providing anchorage and structural requirements, as well as providing a structural support for cell adherence and cell interaction for further recellularization process. Specific methods used for decellularization process play an essential role in the efficacy of cell removal and successful preservation of ultrastructure and biomechanical properties of the tissue. Numerous scaffolding materials and fabrication techniques have been investigated for pancreatic tissue engineering. Techniques of casting, freeze drying, injection molding, and electrospinning have been also used to fabricate scaffolds. Herein, we aim to review the state-of-the-art and the tissue engineering and regenerative medicine technology highlighting the major achievements toward the production of a bioengineered pancreas obtained decellularization techniques and cell-on-scaffold technology.

Keywords

Decellularization · Recellularization · Pancreas · Scaffold · Extracellular matrix · Cell cultures

6.1 History

In 2014, the World Health Organization (WHO) has predicted diabetes as a worldwide disease that afflicts 422 million people with increasing drawbacks for the healthcare system and the economic policies. The prevalence of diabetes mellitus has improved in the as a worldwide disorder which makes it as an epidemic of the twenty-first century. Exogenous insulin treatment and pharmaceutical interventions may fail to restore euglycemia and in a large number of patients, insulin resistance could occur.

Pancreas transplantation has been successfully established for the first time in 1966 by the surgical team driven by Prof. Kelly (1967). In patients with severe diabetic patients, whole pancreas or pancreatic islet transplantation emerged as a fundamental treatment. Although pancreatic islet transplantation is less invasive compared to pancreas transplantation, islets transplantation requires donors and may cause some acute effects comprehend bleeding (Brennan 2004; Villiger 2005), portal vein thrombosis (Brennan 2004) or a transient increase of hepatic

S. Sabetkish · A.-M. Kajbafzadeh (✉)
Pediatric Urology and Regenerative Medicine Research Center, Section of Tissue Engineering and Stem Cells Therapy, Children's Medical Center, Pediatric Center of Excellence, Tehran University of Medical Sciences, No. 62, Dr. Gharib's Street, Keshavarz Boulevard, 1419433151 Tehran, Iran
e-mail: kajbafzd@sina.tums.ac.ir

A. Kajbafzadeh (ed.), *Decellularization Methods of Tissue and Whole Organ in Tissue Engineering*, Advances in Experimental Medicine and Biology 1345,
https://doi.org/10.1007/978-3-030-82735-9_6

inflammatory markers (Ryan et al. 2004). However, regenerative medicine is a possible solution to these problems. Due to the difficulty of expanding adult derived pancreatic β-cells in vitro, many efforts have been made to build up a technique for inducing the self-renewing multipotent stem cells into functional pancreatic β-cells, for further transplantation. The embryonic stem cells, and mesenchymal stem cells (MSCs), are considered as two types of pluripotent cells that are of great interest.

Tissue engineering based on stem cells and scaffolds has been considered as a potential alternative therapy to overcome the adverse effects of the existing pharmacological treatments. Tissue engineering may also serve as a novel treatment foundation for the repair and regeneration of pancreas. By the application of this technique, we may be able to reduce unavoidable consequences of conventional therapies for diabetes and make an immunosuppression free state after transplantation. We may also be able to combine cells such as islets or pancreatic beta cells with scaffolds, which provides mechanical support and appropriate extracellular matrix (ECM) for the survival of cells both in vitro and in vivo.

The ECM has been considered as an important element of the islet microenvironment which may have a significant influence on cellular growth, development, differentiation (Hisaoka et al. 1993; Gittes 1996) proliferation (Hayek 1995), and survival (Hammar 2004; Weber et al. 2008). The ECM in the periphery of the islet has been reported to contain collagen I, III, IV, V, and VI, as well as laminin and fibronectin.

Most recent studies have reported the application of natural acellular pancreatic tissue as a bioscaffold for in vitro recellularization. The suitability of two-dimensional (2D) bioscaffolds in whole-organ engineering was firstly defined in the study of De Carlo et al. (2010). Additionally, whole-organ regeneration of pancreatic tissue has been demonstrated by in vitro techniques, as the first proof-of-concept regeneration of functional islet cells (Napierala 2017).

Our group has extensive experience with regard to the structure of the pancreas (Hashemi 2018a, b), which makes us to our decision to continue with the preparation of pancreas, with the aim of developing and characterizing a new ECM scaffold. The possible applications in experimental and medical settings will be discussed.

6.2 Materials and Methods

Several polymeric biomaterials have been used as scaffold materials for islet tissue engineering, such as heparin, lactic acid-based polymers, polyhydroxyalkanoates, hydrogels, bioploymer films, and biomimetic scaffolds. In particular, using natural scaffolds is crucial for the survival and the right function of both cells and higher organisms.

Cell sheets approach is also included in the tissue-engineered approach. This has been considered as a possible approach for pancreatic islets and β-cell transplantation. Subcutaneous site is one of the new regions of islets injection in animal model (Vériter et al. 2013). It should be also taken into consideration that the most significant differentiation is the absence of the vascular connection between transplanted cells and the blood flow. However, less invasive, the likelihood of repeated procedures in immunological rejection, and the opportunity to safely eliminate transplanted islets are among the advantages of this technique.

6.3 Decellularization Technology

Decellularization techniques are valid in three major types: 1. chemical decellularization, 2. biological decellularization, and 3. physical decellularization.

Acetic and paracetic acids are the most common acids used in chemical decellularization method, which may destroy ECM structure but is capable to remove cellular components (Nouri Barkestani et al. 2021). Calcium hydroxide, sodium sulfide and sodium hydroxide are the most frequent agents used in this method. The application of these bases can eliminate growth factors that enrich the ECM resulting in a loss of bioactivity. Therefore, the application of these agents has been limited. A combination of chemical, physical, and enzymatic protocols is usually applied for complete decellularization of most tissues. However, as the structural properties of pancreas is not complex as compared with other organs such as kidney, liver, or lung, the mildest protocol possible can be applied to achieve a suitable scaffold without disruption of the structural and functional properties of the ECM. Alkaline and acid treatments can remove cellular components and nuclear remnants. However, they destroy ECM components such as GAGs to some extent (Filippo et al. 2002; Falke 2003). Triton X-100, as a nonionic detergent, may affect lipid-lipid and lipid-protein interactions. However, this detergent can keep the proteins within an organ in a functional conformation (Yang 2010). The ionic detergent SDS can be also used for the complete removal of nuclear remnants and cytoplasmic proteins (Rieder 2004). This detergent preserves the structure of the natural tissue while reducing GAG concentration and collagen integrity (Blaudez et al. 2019). Dedicated scientific literature offers many protocol based on the use of SDS for organ or tissue decellularization (Peloso et al. 2015; He 2017). Biological decellularization includes the application of biologic enzymatic and non-enzymatic agents to remove of cell residues. Enzyme such nucleases (DNases or RNases) are able to eliminate nucleotides after cell lysis (Jain 2019). Ethylenediaminetetraacetic acid (EDTA) and ethylene glycol tetraacetic acid (EGTA) are among the non-enzymatic agents. EDTA and EGTA are able to detach cells from ECM. These two agents are added in a multistep protocol as they are not particularly effective if used alone (Tudorache et al. 2007). It has been reported that the use of trypsin for decellularization results in damage to the ECM components (French and Davis 2019).

We recently used a perfusion-based approach using the combination of Triton and SDS for pancreas tissue engineering. The applied method of decellularization completely removed the cells and preserved the ECM structure, while the vasculature remained intact (Hashemi 2018b). In another research, pancreatic bioscaffold of a mouse was created with SDS/Triton that was perfused through the anterior hepatic portal vein. The results showed that the scaffold was capable of mimicking the natural pancreas for pancreatic tissue engineering (Goh 2013). In addition, temperature protocols such as freeze–thaw cycles technology is satisfactory for the decellularization of simple structure (tendon or cartilage base organs). However, the results are hardly applicable for structural complex architectures such as pancreas, kidney, or liver.

6.4 Whole Pancreas Decellularization and Regenerative Medicine

The decellularization technique and the methods of final sterilization are two important points in achieving a suitable scaffold for further recellularization. An ideal decellularization process may be able to produce a suitable scaffold for tissue regeneration. This scaffold should consist of functional ECM components, an intact vascular network, preserved biomechanical properties, for further pancreatic tissue regeneration. Various studies verified that whole-organ perfusion-based decellularization process can result in preservation of ECM, perfusion networks, and complete removal of DNA components, in contrast to other protocols (Soto-Gutierrez 2012). This technique is also naturally designed to permit the delivery of oxygen and nutrients by blood flux and permits the same time-to-contact between detergents and cells in the entire organ, in the attempt of creating a bioengineered pancreas.

In our center, we were able to produce a biocompatible three-dimensional (3D) pancreas scaffold with comparative histological and biomechanical properties of native pancreatic ECM, using a perfusion-based decellularization method (Hashemi 2018a). In one study, the whole pancreas was decellularized using the perfusion-based method from the main pancreatic duct. However, high-dose detergents and enzymes in pancreas dissolve fatty tissue and altered the protein content (Goh 2013).

6.5 Tests to Confirm the Efficacy of the Decellularization Process

Several tests are used to confirm the efficacy of the decellularization process in cell removal as well as preservation of ECM such as histological staining, DNA content, collagens and proteoglycans assessment, and tridimensional imaging technologies (scanning electron microscopy). Residual DNA fragments in the scaffold are able to induce incompatibility and undesirable immunological response (Keane 2012). For that reason, DAPI staining and DNA quantification should be performed to confirm the removal of cellular and nuclear components. Additionally, several studies have confirmed changes in the biomechanical properties of the scaffold after the decellularization process (Haag 2012). Hence, strength testing is of great importance to evaluate the mechanical properties of the obtained scaffold and compared with those of natural pancreas tissue. Moreover, CT angiography is another employed technique to evaluate vascular network competency, for further cell seeding of the scaffolds (Mirmalek-Sani 2013). The above-mentioned investigations can provide crucial data on the effective cellular removal as well as the biological and structural properties of the decellularized matrix intended to seed.

6.6 Cell Seeding

Recellularization strategies play a crucial role for the creation of a functional-bioengineered organoid. The most common cell types used in generating pancreatic islet cells are embryonic stem cells (ESCs), induced pluripotent stem cells (iPSC), pancreatic stem cells, and mesenchymal stem cells (MSC).

Cell–cell adhesion, adhesion signaling, and cell–matrix interactions are indispensable for differentiation. In addition, a number of signals are essential for the metabolism of cells for vital tissue or cell processes such as proliferation and differentiation (Matta and Mobasheri 2014).

The principal feature of ESC contains its ability to self-renew and the potential to differentiate into all embryonic cell types, under in vivo and in vitro conditions. Several studies have tried to produce cells with some level of insulin production from mouse (Soria 2000), monkey (Lester 2004), and human (Assady 2001). However, the in vitro functionality of β-cells has not been confirmed in none of these studies, and the cells were not able to secrete physiologically sufficient amounts of insulin in response to glucose. The reason would be that pancreatic β-cells are not the only cell type that can synthesize and release insulin (Basford 2012). Cultured embryonic stem cells have been also reported to have the capability of differentiation into insulin-positive cells by the application of a cell-trapping strategy (Abdelalim and Emara 2015; Rattananinsruang et al. 2018). These approaches have some limitations including neuronal characteristics in the majority of islet-like structures, a small fraction of insulin-positive cells, loss of Pdx-1 expression which is a specific pancreatic β-cells transcription factor, low insulin content, insufficient insulin response to glucose, and unsatisfactory post-transplantation results (Lumelsky 2001). The results of one in vivo study showed promise results regarding differentiation of the ESC into β-cell precursors, which was characterized by transcription factors Pdx1 (pancreatic and duodenal homeobox 1) and Nkx6.1. Precursor cells were matured into functional β-cells and reverse hyperglycemia after being implanted into immunodeficient mice (Schulz 2012). Although several studies obtained promising results using ESCs, risk of teratocarcinoma formation, immune rejection and social/legal/ethical issues is among the remained obstacles.

Few researches were conducted on deriving insulin-producing cells from iPSC. The Edmonton protocol has been developed for pancreatic islet transplantation in order to replace the lost β-cells (Shapiro 2006). In this method, cadaveric human islets are grafted to patients by applying a minimal, glucocorticoid-free immunosuppressive regimen. However, this protocol provides insulin independence for a limited period of time. Moreover, immune rejection and scarcity of donor islets are the major limitations of this procedure (Shapiro 2000). To overcome these obstacles, insulin-producing cells (IPCs) have been derived from different progenitor sources including embryonic, bone marrow, pancreatic, skin, hepatic duct, or adipose-derived stem cells. In one study in 2012, a combination of activin A and retinoic acid was used to differentiate human-induced pluripotent stem cells (hiPSC) into insulin-producing cells. The outcomes demonstrated that above 10% of the cells became insulin-positive and differentiated cells secreted human C-peptide (Hosoya 2012). The results of another paper showed that diabetes was reversed in immunodeficient mice using murine and rhesus monkey-derived iPSC-differentiated pancreatic precursor cells (He et al. 2014). Since iPSC may activate related oncogenes, the safety issue of these cells should be further investigated before clinical applications.

Regarding the use of pancreatic stem cells, it has been verified that pre-existing β-cells preserved the proliferative capacity which may be the major source of new β-cells in adult life (Dor 2004). However, other papers have found evidence of the participation of pluripotent stem cells in β-cell regeneration.

MSCs may play a significant role in diabetes therapy by islet differentiation and transplantation. In addition, immunoregulatory, anti-inflammatory, and proangiogenic properties of MSCs make these cells beneficial to improve islet engraftment and survival. In one study, MSC derived from mouse and rat bone marrow was cultured under high glucose condition with β-cell-stimulating growth factors. The results depicted capability of expressing pancreatic β-cell genes (insulin, GLUT2, and Pdx1) which reversed hyperglycemia in an animal model of diabetes (Tang 2004). Cotransplantation of syngeneic bone marrow cells with syngeneic/allogeneic MSCs resulted in pancreas tissue repair, normal blood glucose, and serum insulin-level stabilization, which recover the mice suffered from streptozotocin-induced diabetes (Urban 2008). Furthermore, cotransplantation of MSCs with neonatal porcine islets improved the function of the graft in diabetic mice (Hayward 2017). Functional islet-like cells were also differentiated from marrow MSCs, which suggested as a new procedure for clinical diabetes stem-cell therapy to control blood glucose level in diabetic rats (Chen et al. 2004). The effect of rat FM-derived MSCs (rFM-MSCs) and human amnion-derived MSCs (hAMSCs) on the inflammatory reaction in vitro and therapeutic effects in rats with acute and chronic pancreatitis was also investigated. The outcomes verified suppressed inflammatory reaction of acute and chronic pancreatitis in rats (Kawakubo 2016). The therapeutic efficacy of umbilical cord-derived mesenchymal stem cells in patients with type 2 diabetes was also depicted in one study in 2015 (Guan 2015).

Stage-specific differentiation protocol can be also used to aggregate the differentiation of human adipose stem cells to competent functional islet-like cells (Chandra 2011). The availability, autologous source, and large quantity of adipose tissue make it a suitable cell replacement alternative in type 1 diabetes.

Use of pre-seeded or none pre-seeded pancreas scaffolds are crucial for further recellularization process in vivo. Recently, the generation of a whole three-dimensional pancreas scaffold was described in porcine model the effectiveness of which was confirmed by imaging studies (Mirmalek-Sani 2013). Moreover, human stem cells and porcine pancreatic islets were successfully seeded on this scaffold to demonstrate the cellular adhesion and maintenance of cell functions on this bioscaffold. In spite of the successful results of this study, same results may not be obtained in a human-sized model as the best decellularization method has not been determined. Struecker has refined a proof of concept for the repopulation of the decellularized rat pancreas with functional islets of Langerhans (Napierala 2017). Briefly rat pancreas was decellularized via vascular perfusion using Triton X and SDS and then repopulated with islets via the pancreatic duct to test viability and functionality of the islets. TUNEL staining and glucose-stimulated insulin secretion (GSIS) demonstrated the viability and functionality of islets after the injection inside the acellular scaffold. The establishment of circulating stem cells over the transplanted pancreas scaffold could result in functional tissue regeneration, which may activate cell signaling pathways. In our recent study, we applied a surgical technique by transplanting the pancreas scaffold over the host pancreas could the result of which induced islet cell regeneration (Hashemi 2018a).

6.7 Clinical Applications

Regarding the clinical application, we are going to pave the road to customize clinical-relevant size decellularized pancreas that can be regenerated with autologous patient-own cellular population and then transplanted in the same patient. This approach could address the limitations that today affect diabetic treatments. We may also be able to create recellularized scaffolds with functional cells which can be transplanted in the patient whose cells were harvested, without any immunological reactions. This technique may limit the use of immunosuppressive drug regimes in related patients.

6.8 Limitations

With no doubt, the field of stem cell research related to diabetes and pancreatic β-cell development has considerably enhanced. However, further studies are required to address the challenging facing the field. For the application of β-cells for cell replacement therapy, it is essential to differentiate the cells into a pure population of differentiated cells. This may avoid development of teratoma formation after transplantation into patients. In addition, using of hESCs faces a lot of concern, because of the ethical and religious issues in several countries. Moreover, by the application of hESCs for cell therapy, patient may confront with the problem of immune rejection due to the fact that the differentiated cells are not genetically identical to the patients. Future studies are needed to develop our understanding regarding the pathophysiology of different forms of diabetes, the most effect decellularization method of the pancreas organ, as well as providing new treatment strategies. Several fundamental obstacles are not yet overcome and require imperative improvements, above all in vivo short- and long-term functional testing.

6.9 Conclusion

An appropriate and time-efficient decellularization method is essential for a successful regeneration of the pancreas organ through which organ replacement will probably be achieved. As of now, the emergent use of stem cells appears as an exciting and encouraging field to discover. As well, understanding the interaction between cells and the ECM could pave a road in designing an innovative approach for therapeutic applications. In conclusion, it is clear that the decellularized pancreas scaffold is much more similar to native tissue, and appropriate application of this scaffold could be a great assist to patients in need of transplantation.

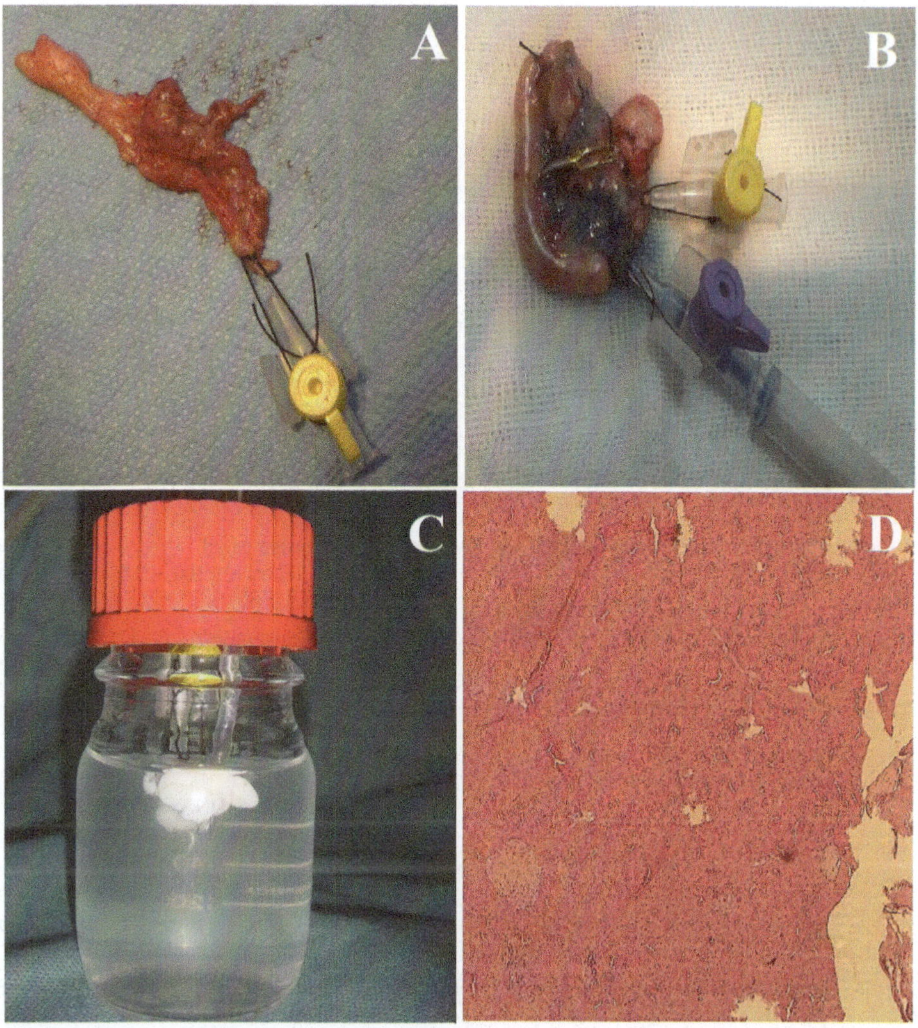

(A) cannulation of pancreas for further decellularization process (B) Injection of methylene blue in decellularized scaffold to demonstrate preservation of structures; the dye did not leak from the scaffold.(C) Decellularized pancreas (D) H&E staining of decellularized scaffold

Reference	Results	Duration	Method of decellularization	Species
1	Complete cells removal + rich ECM scaffold	64 h	Perfusion-based (1% Triton × −100 + 0.1% Ammonia + DNase 2000U/ml + PBS)	Mouse pancreas
2	Complete cells removal + intact vasculature system	72 h	Perfusion-based 0.5% SDS + Milli-Q water containing benzonase + DMEM containing 10% FBS	Mouse pancreas
3	Tissue disruption or incomplete cell removal Intact ECM and proper biomechanical characteristics Intact ECM and proper biomechanical characteristics Intact ECM and proper biomechanical characteristics	16 h 18 h 20 h 20 h	Perfusion-based 0.25% Triton × −100 + 0.5 SDS + PBS 0.25% Triton × −100 + 0.25 SDS + PBS 0.25% Triton × −100 + 0.1 SDS + PBS 0.25% Triton × −100 + 0.05 SDS + PBS	Rat pancreas
4	Complete cells removal + rich ECM scaffold	4 h and 18 min	Perfusion-based 1% Triton × −100 + 0.5% SDS + PBS	Rat pancreas
5	Complete cells removal	26 h	Perfusion-based 1%Triton × −100 + 0.1% ammonium hydroxide + PBS	Rat pancreas
6	Complete cell and DNA clearance, preservation of ECM components, growth factors and stiffness, ability to induce angiogenesis, conservation of the framework of the innate vasculature, and immunogenicity	7 days and 1 h	Perfusion-based 1% Triton × −100 + 0.1% ammonium hydroxide + DNAse + 0.0025% magnesium chloride + PBS	Human pancreas
7	Cellular material removal + preservation of ECM proteins and the native vascular tree	24 h	Perfusion-based 1% Triton × −100 + 0.1% ammonium hydroxide + PBS	Porcine pancreas
8	Production of decellularized scaffold which retained the gross shape of the whole organ	63 h	Perfusion-based 0.05% trypsin + 0.1% Triton × 100 + 0.05% EGTA	Porcine pancreas

References

Abdelalim EM, Emara MM (2015) Advances and challenges in the differentiation of pluripotent stem cells into pancreatic β-cells. World J Stem Cells 7(1):174

Assady S et al (2001) Insulin production by human embryonic stem cells. Diabetes 50(8):1691–1697

Basford CL et al (2012) The functional and molecular characterisation of human embryonic stem cell-derived insulin-positive cells compared with adult pancreatic beta cells. Diabetologia 55(2):358–371

Blaudez F, et al (2019) An overview of decellularisation techniques of native tissues and tissue engineered products for bone, ligament and tendon regeneration. Methods

Brennan DC et al (2004) Portal vein thrombosis complicating islet transplantation in a recipient with the factor V Leiden mutation. Transplantation 78(1): 172–173

Chandra V, et al (2011) Islet-like cell aggregates generated from human adipose tissue derived stem cells ameliorate experimental diabetes in mice. PloS one **6** (6):e20615

Chen L-B, Jiang X-B, Yang L (2004) Differentiation of rat marrow mesenchymal stem cells into pancreatic islet beta-cells. World J Gastroenterol: WJG 10(20): 3016

De Carlo E et al (2010) Pancreatic acellular matrix supports islet survival and function in a synthetic tubular device: in vitro and in vivo studies. Int J Mol Med 25(2):195–202

De Filippo RE, Yoo JJ, Atala A (2002) Urethral replacement using cell seeded tubularized collagen matrices. J Urol 168(4 Part 2):1789–1793

Dor Y et al (2004) Adult pancreatic β-cells are formed by self-duplication rather than stem-cell differentiation. Nature 429(6987):41

Falke G et al (2003) Formation of corporal tissue architecture in vivo using human cavernosal muscle and endothelial cells seeded on collagen matrices. Tissue Eng 9(5):871–879

French KM, Davis ME (2019) Decellularized extracellular matrix-based cardiovascular tissue engineering. Cardiovascular Regenerative Medicine. Springer, pp 35–62

Gittes GK et al (1996) Lineage-specific morphogenesis in the developing pancreas: role of mesenchymal factors. Development 122(2):439–447

Goh S-K et al (2013) Perfusion-decellularized pancreas as a natural 3D scaffold for pancreatic tissue and whole organ engineering. Biomaterials 34(28):6760–6772

Guan LX et al (2015) Therapeutic efficacy of umbilical cord-derived mesenchymal stem cells in patients with type 2 diabetes. Exp Ther Med 9(5):1623–1630

Haag J et al (2012) Biomechanical and angiogenic properties of tissue-engineered rat trachea using genipin cross-linked decellularized tissue. Biomaterials 33(3):780–789

Hammar E et al (2004) Extracellular matrix protects pancreatic β-cells against apoptosis: role of short-and long-term signaling pathways. Diabetes 53(8):2034–2041

Hashemi J et al (2018a) Application of a novel bioreactor for in vivo engineering of pancreas tissue. J Cell Physiol 233(5):3805–3816

Hashemi J et al (2018b) Decellularized pancreas matrix scaffolds for tissue engineering using ductal or arterial catheterization. Cells Tissues Organs 205:72–84

Hayek A et al (1995) Growth factor/matrix-induced proliferation of human adult β-cells. Diabetes 44 (12):1458–1460

Hayward JA et al (2017) Cotransplantation of mesenchymal stem cells with neonatal porcine islets improve graft function in diabetic mice. Diabetes 66(5):1312–1321

He C, Yang Z, Lu N-H (2014) Helicobacter pylori infection and diabetes: is it a myth or fact? World J Gastroenterol: WJG 20(16):4607

He M et al (2017) Optimization of SDS exposure on preservation of ECM characteristics in whole organ decellularization of rat kidneys. J Biomed Mater Res B Appl Biomater 105(6):1352–1360

Hisaoka M, Haratake J, Hashimoto H (1993) Pancreatic morphogenesis and extracellular matrix organization during rat development. Differentiation 53(3):163–172

Hosoya M (2012) Preparation of pancreatic β-cells from human iPS cells with small molecules. Islets 4(3):249–252

Jain A et al (2019) Evaluation of post-transcriptional gene regulation in pancreatic cancer cells: studying RNA binding proteins and their mRNA targets. Pancreatic Cancer. Springer, pp 239–252

Kawakubo K et al (2016) Effect of fetal membrane-derived mesenchymal stem cell transplantation in rats with acute and chronic pancreatitis. Pancreas 45(5): 707–713

Keane TJ et al (2012) Consequences of ineffective decellularization of biologic scaffolds on the host response. Biomaterials 33(6):1771–1781

Kelly WD et al (1967) Allotransplantation of the pancreas and duodenum along with the kidney in diabetic nephropathy. Surgery 61(6):827–837

Lester LB et al (2004) Directed differentiation of rhesus monkey ES cells into pancreatic cell phenotypes. Reprod Biol Endocrinol 2(1):42

Lumelsky N et al (2001) Differentiation of embryonic stem cells to insulin-secreting structures similar to pancreatic islets. Science 292(5520):1389–1394

Matta C, Mobasheri A (2014) Regulation of chondrogenesis by protein kinase C: emerging new roles in calcium signalling. Cell Signal 26(5):979–1000

Mirmalek-Sani S-H et al (2013) Porcine pancreas extracellular matrix as a platform for endocrine pancreas bioengineering. Biomaterials 34(22):5488–5495

Napierala H et al (2017) Engineering an endocrine Neo-Pancreas by repopulation of a decellularized rat pancreas with islets of Langerhans. Sci Rep 7:41777

Nouri Barkestani M, Naserian S, Uzan G, Shamdani S (2021) Postdecellularizationtechniques ameliorate cartilage decellularization process for tissue engineering applications. JTissue Eng 12:2041731420983562

Peloso A, et al (2015) Renal extracellular matrix scaffolds from discarded kidneys maintain glomerular morphometry and vascular resilience and retains critical growth factors. Transplantation 99(9):1807–1816

Rattananinsruang P, Dechsukhum C, Leeanansaksiri W (2018) Establishment of insulin-producing cells from human embryonic stem cells underhypoxic condition for cell based therapy. Front Cell Dev Biol **6**

Rieder E et al (2004) Decellularization protocols of porcine heart valves differ importantly in efficiency of cell removal and susceptibility of the matrix to recellularization with human vascular cells. J Thorac Cardiovasc Surg 127(2):399–405

Ryan EA et al (2004) Risks and side effects of islet transplantation. Curr Diab Rep 4(4):304–309

Schulz TC, et al (2012) A scalable system for production of functional pancreatic progenitors from human embryonic stem cells. PloS one 7(5):e37004

Shapiro AJ et al (2000) Islet transplantation in seven patients with type 1 diabetes mellitus using a glucocorticoid-free immunosuppressive regimen. N Engl J Med 343(4):230–238

Shapiro AJ et al (2006) International trial of the Edmonton protocol for islet transplantation. N Engl J Med 355(13):1318–1330

Soria B et al (2000) Insulin-secreting cells derived from embryonic stem cells normalize glycemia in streptozotocin-induced diabetic mice. Diabetes 49(2): 157–162

Soto-Gutierrez A et al (2012) Perspectives on whole-organ assembly: moving toward transplantation on demand. J Clin Investig 122(11):3817–3823

Tang Q et al (2004) In vitro–expanded antigen-specific regulatory T cells suppress autoimmune diabetes. J Exp Med 199(11):1455–1465
Tudorache I, et al (2007) Tissue engineering of heart valves: biomechanical and morphological properties of decellularized heart valves. J Heart Valve Dis 16(5):567–73. discussion 574
Urban VS et al (2008) Mesenchymal stem cells cooperate with bone marrow cells in therapy of diabetes. Stem Cells 26(1):244–253
Vériter S, Gianello P, Dufrane D (2013) Bioengineered sites for islet cell transplantation. Curr DiabRep 13(5): 745–755
Villiger P et al (2005) Prevention of bleeding after islet transplantation: lessons learned from a multivariate analysis of 132 cases at a single institution. Am J Transplant 5(12):2992–2998
Weber LM, Hayda KN, Anseth KS (2008) Cell–matrix interactions improve β-cell survival and insulin secretion in three-dimensional culture. Tissue Eng Part A 14(12):1959–1968
Yang B et al (2010) Development of a porcine bladder acellular matrix with well-preserved extracellular bioactive factors for tissue engineering. Tissue Eng Part C Methods 16(5):1201–1211

Decellularization of Small Intestinal Submucosa

7

Sahar Jelodari and Esmaeil Sadroddiny

Abstract

Small intestinal submucosa (SIS) is the most studied extracellular matrix (ECM) for repair and regeneration of different organs and tissues. Promising results of SIS-ECM as a vascular graft, led scientists to examine its applicability for repairing other tissues. Overall results indicated that SIS grafts induce tissue regeneration and remodeling to almost native condition. Investigating immunomodulatory effects of SIS is another interesting field of research. SIS can be utilized in different forms for multiple clinical and experimental studies. The aim of this chapter is to investigate the decellularization process of SIS and its common clinical application.

Keywords

Decellularization · Small intestine · Submucosa · ECM · Scaffold

S. Jelodari
Department of Tissue Engineering and Applied Cell Sciences, School of Advanced Technologies in Medicine, Tehran University of Medical Sciences, Tehran, Iran

E. Sadroddiny (✉)
Department of Medical Biotechnology, School of Advanced Technologies in Medicine, Tehran University of Medical Sciences, Tehran, Iran
e-mail: sadroddiny@sina.tums.ac.ir

7.1 History

Small intestinal submucosa (SIS) is the most studied extracellular matrix (ECM) for repair and regeneration of different organs and tissues (Holubec et al. 2015). It is a trilaminar structure consisting of *stratum compactum*, *muscularis mucosa* and *submucosa*. The main components of this structure are various collagen types including I (comprises 90% of SIS), III, IV, V, and VI, fibronectin, glycosaminoglycans such as heparin, growth factors like vascular endothelial growth factor (VEGF), transforming growth factor beta (TGFβ) and basic fibroblast growth factor (bFGF) and proteoglycans (Badylak et al. 1999). Recently, matrix-bound nanovesicles (MBV) which are membranous nanovesicles similar to exosomes, but are bound within collagen fibrils rather than circulating in body fluids, are also introduced as components of different ECM types including SIS. It's been shown that MBVs are responsible for biological effects of ECM bioscaffolds; like modulation of macrophages toward M2 anti-inflammatory phenotype and promoting cell differentiation in vitro (Huleihel et al. 2016).

The very first application of SIS dates back to more than 50 years ago when it was prepared by physical abrasion of inverted small intestine and used as an autograft to replace inferior vena cava in dogs (Matsumoto et al. 1966). A few similar studies were also conducted with inverted small

A. Kajbafzadeh (ed.), *Decellularization Methods of Tissue and Whole Organ in Tissue Engineering*, Advances in Experimental Medicine and Biology 1345,
https://doi.org/10.1007/978-3-030-82735-9_7

intestinal submucosa to replace thoracic aorta and vena cava (Huber et al. 2003; Lawler et al. 1971). Emerging as a new material in the world of regenerative medicine, noninverted SIS also primarily was used to repair small and large diameter vasculatures such as inferarenal aorta (Badylak et al. 1989) and femoral or carotid artery (Lantz et al. 1990). In most of the aforementioned studies the grafts were patent; however, there were also cases of failure due to aneurysm formation (Lawler et al. 1971). In 1992 porcine SIS was studied for the first time and used to substitute carotid artery in dogs. Regarding the results, SIS-ECM xenografts and saphenous vein autografts elicited similar responses, indicating the competence of SIS (Sandusky et al. 1992). Also, it was shown that in comparison to polytetrafluoroethylene (PTFE) grafts, SIS grafts were more resistant to bacterial infection (Badylak et al. 1994). Seemingly, this new material could overcome shortcomings related to vascular and synthetic grafts; it was shown that SIS graft is nonthrombogenic, nonimmunogenic, not infected, not mineralized and also it supports tissue remodeling, neovascularization and endothelial cell growth (Badylak 1993).

From 1994 onward, SIS grafts were intensively investigated to repair defects of bladder (Knapp et al. 1994), ligament (Aiken et al. 1994), tendon (Badylak et al. 1995), skin (Prevel et al. 1995a), abdominal wall (Prevel et al. 1995b), dura mater (Cobb et al. 1996), urethra (Kropp et al. 1998), cartilage (Peel et al. 1998), bone (Suckow et al. 1999), fascia (Dejardin et al. 1999) and corneosclera (Lewin 1999) in animal models. In 1998, Badylak et al. cultured different cell types on SIS-ECM scaffolds and found them suitable (1998). Meanwhile, Surgisis® was marketed as the first FDA approved SIS-derived product. Soon after, other tissue specific SIS products such as Oasis®, Restore® and Cuff-Patch™ were marketed in early 2000s (Badylak 2004). At the same time, clinical applications of these products, which we will discuss later in this chapter, began to come out. Also, examining SIS grafts to repair different new tissue injuries in animal models continued to emerge by the beginning of twenty-first century; namely they had defects in esophagus (Badylak et al. 2000), heart valve (Matheny et al. 2000), small bowel (Chen and Badylak 2001), ureter (Jaffe et al. 2001; Sofer et al. 2002), meniscus (Welch et al. 2002), heart ventricle (Badylak et al. 2003), larynx (Huber et al. 2003) and tympanic membrane (Spiegel and Kessler 2005). Although most studies demonstrated compelling results, there were also failures and shortcomings using SIS grafts in preclinical and clinical studies; which will be mentioned later in this chapter.

The most common type of SIS in preclinical and clinical settings are single or multilayered sheets, but powders (Khang et al. 2002) and hydrogels (Keane et al. 2017) are also prepared from this biomaterial (Fig. 7.1). For example, SIS hydrogel was examined to treat ulcerative colitis in a study by Keane et al. The hydrogel was adhesive to colonic tissue and could reduce the signs of the disease in a rodent model (2017). Cell seeding is another way to produce a SIS scaffold with more desirable properties in some applications. On the other hand, SIS-ECM modifications are also investigated in some studies; although primarily it was emphasized that these changes, such as crosslinking, can reduce its impact on tissue remodeling and regeneration by inducing foreign body reaction and decreasing of SIS degradation rate (Badylak 2004). The purpose of these modifications is to obtain scaffolds with greater strength and homogeneity or SIS-derived improvement of biological properties. In one study, SIS was modified by poly(lactic-co-glycolic) acid nanoparticles (PLGA NPs) which resulted in a more homogenous SIS and also enhanced growth of endothelial cells in vitro (Mondalek et al. 2008). It's been shown that Electrospun poly(ε-caprolactone) (PCL)/SIS powder conduits have enhanced hydrophilicity and caused more favorable neural cell attachment in comparison to PCL (Hong and Kim 2010). Also, injectable SIS biomaterial was successfully applied for repair of ischemic myocardium in murine models (Okada et al. 2010) or as drug depot to treat rheumatoid arthritis (Kim et al. 2016). Also, Adipose-derived stem cells on SIS microcarriers accelerated

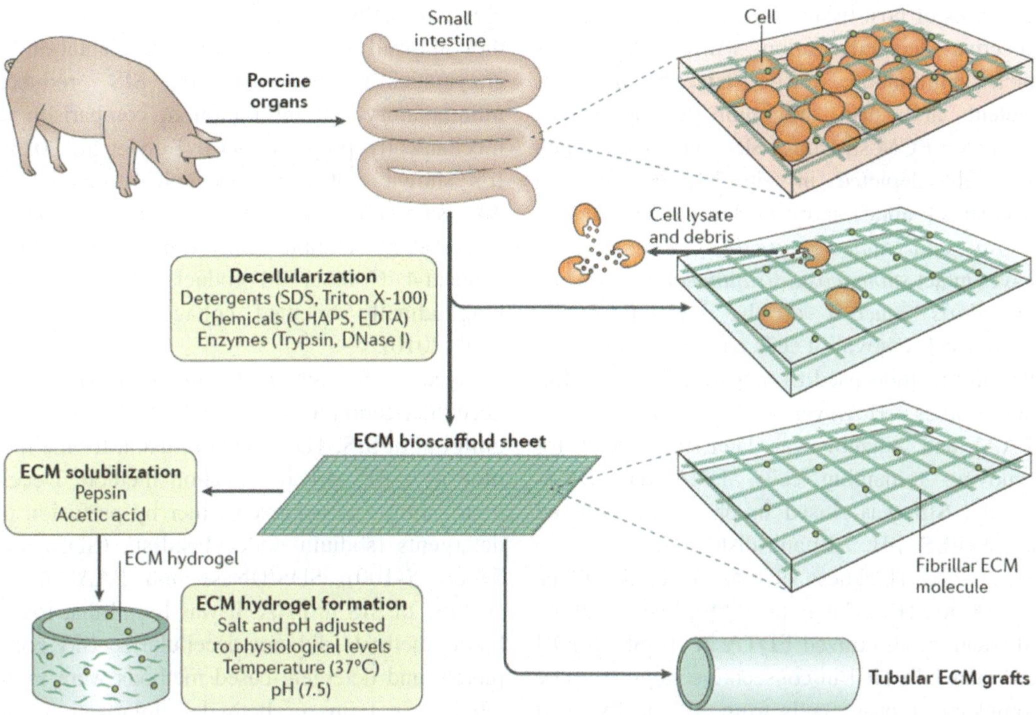

Fig. 7.1 Preparation of bioscaffolds from small intestinal submucosa (SIS) requires mechanical, detergent-based (such as sodium dodecyl sulfate (SDS)), chemical (such as EDTA) and enzymatic (such as DNase and trypsin) treatment of the tissue to remove cellular components. After decellularization, the SIS sheet can also be utilized in tubular form for specific applications like gastrointestinal tract reconstruction and also as hydrogels. Pepsin is usually the solubilizing enzyme to obtain the precursor material for producing hydrogel out of SIS (Hussey et al. 2017). Reprinted from Hussey et al. (2017) with permission from Springer Nature

wound closure in vivo (Zhou et al. 2011). Improved mechanical properties and better ECM production in hybrid PLGA-SIS scaffolds has also been reported; respectively due to PLGA and SIS (Kim et al. 2014). According to Da et al. chemically crosslinked polyurethane (PU)/SIS scaffolds are suitable for soft tissue engineering because of their superior resilience and improved cell viability (2017). FGF2-impregnated gelatin hydrogel sheet was embedded in SIS sheet and utilized successfully as a cardiac patch with FGF2 controlled release in pigs (Tanaka et al. 2015).

Investigating immunomodulatory effects of SIS is another interesting field of research. SIS elicits Th2-mediated immune response which can promote graft remodeling by attenuating pro-inflammatory cytokine induction of Th1 pathway. SIS enhanced immunity against prostate tumor by evoking cell-mediated immunity (Suckow et al. 2008). Adjuvanticity of commercial SIS was also evaluated in a mouse model and compared with alum. In contrast to alum, SIS did not provoke pro-inflammatory cytokines; but SIS-mediated Th2 antibody response was as effective as alum (Aachoui and Ghosh 2011).

In the following sections, we will discuss different methods of decellularization and SIS preparation protocols. Also seeding of different cell types, various clinical applications and possible limitations will be mentioned separately.

7.2 Materials and Methods

SIS which is commonly derived from jejunum of porcine small intestine has to be decellularized due to the inflammatory and immunological responses of host tissues to cellular components. Effective decellularization will result in the nuclei elimination, drastic reduction of DNA content and also minimum disruption of remaining ECM structure. SIS preparation which is nicely depicted in Fig. 7.1, begins with mechanical abrasion of *tunica mucosa*, *tunica serosa*, and *tunica muscularis externa*; and leaves intact *submucosa*, *stratum compactum* and *muscularis mucosa* at the end. Primarily, mechanical removal of cells and saline wash was the only method used to prepare SIS in preclinical studies. However, 0.1% peracetic acid (PAA) and ethanol was later introduced for removing remaining cells, DNA and RNA. Finally, SIS was rinsed in phosphate buffered saline (PBS, pH: 7.0) and distilled water extensively. The Thickness of final material will be 80–100 μm (Badylak et al. 1999). Later in 2000, Abraham et al. utilized EDTA in 10 mM NaOH and then 1 M HCl in consecutive steps, to more completely remove cells from SIS. In the next step, it was washed with PBS and degreased by methylene chloride (2000). In another study, enzymatic digestion and detergent treatment after mechanical abrasion and lipid removal helped to eliminate cellular components (Fig. 7.2). Also, they showed that subcutaneous implantation of enzymatically decellularized SIS reduced inflammatory cell infiltration in comparison to mechanically prepared ones (Luo et al. 2011). It's shown that after mechanical removal of mucosa and serosa, further decellularization was achieved by sodium deoxycholate and sodium azide at 4 °C. Final SIS product was durable with longitudinally oriented collagen fibers (Hata et al. 2010).

Syed et al. compared different protocols of decellularization to produce a tubular esophageal graft out of SIS. They used perfusion for the first time as a SIS decellularization method. Decellularization was achieved either by perfusion of detergents (sodium dodecyl sulfate (SDS) and Triton X-100), SDS/DNase and PAA or by shaking of SIS in PAA. It was found that PAA-based methods did not decellularize SIS completely and detergent-based methods were more efficient in terms of both decellularization and retaining mechanical properties. It was assumed that the tubular SIS or the method caused this insufficient decellularization rather than the PAA itself. On the other hand, biocompatibility of

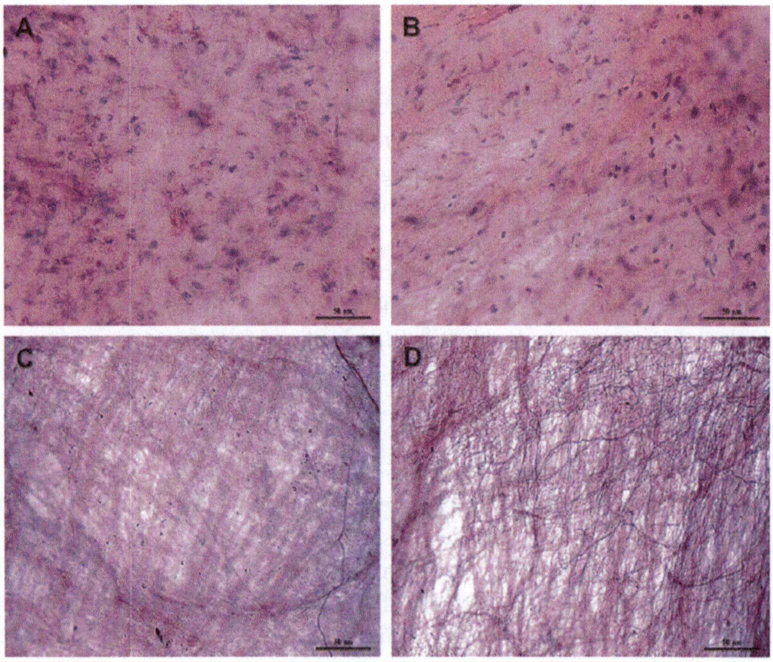

Fig. 7.2 H&E staining of SIS sample after decellularization by **a** mechanical treatment, **b** mechanical treatment and defatting, **c** Enzymatic digestion and **d** Detergent treatment. Cells are present in a and b (Luo et al. 2011). Reprinted from Luo et al. (2011) with permission from Elsevier

PAA treated SIS was far better than SDS/Triton X-100 treated ones; but lower than that of SDS/DNase treated samples. As a result, it is concluded that SDS/DNase method would be suitable in all aspects if DNase elimination step will be added (2014). In 2018, Ji et al. evaluated four decellularization methods; namely simple mechanical abrasion (SIS-N) or PAA plus ethanol treatment (SIS-A), acid-base treatment (SIS-B) and enzyme-detergent treatment (SIS-C) after mechanical delamination. The level of decellularization was similar in SIS-A, B and C and much more than SIS-N. Cell proliferation was also achieved at the same level in all processed tissues. However, microstructure of SIS-C was damaged the most with obvious collagen fibers breakage. SIS-C also had the minimum types of proteins between these groups. At the end, it was stated that since SIS preparation by badylak method (SIS-A) underwent brief decellularization with optimum results and retaining of active components and native microstructure; it could be the effective way for SIS decellularization (2019).

Other sources of SIS were also examined to evaluate their advantages or disadvantages over porcine SIS. Rashtbar et al. reported the decellularization of ovine SIS, which is thinner than porcine SIS, with lower concentration of detergents in comparison to common protocols and so resulted in less harmful effects of detergent to the Scaffold (2018). In another study, bovine SIS was subjected to freeze-thaw cycles for better disruption of cell membrane, before PAA treatment (Parmaksiz et al. 2018). It is worth mentioning that small bowel decellularization as a whole was also studied in order to retain fine structure of small intestine which could replace properly the defected intestinal tissue (Nowocin et al. 2016). In a well-established study, Totonelli et al. decellularized rat small intestine by the combination of detergent (sodium deoxycholate) and enzymatic (DNaseI) treatment in a continuous peristaltic delivery setting via both lumen and vasculature. This method resulted in decellularized small intestine with preserved villus-crypt architecture (2012).

After decellularization, the common procedure is lyophilization and then sterilization of SIS with ethylene oxide, gamma or e-beam irradiation. It's been shown that hydrolytic degradation of SIS would accelerate after sterilization and unsterilized SIS has more prolonged effect on protein expression and cell growth (Grimes et al. 2005). Retention of growth factors like TGFβ, VEGF and bFGF after decellularization and sterilization of SIS has been confirmed (McDevitt et al. 2003; Azzarello et al. 2007); so their beneficial effects on cell growth and proliferation would not be eliminated by processing. Dehydration of the product by means of lyophilization renders a more durable product with prolonged shelf life. Also, multilayered SIS was usually produced to bring mechanical strength for different applications. Vacuum pressing is a physical procedure that is established to produce laminated SIS (applied to obtain 10-layered Restore® product). Carbodiimide can also be used as a crosslinker to laminate SIS, which does not inhibit cell growth in contrast to glutaraldehyde (Abraham et al. 2000). It is noteworthy mentioning that since SIS has high rate of degradation and bioinductivity, mechanical properties will change rapidly after implantation.

SIS hydrogel preparation has also been examined in different in vitro and in vivo studies. Usually, after decellularization and lyophilization, ECM is pulverized and then solubilized by pepsin and different acidic condition or variation in treatment duration (Fig. 7.1). Then temperature-mediated gelation is achieved by assembly of collagen fibers at 37 °C. Possibility of applying in minimally invasive procedures while retaining biochemical properties is the advantage of the SIS hydrogels. ECM hydrogel preparation methods and characterization of these hydrogels is comprehensively reviewed by Saldin et al. (2017).

7.3 Cell Seeding

Suitable SIS microstructure and therapeutic importance of cellular component encouraged scientists to culture different cell types on SIS scaffolds and assess their capability in various tissue regeneration applications either by in vitro or in vivo studies. In most of these studies SIS-ECM supported the growth, proliferation, migration and sometime differentiation of different cell types. Primarily, Badylak et al. cultured six different cell types including three types of fibroblasts, human primary keratinocytes, human microvascular endothelial cells (HMECs) and an osteosarcoma cell line. All of these cell types proliferated on SIS substrate similar to their in vivo environment (1998). This distinctive spatial orientation and behavior regarding cell type was confirmed by co-culturing fibroblasts and epidermal cells on SIS. Presence of epidermal cells facilitated fibroblast migration and invasiveness (Lindberg and Badylak 2001). In another study, different cell types were cultured on SIS and SIS-derived gel. The ability of both substrata to support specific morphology, growth and differentiation of these cell types were equivalent or superior in comparison to Matrigel and Vitrogen (Voytik-Harbin et al. 1998). It's been shown by Badylak et al. that HMECs adhere to hydrated SIS in a greater degree than plastic dishes and dehydrated or rehydrated SIS. Also, collagen type IV and fibronectin are not the only proteins which cause cell attachment to SIS; and RGD peptide is not critical for HMECs binding to SIS (1999). It's been shown that culture of pancreatic islets on SIS would improve their in vitro function and help them retain their morphology (Woods et al. 2004). Another in vitro study successfully recruited muscle-derived cells to remodel SIS to the state of similar compliance to bladder wall (Lu et al. 2005). Growth and differentiation of bone marrow derived-mesenchymal stem cells (MSC) (Zhang et al. 2005) and migration and growth of embryonic stem cells (ESCs) (Lakshmanan et al. 2005) is also supported by SIS-ECM. Biomechanical properties of cell-seeded SIS after cyclic loading was also increased in comparison to non-loaded and non-seeded SIS; so made it suitable for tendon repair (Androjna et al. 2007). In cardiovascular field, Thick contracting myocardial graft was generated from cardiomyocyte (CM)-seeded SIS accompanied by stacking of one or three-layered CM cell sheets (Hata et al. 2010). Human crypt-derived primary intestinal epithelial (IECs) cells from small intestinal organoids were seeded as single cells on SIS and cocultured with fibroblasts in transwell-like setup to establish small intestine model. This led to emergence of differentiated intestinal cells and formation of intact epithelial barrier (Schweinlin et al. 2016). Sun et al. examined the osteogenic and angiogenic capacity of composite of SIS and mesoporous bioactive glass (MBG). Osteogenic and angiogenic gene expression of rat bone marrow stem cells (rBMSCs) and human umbilical vein endothelial cells (HUVECs) were respectively upregulated in composite scaffolds in comparison to SIS and SIS/non-mesoporous BG (2018). However, generally speaking, it's believed that tissue-specific ECM can profoundly contribute to maintenance of cell phenotype and promotion of cell proliferation and differentiation in tissue-specific manner (Hussey et al. 2017).

There are also in vivo experiments utilizing cell-seeded SIS-ECM; not only to improve mechanical and biological properties of grafts, but also to establish more suitable niches for cell growth and differentiation. In an early example, chondrocytes were cultured on SIS and implanted to repair articular cartilage in rabbits. Overall results suggested that healing in SIS-cartilaginous transplanted defects and unfilled self-repaired ones was similar. It seemed that SIS alone inhibited host tissue formation in vivo (Peel et al. 1998). Bladder augmentation in a mouse model with smooth muscle and urothelial cell-seeded SIS was also conducted successfully and organized bladder regeneration was achieved; although the number of cells were significantly reduced after implantation (Zhang et al. 2004). MSC-seeded SIS were applied to the infarct area in a rabbit model. SIS sheets were incorporated into infarcted host tissue with mild inflammatory reaction and no rejection response. Improved left ventricle contractile function and angiogenesis were observed in the patch implantation region (Tan et al. 2009). $Isl1^+$ embryonic cardiac progenitor cells (CPCs) were viable on SIS-ECM scaffolds and after 7 days differentiated in to cardiomyocytes and endothelial cells. These cardiomyocytes started contraction spontaneously and improved mouse cardiac function after myocardial infarction (Wang et al. 2017). Tongue regeneration of a rat model was also investigated by applying human gingiva-derived mesenchymal stem cells (GMSCs) seeded on SIS-ECM. Soft tissue healing and expression of myogenic transcription factor were promoted in GMSC/SIS-ECM group in comparison to control groups (Xu et al. 2017). Kim et al. studied the SIS sponge seeded with rBMSCs for repairing cranial bone defect in a rat model. Bone formation was significant in rBMSC-seeded SIS scaffolds in comparison to SIS itself and other control groups (2010). In order to achieve more controllable mechanical stability and degradation, SIS was also modified by four-arm polyethylene glycol (PEG) crosslinker. The bioactive sponge was successfully utilized to deliver skin cells to the wound site and accelerate healing (Dong et al. 2019). For further reading on older in vitro and in vivo applications of cell-seeded SIS we refer the readers to a review article by Andree et al. (2013).

7.4 Clinical Applications

Different SIS-based products such as Surgisis®, Durasis®, stratasis®, CuffPatch™, Oasis®, Restore®, CorMatrix® and Dynamatrix® are available for specific clinical applications. It is estimated that more than one million patients were subjected to SIS grafting in clinical basis (Mosala Nezhad et al. 2016; Badylak 2007). The popularity of ECM scaffolds in general and SIS scaffolds in particular, which led them to make their way toward clinic, is rooted in their bioinductive and biomechanical properties, modest immune response and reasonable degradation rate (Badylak 2007). Most of the marketed SIS products are available as patches and rehydration before application is necessary. Among these products, here we summarize the results related to those examined in several studies and entered to clinical trials and applications.

Surgisis® (Cook Biotech Inc., West Lafayette, IN, USA) is a 4 or 6-ply SIS sheet and mostly substituted to repair soft tissues like hiatal (Oelschlager et al. 2012; Watson et al. 2015), diaphragmatic (Smith et al. 2004; Romao et al. 2012) and inguinal hernia (Ravo and Falasco 2019). Although in most cases SIS grafts were support tissue repair with less pain and complications, hernia recurrence in long term is an obvious shortcoming of the patch in comparison to suturing and also evidence in favor of SIS over synthetic meshes is not sufficient (Antoniou et al. 2015; Grethel et al. 2006; Sarr et al. 2014). Synthetic mesh and 4 or 6-ply SIS grafting in patients with gastroesophageal reflux disease who underwent crura reinforcement showed that with respect to reduction of hernia recurrence and patient satisfaction, 6-ply SIS is a promising substitute (Wang et al. 2016). There are also studies which grafted SIS for vaginal prolapse treatment. Overall results were favorable with reduced recurrence rate and improvement of disease condition (Armitage et al. 2012; Feldner et al. 2010). SIS-assisted bladder augmentation in a small group of patients was not satisfactory due to insufficient increase in bladder compliance (Schaefer et al. 2013). Peyronie's disease (PD) is an unusual curvature in penis which is sometimes accompanied by erectile dysfunction (ED). Inflammation with subsequent plaque formation in the tunica albuginea is the cause of this abnormality. Patients with PD were also subjected to SIS grafting in several studies. In most of these cases ED was reported after surgery. However, except one study that reported curvature recurrence (John et al. 2006), the overall outcome of these studies was reported as satisfactory and with a low risk of complication (Cosentino et al. 2016; Rosenhammer et al. 2018). Treatment of anal fistula by SIS grafts was also examined in clinical studies; with the reported success rate ranging from 20 to 94% (Garg et al. 2010; O'Riordan et al. 2012). There are also two studies with small sample size that compared Surgisis® and Gore synthetic plug in anal fistula repair. In one study, healing rate was 12.5% for SIS compared to 54.4% in Gore plug (Buchberg et al. 2010); and 39% for both plugs in another study (Hansen et al. 2019). Chest wall reconstruction after Ewing's tumor resection is also case-reported. Superior wound healing and symmetric growth of chest wall led to the conclusion that SIS utilization is a reasonable path toward chest wall reconstruction. However, the authors mentioned that the follow-up period was short (Murphy and Corbally 2007). Tubular SIS grafts were successfully transplanted in five patients with esophageal adenocarcinoma. The follow-up showed restitution of normal esophagus without recalcitrant stricture formation and recurrence of neoplasia (Badylak et al. 2011).

Treatment of different types of skin wounds was also clinically evaluated by SIS grafting. Oasis® wound matrix (Smith & Nephew, Inc., Andover, MA, USA) has been evaluated for treatment of chronic leg ulcers (Mostow et al. 2005), diabetic ulcers (Niezgoda et al. 2005; Landsman et al. 2008) and pressure ulcers (Brown-Etris et al. 2019). All of these randomized studies suggested that SIS is a suitable wound care product; the clinical outcomes are comparable to other available products and procedures like Regranex Gel (Niezgoda et al. 2005), living skin equivalent (Dermagraft®, Advanced BioHealing, La Jolla, Calif) (Landsman et al. 2008), compression therapy alone (Mostow et al. 2005) and standard care alone (Brown-Etris et al. 2019). Treatment of mixed arterial/venous (A/V) ulcers by SIS were also studied and compared to Hyaloskin® (Romanelli et al. 2007) and a moist wound dressing (Romanelli et al. 2010). In both cases, application of SIS-ECM resulted in higher number of ulcer closure.

Clinical application of 4-ply SIS CorMatrix® (CorMatrix Cardiovascular, Inc., Roswell, GA, USA) which is the most widely used SIS product in cardiovascular surgery, started almost 10 years ago; after multitude of preclinical studies (Mosala Nezhad et al. 2016). It is a strong and durable construct which makes it suitable in cardiovascular applications. Pericardial reconstruction by means of SIS patch was the first clinical report using this product. Although there were limitations in this study, relative risk of atrial fibrillation was reduced by 54%; which was compelling to consider this product as a suitable substitute in cardiac clinical applications (Boyd et al. 2010). Thereafter, CorMatrix® was applied to reconstruct pediatric and adult valve and vascular defects, congenital cardiac defects, endocarditis and infarcted myocardium; which have been comprehensively reviewed by Mosala Nezhad et al. (2016). Although most of the studies supported suitability of the patch, severe chronic inflammation (Zaidi et al. 2014; Rosario-Quinones et al. 2015), calcification (Stelly and Stelly 2013) and stenosis (Witt et al. 2013; Poulin et al. 2013) are major complications occurred in clinical settings and indicating the importance of further evaluation of CorMatrix® in randomized studies with large number of patients and longer follow-up periods (Mosala Nezhad et al. 2016). For further reading on preclinical studies using this SIS-ECM patch and other ECM based products, the readers are referred to a recent article by Iop et al. (2018).

7.5 Limitations

Although SIS products are widely examined in preclinical and clinical applications, reports of adverse or no beneficial effects should not be ignored. In clinical applications, failure of rotator cuff regeneration is an infamous example. In this regard; unimproved regeneration, retearing and low healing rate in comparison to control group were common complications (Sclamberg et al. 2004; Walton et al. 2007; Iannotti et al. 2006). In the case of stress urinary incontinence treatment by SIS, inflammatory response and failed procedure has been reported (John et al. 2008; Siracusano et al. 2011). It seems that matching of mechanical properties between target tissue and the scaffold, more importantly in under stress tissues, has an impact in success rate. Another issue to be considered is the differential regenerative impact of distinct segments of SIS. It's been shown that distal ileum resulted in a more robust regeneration and less adverse effects than proximal jejunum (Kropp et al. 2004). Insufficient nourishment of cells in 3D constructs of multilayer SIS is another limitation for establishing cell-seeded SIS to reinforce beneficial aspects of this natural scaffold. Chemical manipulation of SIS by means of other synthetic or natural materials and also crosslinkers are strategies of improving mechanical properties or to better control degradation rate of SIS; however, mostly they lead to weaker tissue remodeling. As a result, proper evaluation of the situation and considering all the benefits and limitations of SIS scaffolds and its variations would help to improve final outcome of tissue regeneration and lessen the risk of failure.

7.6 Conclusion

From an inverted autograft for vena cava replacement in dogs to several marketed products, SIS has come a long way before being established as a reliable substitute in regenerative medicine applications. Although whole small bowel decellularization was also reported, its applications will be limited to replacement of small intestinal segments and it can cause more pronounced immune response. The natural structure and properties of SIS-ECM scaffolds make them suitable substitutes in regenerative medicine and expand their application by their emergence. The growth factors like VEGF, TGFβ and bFGF which are preserved after SIS processing, enable tissue remodeling and regeneration. On the other hand, this scaffold promotes differentiation of monocytes toward M2 macrophages rather than M1 and accordingly facilitates tissue remodeling by releasing anti-inflammatory cytokines. Although some limitations and failures were also reported in preclinical and clinical applications, the beneficial properties of SIS-ECM predominated and opened up lots of opportunities for marketing SIS-based products. Obviously, like other synthetic and natural scaffolds, improving mechanical and functional properties of SIS-based scaffolds would enlarge its applicability in different fields of regenerative medicine.

References

Aachoui Y, Ghosh SK (2011) Extracellular matrix from porcine small intestinal submucosa (SIS) as immune adjuvants. PLoS One 6(11):e27083.https://doi.org/10.1371/journal.pone.0027083

Abraham GA, Murray J, Billiar K, Sullivan SJ (2000) Evaluation of the porcine intestinal collagen layer as a biomaterial. J Biomed Mater Res 51(3):442–452

Aiken SW, Badylak SF, Toombs JP, Shelbourne KD, Hiles MC, Lantz GC, Van Sickle D (1994) Small intestinal submucosa as an intra-articular ligamentous graft material: a pilot study in dogs. Vet Comp Orthop Traumatol 07(03):124–128. https://doi.org/10.1055/s-0038-1633133

Andree B, Bar A, Haverich A, Hilfiker A (2013) Small intestinal submucosa segments as matrix for tissue engineering: review. Tissue Eng Part B Rev 19 (4):279–291. https://doi.org/10.1089/ten.TEB.2012.0583

Androjna C, Spragg RK, Derwin KA (2007) Mechanical conditioning of cell-seeded small intestine submucosa: a potential tissue-engineering strategy for tendon repair. Tissue Eng 13(2):233–243. https://doi.org/10.1089/ten.2006.0050

Antoniou SA, Pointner R, Granderath FA, Kockerling F (2015) The use of biological meshes in diaphragmatic defects—an evidence-based review of the literature. Front Surg 2:56. https://doi.org/10.3389/fsurg.2015.00056

Armitage S, Seman EI, Keirse MJNC (2012) Use of surgisis for treatment of anterior and posterior vaginal prolapse. Obstet Gynecol Int 2012:376251–376251. https://doi.org/10.1155/2012/376251

Azzarello J, Ihnat MA, Kropp BP, Warnke LA, Lin HK (2007) Assessment of angiogenic properties of biomaterials using the chicken embryo chorioallantoic membrane assay. Biomed Mater (bristol, England) 2 (2):55–61. https://doi.org/10.1088/1748-6041/2/2/001

Badylak SF (2004) Xenogeneic extracellular matrix as a scaffold for tissue reconstruction. Transpl Immunol 12(3–4):367–377. https://doi.org/10.1016/j.trim.2003.12.016

Badylak SF (2007) The extracellular matrix as a biologic scaffold material. Biomaterials 28(25):3587–3593. https://doi.org/10.1016/j.biomaterials.2007.04.043

Badylak SF, Lantz GC, Coffey A, Geddes LA (1989) Small intestinal submucosa as a large diameter vascular graft in the dog. J Surg Res 47(1):74–80

Badylak SF (1993) Small intestinal submucosa (SIS): a biomaterial conducive to smart tissue remodeling. In: Bell E (ed) Tissue engineering: current perspectives. Birkhäuser Boston, Boston, MA, pp 179–189. https://doi.org/10.1007/978-1-4615-8186-4_17

Badylak SF, Coffey AC, Lantz GC, Tacker WA, Geddes LA (1994) Comparison of the resistance to infection of intestinal submucosa arterial autografts versus polytetrafluoroethylene arterial prostheses in a dog model. J Vasc Surg 19(3):465–472

Badylak SF, Tullius R, Kokini K, Shelbourne KD, Klootwyk T, Voytik SL, Kraine MR, Simmons C (1995) The use of xenogeneic small intestinal submucosa as a biomaterial for Achilles tendon repair in a dog model. J Biomed Mater Res 29(8):977–985. https://doi.org/10.1002/jbm.820290809

Badylak SF, Record R, Lindberg K, Hodde J, Park K (1998) Small intestinal submucosa: a substrate for in vitro cell growth. J Biomater Sci Polym Ed 9 (8):863–878

Badylak S, Liang A, Record R, Tullius R, Hodde J (1999) Endothelial cell adherence to small intestinal submucosa: an acellular bioscaffold. Biomaterials 20(23–24):2257–2263

Badylak S, Meurling S, Chen M, Spievack A, Simmons-Byrd A (2000) Resorbable bioscaffold for esophageal repair in a dog model. J Pediatr Surg 35(7):1097–1103. https://doi.org/10.1053/jpsu.2000.7834

Badylak S, Obermiller J, Geddes L, Matheny R (2003) Extracellular matrix for myocardial repair. Heart Surg Forum 6(2):E20-26

Badylak SF, Hoppo T, Nieponice A, Gilbert TW, Davison JM, Jobe BA (2011) Esophageal preservation in five male patients after endoscopic inner-layer circumferential resection in the setting of superficial cancer: a regenerative medicine approach with a biologic scaffold. Tissue Eng Part A 17(11–12):1643–1650. https://doi.org/10.1089/ten.tea.2010.0739

Boyd WD, Johnson WE 3rd, Sultan PK, Deering TF, Matheny RG (2010) Pericardial reconstruction using an extracellular matrix implant correlates with reduced risk of postoperative atrial fibrillation in coronary artery bypass surgery patients. Heart Surg Forum 13 (5):E311-316. https://doi.org/10.1532/hsf98.20091184

Brown-Etris M, Milne CT, Hodde JP (2019) An extracellular matrix graft (Oasis((R)) wound matrix) for treating full-thickness pressure ulcers: a randomized clinical trial. J Tissue Viability 28(1):21–26. https://doi.org/10.1016/j.jtv.2018.11.001

Buchberg B, Masoomi H, Choi J, Bergman H, Mills S, Stamos MJ (2010) A tale of two (anal fistula) plugs: is there a difference in short-term outcomes? Am Surg 76 (10):1150–1153

Chen MK, Badylak SF (2001) Small bowel tissue engineering using small intestinal submucosa as a scaffold. J Surg Res 99(2):352–358. https://doi.org/10.1006/jsre.2001.6199

Cobb MA, Badylak SF, Janas W, Boop FA (1996) Histology after dural grafting with small intestinal submucosa. Surg Neurol 46(4):389–393, discussion 393–384

Cosentino M, Kanashiro A, Vives A, Sanchez J, Peraza MF, Moreno D, Perona J, De Marco V, Ruiz-Castane E, Sarquella J (2016) Surgical treatment of Peyronie's disease with small intestinal submucosa graft patch. Int J Impot Res 28(3):106–109. https://doi.org/10.1038/ijir.2016.10

Da L, Gong M, Chen A, Zhang Y, Huang Y, Guo Z, Li S, Li-Ling J, Zhang L, Xie H (2017) Composite elastomeric polyurethane scaffolds incorporating small intestinal submucosa for soft tissue engineering. Acta Biomater 59:45–57. https://doi.org/10.1016/j.actbio.2017.05.041

Dejardin LM, Arnoczky SP, Clarke RB (1999) Use of small intestinal submucosal implants for regeneration of large fascial defects: an experimental study in dogs. J Biomed Mater Res 46(2):203–211

Dong Y, Liu Z, Qi F, Jin L, Zhang L, Zhu N (2019) Polyethylene-glycol-ornamented small intestinal submucosa biosponge for skin tissue engineering. ACS Biomater Sci Eng 5(5):2457–2465. https://doi.org/10.1021/acsbiomaterials.8b01592

Feldner PC Jr, Castro RA, Cipolotti LA, Delroy CA, Sartori MG, Girao MJ (2010) Anterior vaginal wall prolapse: a randomized controlled trial of SIS graft versus traditional colporrhaphy. Int Urogynecol J 21 (9):1057–1063. https://doi.org/10.1007/s00192-010-1163-8

Garg P, Song J, Bhatia A, Kalia H, Menon GR (2010) The efficacy of anal fistula plug in fistula-in-ano: a systematic review. Colorectal Dis: the official journal of the Association of Coloproctology of Great Britain and Ireland 12(10):965-970.https://doi.org/10.1111/j.1463-1318.2009.01933.x

Grethel EJ, Cortes RA, Wagner AJ, Clifton MS, Lee H, Farmer DL, Harrison MR, Keller RL, Nobuhara KK (2006) Prosthetic patches for congenital diaphragmatic hernia repair: surgisis vs gore-tex. J Pediatr Surg 41 (1):29–33. https://doi.org/10.1016/j.jpedsurg.2005.10.005

Grimes M, Pembroke JT, McGloughlin T (2005) The effect of choice of sterilisation method on the biocompatibility and biodegradability of SIS (small intestinal submucosa). Bio-Med Mater Eng 15(1–2):65–71

Hansen MS, Kjaer ML, Andersen J (2019) Efficacy of plug treatment for complex anorectal fistulae: long-term Danish results. Annal Coloproctol. https://doi.org/10.3393/ac.2018.07.14

Hata H, Bar A, Dorfman S, Vukadinovic Z, Sawa Y, Haverich A, Hilfiker A (2010) Engineering a novel three-dimensional contractile myocardial patch with cell sheets and decellularised matrix. Eur J Cardio-thoracic Surg: official journal of the European Association for Cardio-thoracic Surgery 38(4):450–455. https://doi.org/10.1016/j.ejcts.2010.02.009

Holubec T, Caliskan E, Sundermann SH, Starck CT, Plass A, Bettex D, Falk V, Maisano F (2015) Use of extracellular matrix patches in cardiac surgery. J Card Surg 30(2):145–148. https://doi.org/10.1111/jocs.12494

Hong S, Kim G (2010) Electrospun micro/nanofibrous conduits composed of poly(epsilon-caprolactone) and small intestine submucosa powder for nerve tissue regeneration. J Biomed Mater Res B Appl Biomater 94(2):421–428. https://doi.org/10.1002/jbm.b.31670

Huber JE, Spievack A, Simmons-Byrd A, Ringel RL, Badylak S (2003) Extracellular matrix as a scaffold for laryngeal reconstruction. Ann Otol Rhinol Laryngol 112(5):428–433. https://doi.org/10.1177/000348940311200508

Huleihel L, Hussey GS, Naranjo JD, Zhang L, Dziki JL, Turner NJ, Stolz DB, Badylak SF (2016) Matrix-bound nanovesicles within ECM bioscaffolds. Sci Adv 2(6):e1600502.https://doi.org/10.1126/sciadv.1600502

Hussey GS, Keane TJ, Badylak SF (2017) The extracellular matrix of the gastrointestinal tract: a regenerative medicine platform. Nat Rev Gastroenterol Hepatol 14 (9):540–552. https://doi.org/10.1038/nrgastro.2017.76

Iannotti JP, Codsi MJ, Kwon YW, Derwin K, Ciccone J, Brems JJ (2006) Porcine small intestine submucosa augmentation of surgical repair of chronic two-tendon rotator cuff tears. A randomized, controlled trial. Journal Bone Joint Surg Am 88(6):1238–1244. https://doi.org/10.2106/jbjs.e.00524

Iop L, Palmosi T, Dal Sasso E, Gerosa G (2018) Bioengineered tissue solutions for repair, correction and reconstruction in cardiovascular surgery. J Thorac Dis 10(Suppl 20):S2390-s2411. https://doi.org/10.21037/jtd.2018.04.27

Jaffe JS, Ginsberg PC, Yanoshak SJ, Costa LE, Jr., Ogbolu FN, Moyer CP, Greene CH, Finkelstein LH, Harkaway RC (2001) Ureteral segment replacement using a circumferential small-intestinal submucosa xenogenic graft. J Invest Surg: the official journal of the Academy of Surgical Research 14(5):259–265

Ji Y, Zhou J, Sun T, Tang K, Xiong Z, Ren Z, Yao S, Chen K, Yang F, Zhu F, Guo X (2019) Diverse preparation methods for small intestinal submucosa (SIS): decellularization, components, and structure. J Biomed Mater Res, Part A 107(3):689–697. https://doi.org/10.1002/jbm.a.36582

John TT, Aggarwal N, Singla AK, Santucci RA (2008) Intense inflammatory reaction with porcine small intestine submucosa pubovaginal sling or tape for stress urinary incontinence. Urology 72(5):1036–1039. https://doi.org/10.1016/j.urology.2008.07.043

John T, Bandi G, Santucci R (2006) Porcine small intestinal submucosa is not an ideal graft material for Peyronie's disease surgery. J Urol 176(3):1025–1028; discussion 1029. https://doi.org/10.1016/j.juro.2006.04.039

Keane TJ, Dziki J, Sobieski E, Smoulder A, Castleton A, Turner N, White LJ, Badylak SF (2017) Restoring mucosal barrier function and modifying macrophage phenotype with an extracellular matrix hydrogel: potential therapy for ulcerative colitis. J Crohns Colitis 11 (3):360–368. https://doi.org/10.1093/ecco-jcc/jjw149

Khang G, Rhee JM, Shin P, Kim IY, Lee B, Lee SJ, Lee YM, Lee HB, Lee I (2002) Preparation and characterization of small intestine submucosa powder impregnated poly(L-lactide) scaffolds: the application for tissue engineered bone and cartilage. Macromol Res 10 (3):158–167. https://doi.org/10.1007/BF03218266

Kim KS, Lee JY, Kang YM, Kim ES, Kim GH, Rhee SD, Cheon HG, Kim JH, Min BH, Lee HB, Kim MS (2010) Small intestine submucosa sponge for in vivo support of tissue-engineered bone formation in the presence of rat bone marrow stem cells. Biomaterials 31(6):1104–1113. https://doi.org/10.1016/j.biomaterials.2009.10.020

Kim SH, Song JE, Lee D, Khang G (2014) Development of poly(lactide-co-glycolide) scaffold-impregnated small intestinal submucosa with pores that stimulate extracellular matrix production in disc regeneration. J Tissue Eng Regen Med 8(4):279–290. https://doi.org/10.1002/term.1520

Kim K, Park JH, Park SH, Lee HY, Kim JH, Kim MS (2016) An injectable, click-cross-linked small intestinal submucosa drug depot for the treatment of rheumatoid arthritis. Adv Healthcare Mater 5(24):3105–3117. https://doi.org/10.1002/adhm.201601040

Knapp PM, Lingeman JE, Siegel YI, Badylak SF, Demeter RJ (1994) Biocompatibility of small-intestinal submucosa in urinary tract as augmentation cystoplasty graft and injectable suspension. J Endourol 8(2):125–130. https://doi.org/10.1089/end.1994.8.125

Kropp BP, Ludlow JK, Spicer D, Rippy MK, Badylak SF, Adams MC, Keating MA, Rink RC, Birhle R, Thor KB (1998) Rabbit urethral regeneration using small intestinal submucosa onlay grafts. Urology 52 (1):138–142. https://doi.org/10.1016/S0090-4295(98)00114-9

Kropp BP, Cheng EY, Lin HK, Zhang Y (2004) Reliable and reproducible bladder regeneration using unseeded distal small intestinal submucosa. J Urol 172(4 Pt 2):1710–1713

Lakshmanan Y, Frimberger D, Gearhart JD, Gearhart JP (2005) Human embryoid body-derived stem cells in co-culture with bladder smooth muscle and urothelium. Urology 65(4):821–826. https://doi.org/10.1016/j.urology.2004.11.022

Landsman A, Roukis TS, DeFronzo DJ, Agnew P, Petranto RD, Surprenant M (2008) Living cells or collagen matrix: which is more beneficial in the treatment of diabetic foot ulcers? Wounds Compendium Clin Res Pract 20(5):111–116

Lantz GC, Badylak SF, Coffey AC, Geddes LA, Blevins WE (1990) Small intestinal submucosa as a small-diameter arterial graft in the dog. J Invest Surg: the official journal of the Academy of Surgical Research 3(3):217–227

Lawler MR Jr, Foster JH, Scott HW Jr (1971) Evaluation of canine intestinal submucosa as a vascular substitute. Am J Surg 122(4):517–519

Lewin GA (1999) Repair of a full thickness corneoscleral defect in a German shepherd dog using porcine small intestinal submucosa. J Small Anim Pract 40(7):340–342

Lindberg K, Badylak SF (2001) Porcine small intestinal submucosa (SIS): a bioscaffold supporting in vitro primary human epidermal cell differentiation and

synthesis of basement membrane proteins. Burns J Int Soc Burn Injuries 27(3):254–266
Lu S-H, Sacks MS, Chung SY, Gloeckner DC, Pruchnic R, Huard J, de Groat WC, Chancellor MB (2005) Biaxial mechanical properties of muscle-derived cell seeded small intestinal submucosa for bladder wall reconstitution. Biomaterials 26(4):443–449. https://doi.org/10.1016/j.biomaterials.2004.05.006
Luo JC, Chen W, Chen XH, Qin TW, Huang YC, Xie HQ, Li XQ, Qian ZY, Yang ZM (2011) A multistep method for preparation of porcine small intestinal submucosa (SIS). Biomaterials 32(3):706–713. https://doi.org/10.1016/j.biomaterials.2010.09.017
Matheny RG, Hutchison ML, Dryden PE, Hiles MD, Shaar CJ (2000) Porcine small intestine submucosa as a pulmonary valve leaflet substitute. J Heart Valve Dis 9(6):769–774, discussion 774–765
Matsumoto T, Holmes RH, Burdick CO, Heisterkamp CA 3rd, O'Connell TJ Jr (1966) Replacement of large veins with free inverted segments of small bowel: autografts of submucosal membrane in dogs and clinical use. Ann Surg 164(5):845–848
McDevitt CA, Wildey GM, Cutrone RM (2003) Transforming growth factor-beta1 in a sterilized tissue derived from the pig small intestine submucosa. J Biomed Mater Res, Part A 67(2):637–640. https://doi.org/10.1002/jbm.a.10144
Mondalek FG, Lawrence BJ, Kropp BP, Grady BP, Fung KM, Madihally SV, Lin HK (2008) The incorporation of poly(lactic-co-glycolic) acid nanoparticles into porcine small intestinal submucosa biomaterials. Biomaterials 29(9):1159–1166. https://doi.org/10.1016/j.biomaterials.2007.11.020
Mosala Nezhad Z, Poncelet A, de Kerchove L, Gianello P, Fervaille C, El Khoury G (2016) Small intestinal submucosa extracellular matrix (CorMatrix (R)) in cardiovascular surgery: a systematic review. Interact Cardiovasc Thorac Surg 22(6):839–850. https://doi.org/10.1093/icvts/ivw020
Mostow EN, Haraway GD, Dalsing M, Hodde JP, King D (2005) Effectiveness of an extracellular matrix graft (OASIS Wound Matrix) in the treatment of chronic leg ulcers: a randomized clinical trial. J Vasc Surg 41(5):837–843. https://doi.org/10.1016/j.jvs.2005.01.042
Murphy F, Corbally MT (2007) The novel use of small intestinal submucosal matrix for chest wall reconstruction following Ewing's tumour resection. Pediatr Surg Int 23(4):353–356. https://doi.org/10.1007/s00383-007-1882-1
Niezgoda JA, Van Gils CC, Frykberg RG, Hodde JP, Group ODUS (2005) Randomized Clinical Trial Comparing OASIS Wound Matrix to Regranex Gel for Diabetic Ulcers. Adv Skin Wound Care 18(5):258–266
Nowocin AK, Southgate A, Gabe SM, Ansari T (2016) Biocompatibility and potential of decellularized porcine small intestine to support cellular attachment and growth. J Tissue Eng Regen Med 10(1):E23-33. https://doi.org/10.1002/term.1750
Oelschlager BK, Petersen RP, Brunt LM, Soper NJ, Sheppard BC, Mitsumori L, Rohrmann C, Swanstrom LL, Pellegrini CA (2012) Laparoscopic paraesophageal hernia repair: defining long-term clinical and anatomic outcomes. J Gastrointest Surg: official journal of the Society for Surgery of the Alimentary Tract 16(3):453–459.https://doi.org/10.1007/s11605-011-1743-z
Okada M, Payne TR, Oshima H, Momoi N, Tobita K, Huard J (2010) Differential efficacy of gels derived from small intestinal submucosa as an injectable biomaterial for myocardial infarct repair. Biomaterials 31(30):7678–7683. https://doi.org/10.1016/j.biomaterials.2010.06.056
O'Riordan JM, Datta I, Johnston C, Baxter NN (2012) A systematic review of the anal fistula plug for patients with Crohn's and non-Crohn's related fistula-in-ano. Dis Colon Rectum 55(3):351–358. https://doi.org/10.1097/DCR.0b013e318239d1e4
Parmaksiz M, Elcin AE, Elcin YM (2018) Decellularization of bovine small intestinal submucosa. Methods in Mol Biol (clifton, NJ) 1577:129–138. https://doi.org/10.1007/7651_2017_33
Peel SAF, Chen H, Renlund R, Badylak SF, Kandel RA (1998) Formation of a SIS–cartilage composite graft in vitro and its use in the repair of articular cartilage defects. Tissue Eng 4(2):143–155. https://doi.org/10.1089/ten.1998.4.143
Poulin F, Horlick EM, David T, Woo A, Thavendiranathan P (2013) 3-Dimensional transesophageal echocardiography-guided closure of a Gerbode shunt due to CorMatrix patch dehiscence. J Am Coll Cardiol 62(3):e5.https://doi.org/10.1016/j.jacc.2013.02.090
Prevel CD, Eppley BL, Summerlin DJ, Sidner R, Jackson JR, McCarty M, Badylak SF (1995a) Small intestinal submucosa: utilization as a wound dressing in full-thickness rodent wounds. Ann Plast Surg 35(4):381–388
Prevel CD, Eppley BL, Summerlin DJ, Jackson JR, McCarty M, Badylak SF (1995b) Small intestinal submucosa: utilization for repair of rodent abdominal wall defects. Ann Plast Surg 35(4):374–380
Rashtbar M, Hadjati J, Ai J, Jahanzad I, Azami M, Shirian S, Ebrahimi-Barough S, Sadroddiny E (2018) Characterization of decellularized ovine small intestine submucosal layer as extracellular matrix-based scaffold for tissue engineering. J Biomed Mater Res B Appl Biomater 106(3):933–944. https://doi.org/10.1002/jbm.b.33899
Ravo B, Falasco G (2019) Pure tissue inguinal hernia repair with the use of biological mesh: a 10-year follows up. A prospective study. Hernia: the journal of hernias and abdominal wall surgery.https://doi.org/10.1007/s10029-019-01976-y
Romanelli M, Dini V, Bertone M, Barbanera S, Brilli C (2007) OASIS® wound matrix versus Hyaloskin® in the treatment of difficult-to-heal wounds of mixed arterial/venous aetiology. Int Wound J 4(1):3–7. https://doi.org/10.1111/j.1742-481X.2007.00300.x

Romanelli M, Dini V, Bertone MS (2010) Randomized comparison of OASIS wound matrix versus moist wound dressing in the treatment of difficult-to-heal wounds of mixed arterial/venous etiology. Adv Skin Wound Care 23(1):34–38. https://doi.org/10.1097/01.asw.0000363485.17224.26

Romao RLP, Nasr A, Chiu PPL, Langer JC (2012) What is the best prosthetic material for patch repair of congenital diaphragmatic hernia? Comparison and meta-analysis of porcine small intestinal submucosa and polytetrafluoroethylene. J Pediatr Surg 47(8):1496–1500. https://doi.org/10.1016/j.jpedsurg.2012.01.009

Rosario-Quinones F, Magid MS, Yau J, Pawale A, Nguyen K (2015) Tissue reaction to porcine intestinal Submucosa (CorMatrix) implants in pediatric cardiac patients: a single-center experience. Ann Thorac Surg 99(4):1373–1377. https://doi.org/10.1016/j.athoracsur.2014.11.064

Rosenhammer B, Sayedahmed K, Fritsche HM, Burger M, Kübler H, Hatzichristodoulou G (2018) Long-term outcome after grafting with small intestinal submucosa and collagen fleece in patients with Peyronie's disease: a matched pair analysis. Int J Impot Res. https://doi.org/10.1038/s41443-018-0071-1

Saldin LT, Cramer MC, Velankar SS, White LJ, Badylak SF (2017) Extracellular matrix hydrogels from decellularized tissues: structure and function. Acta Biomater 49:1–15. https://doi.org/10.1016/j.actbio.2016.11.068

Sandusky GE Jr, Badylak SF, Morff RJ, Johnson WD, Lantz G (1992) Histologic findings after in vivo placement of small intestine submucosal vascular grafts and saphenous vein grafts in the carotid artery in dogs. Am J Pathol 140(2):317–324

Sarr MG, Hutcher NE, Snyder S, Hodde J, Carmody B (2014) A prospective, randomized, multicenter trial of surgisis Gold, a biologic prosthetic, as a sublay reinforcement of the fascial closure after open bariatric surgery. Surgery 156(4):902–909. https://doi.org/10.1016/j.surg.2014.06.022

Schaefer M, Kaiser A, Stehr M, Beyer HJ (2013) Bladder augmentation with small intestinal submucosa leads to unsatisfactory long-term results. J Pediatr Urol 9(6 Pt A):878–883. https://doi.org/10.1016/j.jpurol.2012.12.001

Schweinlin M, Wilhelm S, Schwedhelm I, Hansmann J, Rietscher R, Jurowich C, Walles H, Metzger M (2016) Development of an advanced primary human in vitro model of the small intestine. Tissue Eng Part C Methods 22(9):873–883. https://doi.org/10.1089/ten.tec.2016.0101

Sclamberg SG, Tibone JE, Itamura JM, Kasraeian S (2004) Six-month magnetic resonance imaging follow-up of large and massive rotator cuff repairs reinforced with porcine small intestinal submucosa. J Shoulder Elbow Surg 13(5):538–541. https://doi.org/10.1016/s1058274604001193

Siracusano S, Ciciliato S, Lampropoulou N, Cucchi A, Visalli F, Talamini R (2011) Porcine small intestinal submucosa implant in pubovaginal sling procedure on 48 consecutive patients: long-term results. Eur J Obstet Gynecol Reprod Biol 158(2):350–353. https://doi.org/10.1016/j.ejogrb.2011.06.012

Smith MJ, Paran TS, Quinn F, Corbally MT (2004) The SIS extracellular matrix scaffold-preliminary results of use in congenital diaphragmatic hernia (CDH) repair. Pediatr Surg Int 20(11–12):859–862. https://doi.org/10.1007/s00383-004-1298-0

Sofer M, Rowe E, Forder DM, Denstedt JD (2002) Ureteral segmental replacement using multilayer porcine small-intestinal submucosa. J Endourol 16(1):27–31. https://doi.org/10.1089/089277902753483682

Spiegel JH, Kessler JL (2005) Tympanic membrane perforation repair with acellular porcine submucosa. Otol Neurotology: official publication of the American Otological Society, American Neurotology Society [and] European Academy of Otology and Neurotology 26(4):563–566

Stelly M, Stelly TC (2013) Histology of CorMatrix bioscaffold 5 years after pericardial closure. Ann Thorac Surg 96(5):e127-129. https://doi.org/10.1016/j.athoracsur.2013.06.114

Suckow MA, Hall P, Wolter W, Sailes V, Hiles MC (2008) Use of an extracellular matrix material as a vaccine carrier and adjuvant. Anticancer Res 28(5a):2529–2534

Suckow MA, Voytik-Harbin SL, Terril LA, Badylak SF (1999) Enhanced bone regeneration using porcine small intestinal submucosa. J Invest Surg: the official journal of the Academy of Surgical Research 12(5):277–287

Sun T, Liu M, Yao S, Ji Y, Xiong Z, Tang K, Chen K, Yang H, Guo X (2018) Biomimetic composite scaffold containing small intestinal submucosa and mesoporous bioactive glass exhibits high osteogenic and angiogenic capacity. Tissue Eng Part A 24(13–14):1044–1056. https://doi.org/10.1089/ten.tea.2017.0398

Syed O, Walters NJ, Day RM, Kim HW, Knowles JC (2014) Evaluation of decellularization protocols for production of tubular small intestine submucosa scaffolds for use in oesophageal tissue engineering. Acta Biomater 10(12):5043–5054. https://doi.org/10.1016/j.actbio.2014.08.024

Tan MY, Zhi W, Wei RQ, Huang YC, Zhou KP, Tan B, Deng L, Luo JC, Li XQ, Xie HQ, Yang ZM (2009) Repair of infarcted myocardium using mesenchymal stem cell seeded small intestinal submucosa in rabbits. Biomaterials 30(19):3234–3240. https://doi.org/10.1016/j.biomaterials.2009.02.013

Tanaka A, Kawaji K, Patel AR, Tabata Y, Burke MC, Gupta MP, Ota T (2015) In situ constructive myocardial remodeling of extracellular matrix patch enhanced with controlled growth factor release. J Thorac Cardiovasc Surg 150(5):1280-1290.e1282. https://doi.org/10.1016/j.jtcvs.2015.07.073

Totonelli G, Maghsoudlou P, Garriboli M, Riegler J, Orlando G, Burns AJ, Sebire NJ, Smith VV, Fishman JM, Ghionzoli M, Turmaine M, Birchall MA, Atala A,

Soker S, Lythgoe MF, Seifalian A, Pierro A, Eaton S, De Coppi P (2012) A rat decellularized small bowel scaffold that preserves villus-crypt architecture for intestinal regeneration. Biomaterials 33(12):3401–3410. https://doi.org/10.1016/j.biomaterials.2012.01.012

Voytik-Harbin SL, Brightman AO, Waisner BZ, Robinson JP, Lamar CH (1998) Small intestinal submucosa: a tissue-derived extracellular matrix that promotes tissue-specific growth and differentiation of cells in vitro. Tissue Eng 4(2):157–174. https://doi.org/10.1089/ten.1998.4.157

Walton JR, Bowman NK, Khatib Y, Linklater J, Murrell GA (2007) Restore orthobiologic implant: not recommended for augmentation of rotator cuff repairs. J Bone Joint Surg Am 89(4):786–791. https://doi.org/10.2106/jbjs.f.00315

Wang B, Zhang W, Shan CX, Liu S, Jiang ZG, Qiu M (2016) Long-term outcomes of cruroplasty reinforcement with composite versus biologic mesh for gastroesophageal reflux disease. Surg Endosc 30(7):2865–2872. https://doi.org/10.1007/s00464-015-4570-6

Wang L, Meier EM, Tian S, Lei I, Liu L, Xian S, Lam MT, Wang Z (2017) Transplantation of Isl1(+) cardiac progenitor cells in small intestinal submucosa improves infarcted heart function. Stem Cell Res Ther 8(1):230–230. https://doi.org/10.1186/s13287-017-0675-2

Watson DI, Thompson SK, Devitt PG, Smith L, Woods SD, Aly A, Gan S, Game PA, Jamieson GG (2015) Laparoscopic repair of very large hiatus hernia with sutures versus absorbable mesh versus nonabsorbable mesh: a randomized controlled trial. Ann Surg 261(2):282–289. https://doi.org/10.1097/sla.0000000000000842

Welch JA, Montgomery RD, Lenz SD, Plouhar P, Shelton WR (2002) Evaluation of small-intestinal submucosa implants for repair of meniscal defects in dogs. Am J Vet Res 63(3):427–431

Witt RG, Raff G, Van Gundy J, Rodgers-Ohlau M, Si MS (2013) Short-term experience of porcine small intestinal submucosa patches in paediatric cardiovascular surgery. Eur J Cardio-thoracic Surg: official journal of the European Association for Cardio-thoracic Surgery 44(1):72–76.https://doi.org/10.1093/ejcts/ezs638

Woods EJ, Walsh CM, Sidner RA, Zieger MA, Lakey JR, Ricordi C, Critser JK (2004) Improved in vitro function of islets using small intestinal submucosa. Transpl Proc 36(4):1175–1177. https://doi.org/10.1016/j.transproceed.2004.04.042

Xu Q, Shanti RM, Zhang Q, Cannady SB, O'Malley BW Jr, Le AD (2017) A gingiva-derived mesenchymal stem cell-laden porcine small intestinal submucosa extracellular matrix construct promotes myomucosal regeneration of the tongue. Tissue Eng Part A 23(7–8):301–312. https://doi.org/10.1089/ten.TEA.2016.0342

Zaidi AH, Nathan M, Emani S, Baird C, del Nido PJ, Gauvreau K, Harris M, Sanders SP, Padera RF (2014) Preliminary experience with porcine intestinal submucosa (CorMatrix) for valve reconstruction in congenital heart disease: histologic evaluation of explanted valves. J Thoracic Cardiovasc Surg 148(5):2216–2214, 2225.e2211. https://doi.org/10.1016/j.jtcvs.2014.02.081

Zhang Y, Kropp BP, Lin HK, Cowan R, Cheng EY (2004) Bladder regeneration with cell-seeded small intestinal submucosa. Tissue Eng 10(1–2):181–187. https://doi.org/10.1089/107632704322791835

Zhang Y, Lin HK, Frimberger D, Epstein RB, Kropp BP (2005) Growth of bone marrow stromal cells on small intestinal submucosa: an alternative cell source for tissue engineered bladder. BJU Int 96(7):1120–1125. https://doi.org/10.1111/j.1464-410X.2005.05741.x

Zhou Y, Yan Z, Zhang H, Lu W, Liu S, Huang X, Luo H, Jin Y (2011) Expansion and delivery of adipose-derived mesenchymal stem cells on three microcarriers for soft tissue regeneration. Tissue Eng Part A 17 (23–24):2981–2997. https://doi.org/10.1089/ten.tea.2010.0707

The Isolated Mucosa of the Rat Colon Decellularization, Microscopy and Cell Cultures

8

Pedro Mestres-Ventura, Laura López Gómez, Gilberto del Rosario Hernández, and Teresa Núñez López

Abstract

This project presents the most important findings of the studies, which we carried out in our laboratory on the decellularization of the rat isolated colonic mucosa. We have also included some details of the experiences gathered with the muscle layer as well as the whole wall of the colon. The question of the cytocompatibility of this new substrate has been addressed with the application of primary cultures of human cells and well-established cell lines. The possible applications in experimental and medical settings will be discussed.

P. Mestres-Ventura (✉)
Department of Anatomy and Cell Biology, Faculty of Medicine, Saarland University, 66421 Homburg Saar, Germany
e-mail: pedro.mestres@uks.eu

L. López Gómez
Department of Basic Health Sciences, Faculty for Health Sciences. University Rey Juan Carlos, Av. de Atenas s/n, 28922 Alcorcón, Madrid, Spain

G. del Rosario Hernández
Laboratory of Electron Microscopy, Center for Technical Support (CAT), University Rey Juan Carlos, 28933 Mostoles, Madrid, Spain

T. Núñez López
Center for Technical Support (CAT), Faculty of Health Sciences, University Rey Juan Carlos, Av. de Atenas s/n, 28922 Alcorcon, Madrid, Spain

Keywords

Decellularization · Colon · Rat · ECM scaffold · Cell cultures · Microscopy

8.1 History

We began to work with the isolated mucosa of the rat colon in a project in which this structure was investigated in vitro using the Ussing chamber (Böhme et al. 1992). The project examined the secretory mechanisms of colon enterocytes under the influence of the mucosal nerve plexus as well as under the effect of mediators (Diener et al. 1989).

In our model, the basement membrane (BM) was a support par excellence for the enterocytes, to which they are attached through cell membrane adhesions and specialized contacts, providing them with mechanical support (Mestres et al. 1991). An important aspect in this context was the possible barrier function of the basement membrane; which is why ultrastructural studies were performed (Mestres et al. 1991, 2014).

In order to examine the surface of the basement membrane, procedures were applied to remove the attached enterocytes with minimal damage to the BM while, at the same time, preserving vitality and viability of the cells (Andres et al. 1985; Diener et al. 1989). The

A. Kajbafzadeh (ed.), *Decellularization Methods of Tissue and Whole Organ in Tissue Engineering*, Advances in Experimental Medicine and Biology 1345,
https://doi.org/10.1007/978-3-030-82735-9_8

decellularization procedure selected in these experiments only relates to the decellularization of the superficial epithelium, and not to the innermost layers of the mucosa, such as the lamina propria with proper neural elements and muscular layer of the mucosa (Mestres et al. 1992a, b). Our first observations after this treatment were that BM had been preserved and presented a discretely globular surface with fenestrations. The latter is related to the passage of delicate prolongations of the basal pole of enterocytes through the BM as well as macrophages and other cells able to move through the mucosa compartments (Mestres et al. 1991).

The superficial epithelium of the colon was dislodged using EDTA and controlled mechanical vibration (Diener et al. 1989). However, the degree of decellularization of the mucosa was, on the whole, not satisfactory, as the large number of cells which remained in the network of the extracellular matrix represent a disruptive factor. Therefore, decellularization was intensified with the application of a protocol based on detergents, which will be described in detail later (see Lopez Gomez et al. 2018).

There are many decellularization strategies, which have been applied with varying success, depending on the tissues and organs selected (Crapo et al. 2011). Hollow organs have received particular attention as they permit the obtain of scaffolds of adequate size and easy handling. Good examples of this are the esophagus (Ozeki et al. 2006; Keane et al. 2013), the small intestine (Maghsoudlou et al. 2013; Oliveira et al. 2013; Denost et al. 2015; Massie et al. 2017), and the urinary bladder (Barnes et al. 2011). Even tumor extracellular matrix (ECM) has been isolated in order to study the interactions of this particular ECM with diverse cell types (Nietzer et al. 2016).

Studies of the relevant literature reveal that the wall of the colon has rarely aroused the interest of researchers and merely a few studies of this organ and its walls, especially in the context of its uses in tissue engineering, have been carried out. One exception is a study using sheep colon, in which the entire wall was decellularized (Kajbafzadeh et al. 2014). Another more recent study deals with the mucosa of the colon, studying the host responses to the ECM derived from this organ (Keane et al. 2017). As previously mentioned, our group has extensive experience with regard to the structure of the rat colon (Mestres et al. 2014; Lopez Gomez et al. 2018). This fact, together with the circumstances previously explained, led to our decision to continue with the preparation of this organ, with the aim of developing and characterizing a new ECM scaffold derived from the colon wall.

8.2 Material and Methods

8.2.1 Animal Care

Adult Wistar and Sprague–Dawley rats of both sexes (200–300 gr body weight) were kept in animal facilities under controlled light, temperature, humidity, and food conditions. The animals were euthanized with carbon dioxide and in accordance with the relevant Spanish and EU regulations. The project was also examined and authorized by the ethical commission of the Rey Juan Carlos University (Madrid, Spain).

8.3 Organ Dissection and Tissue Preparation

After the opening of the abdominal wall, the entire colon was removed and carefully separated from its peritoneal adhesions. The colon was then washed to free it of fecal contents, for which a physiological solution was used at pH 7.2 and 4 °C (Parsons and Paterson 1965).

To separate the mucosa from the muscular layer, a cylindrical rod of 5 mm caliber was introduced into the lumen of the colon, and a circular section was made in one of the ends in order to lift the muscular layer slightly and separate it from the submucosa and mucosa, which remained on the cylinder. By carrying out a longitudinal section, the luminal surface of the colon was exposed (Bridges et al. 1986; Mestres et al. 1991). The isolated mucosa was spread onto a piece of smooth and uncreased filter paper, enabling simple and safe handling of the samples. Generally, two or three relatively large samples can be obtained from each colon (Lopez Gomez 2015).

With this preparation, the muscle layer (tunica muscularis) of the colon can also be separated, forming a rather large specimen somewhat rougher than the mucosa. As the muscle layer is also susceptible to decellularization, our procedure provided a substrate with a surface constituted mainly of deep-running connective tissue fibers, which are in contact with the sheaths of the basement membrane of the smooth muscle cells of the muscularis mucosae.

8.4 Decellularization

The mucosa, still attached to the filter paper, was divided into squares of approx. 2 cm side length and attached to plastic holders by means of an acrylic glue used for surgery (Histoacryl®, Braun-Aeskulap, Germany). The holders (Fig. 8.1) have a hole of 1.5 cm in diameter that is covered by the mucosa.

In a first decellularization step, the holders with the mucosa were plunged into a Parsons and Paterson solution containing 10 mM EDTA at pH 7.4, but without Ca^{2+} or Mg^{2+}, for 5 min and vibrated (frequency 50 Hz) for 15 s. This treatment was repeated for 45 min, resulting in the complete removal of surface mucosal epithelium. To restore ion levels, the tissue was washed with Parsons and Paterson solution containing Ca^{2+} and Mg^{2+}.

In a second decellularization step, two detergents were used: First Triton X-100 (Sigma-Aldrich) and then SDS (Sigma-Aldrich). They were applied separately at a temperature of 4 °C under constant stirring.

The two following procedures were applied: *A short treatment*: Triton X-100 (0.5% in PBS) for 30 min, followed by SDS (0.5% in PBS) for 10 min and 15 min with DNase I Invitrogen (1U/µl) and a *long treatment*: Triton X-100 (0.5% in PBS) for 2 h, followed by SDS (0.5% in PBS) for 30 min and then 15 min with DNase I, with the washing in Parsons and Paterson solution between each step (Lopez Gomez et al. 2018). The treatment with DNase I, which was carried out at 37 °C, was especially required after SDS treatment in order to remove free-spreading DNA, which was visualized on the HRSEM pictures (see Fig. 8.4d). In order to eliminate detergent and enzyme residues and re-establish the ionic medium, the samples were thoroughly washed with the Parsons and Peterson solution.

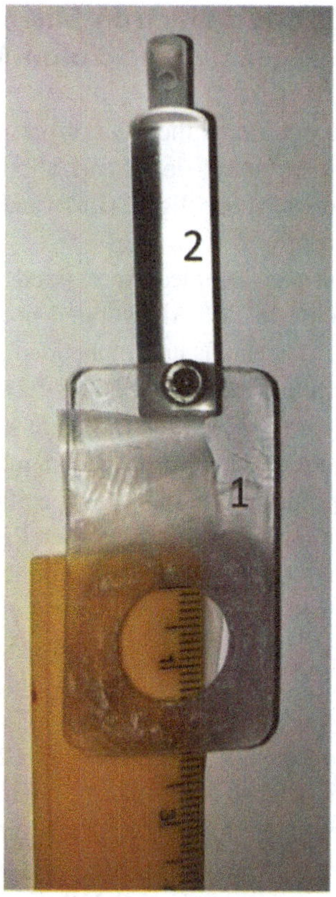

Fig. 8.1 Plastic holder to which the isolated colonic mucosa is glued with Histoacryl®. 1: holder with 1.5 cm diameter hole; 2: rod for attachment to the vibration equipment

The reaction time of both detergents was tested in experiments carried out in one of our department's other laboratories, using our protocol but with prolonged application times: Triton X-100 overnight at 4 °C and SDS at room temperature for one hour. The results were good in both cases (Pötschke 2012).

8.5 Light and Electron Microscopy of Decellularized Colon Mucosa

In order to determine the state of preservation of colonic ECM after decellularization, samples were processed for light (LM) and electron (EM) microscopy.

For LM the samples were fixed with 10% formaldehyde in PBS, dehydrated in an ascending series of ethanol, and embedded in paraffin (Panreac Quimica SA, MP 56-58 °C). Histological sections of 5 µm thickness were stained with hematoxyline-eosin (general ECM morphology) or with propidium iodide (a specific staining agent for nuclear nucleic acids, see Riccardi and Nicoletti 2006). Figure 8.2 shows the effects of our treatments on the ECM and on cell nuclei of the isolated colonic mucosa (Lopez Gomez et al. 2018). The treatments with detergents practically dissected the structures of the ECM, leaving them free of cellular elements (Fig. 8.2). In order to have a negative control of basement membrane preservation and, at the same time, to visualize the ECM of the lamina propria, some samples were incubated in 0.25% trypsin-EDTA in PBS at pH 7.4 for 2 and 24 h at 4 °C with constant stirring. The basement membrane was already eliminated after the 2 h treatment (Figs. 8.2d–g). The samples were then washed in the Parsons and Paterson solution in order to remove traces of the protease and to restore the ionic medium.

A particular compartment of the ECM is the basement membrane and to determine its persistence after decellularization some of its specific molecular components were detected by immunohistochemistry (ICC). The following antibodies were used: anti-collagen IV (rabbit polyclonal, Abcam, ab6586), anti-laminin (rabbit polyclonal laminin 1 + 2, Abcam, ab7463), and anti-perlecan (rabbit polyclonal, Santa Cruz Biotechnology, H-300). The histological sections were de-waxed, rehydrated, and then treated with pepsin (pepsin crystalline, Sigma-Aldrich, Germany, P6887) to unmask collagen IV and laminin antigens, while for perlecan this pretreatment was unnecessary. Endogenous peroxidase was inhibited with H_2O_2 for 15 min at room temperature, and unspecific binding sites were blocked with 10% FCS in PBS for 30 min in a humid chamber at room temperature. The primary antibodies against collagen IV (dilution 1/1000), laminin (dilution 1/1500), and perlecan (dilution 1/1000) were incubated overnight in a humid chamber at 4 °C. The binding of the primary antibodies was detected using Histofine Simple Stain MAX PO (Nichirei Biosciences, Tokyo, Japan).

For LM an Axioplan 2 Optical Microscope (Zeiss, Oberkochen) equipped with an HCR Axiocam camera (Zeiss, Oberkochen) was used and digital images were obtained and stored in JPEG and TIFF format. Figure 8.3 shows the results of the three BM-proteins mentioned above obtained by immunohistochemistry. While collagen and laminin appear in BMs, perlecan also appears weak in other locations of the ECM. Together they demonstrate that the BM was preserved.

Electron microscopy allowed us not only to examine details of the fine structure of ECM and basement membrane but also to see how cells cultured on this scaffold attach and migrate inside it. Thus, scanning (SEM) as well as the new technique called block-face SEM (BFSEM) and transmission electron microscopy (TEM) have proven very useful.

For EM the samples were processed as follows: Samples attached to plastic holders (Fig. 8.1) were fixed in 2.5% glutaraldehyde (Fluka, Switzerland) buffered with cacodylate 0.12 M, at pH 7.4, post-fixed in buffered 1% osmium tetroxide (Electron Microscopy Sciences, EMS) for 2 h at 4 °C and then dehydrated in an increasing alcohol series. The specimen remained attached to the holder during all these steps.

For SEM the fixed and dehydrated samples were transferred to 100% acetone and dried according to the critical point method (CPD) using CO_2 as an intermediary (Mestres et al. 2014). The drying procedure is an important step as otherwise, the 3D structure of the sample would collapse and the original tissue architecture

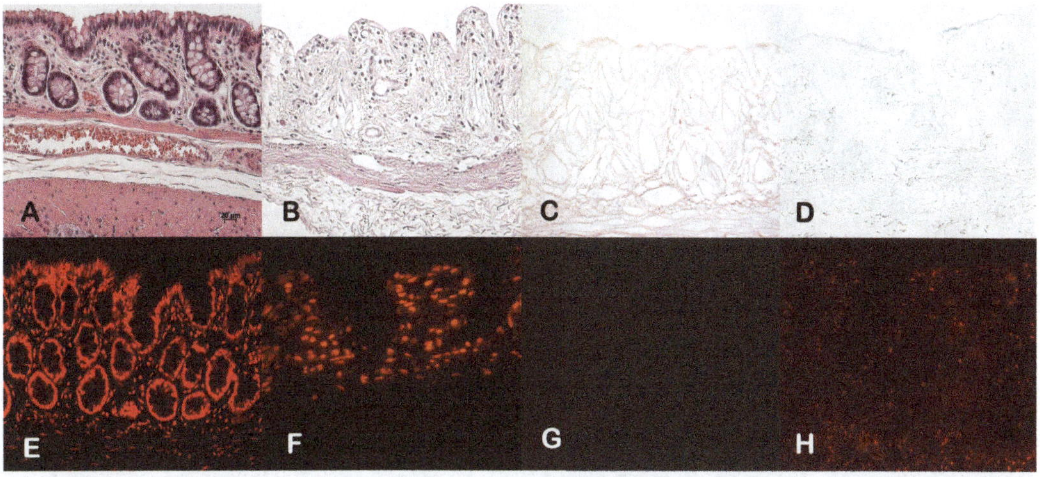

Fig. 8.2 Light microscopy (reproduced with permission of reference Lopez Gomez et al. 2018). **a** Partial view of the colon wall (control). H/E staining. 1: Surface epithelium; 2: Lamina propria; 3: Crypts in cross and length section; 4: Tunica muscularis mucosae; 5: Submucosa with a large blood vessel; 6: Partial view of tunica muscularis. In the following light microscopy images, structural features are indicated as in Fig. 8.2a. **b** Isolated colonic mucosa after decellularization after EDTA/vibration treatment. H/E staining. The surface epithelium has been completely removed. **c** Isolated colonic mucosa decellularized after the combined treatment (see diagram on Fig. 8.2). H/E staining. The surface epithelium has been completely removed. **d** Isolated colonic mucosa after trypsin treatment. H/E staining. The surface epithelium has been completely removed (arrow). **e** Untreated colon wall stained with PI. Note the cell nuclei in red. **f** Isolated colonic mucosa treated with EDTA/vibration and stained with PI. The surface epithelium has been completely removed (arrow). **g** Isolated colonic mucosa after combined short treatment (see Fig. 8.1) and stained with PI. Cell nuclei are no longer visible. **h** Isolated colonic mucosa after trypsin treatment. Particles stained with PI are still visible and correspond to fragments of cell nuclei trapped in the ECM structure. The surface epithelium has been completely removed (arrow)

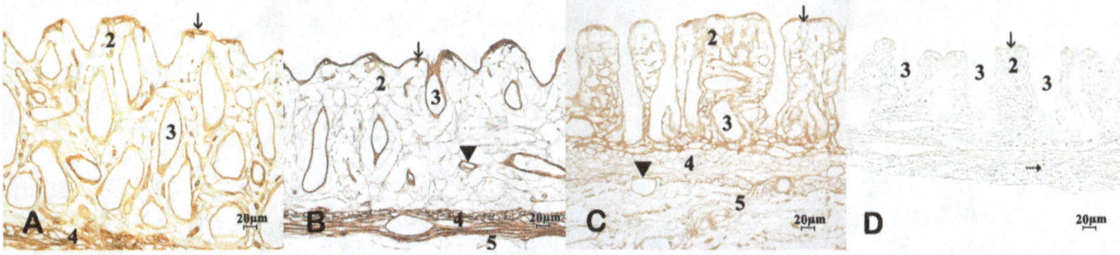

Fig. 8.3 Immunohistochemical detection of marker proteins of BM after decellularization of isolated colonic mucosa (short treatment). (Reproduced with permission of reference: Lopez Gomez et al. 2018). **a** Collagen IV. The luminal side of the colonic mucosa appears cell free with a sharp line corresponding to the BM (arrow). The BM is also clearly stained at the crypts (3) around blood vessels and smooth muscle of the muscularis mucosae layer (4); **b** Laminin. The BM membrane of the surface epithelium (arrow), blood vessels in the lamina propria, and in the smooth muscle of lamina muscularis mucosae layer, are well stained (arrow head). In contrast, the submucosa (5) remains unstained; **c** Perlecan. The BM membranes appear stained but there is also a weak staining of extracellular matrix (ECM) in other parts of the colon wall; **d** Negative control in which the primary antibody was omitted

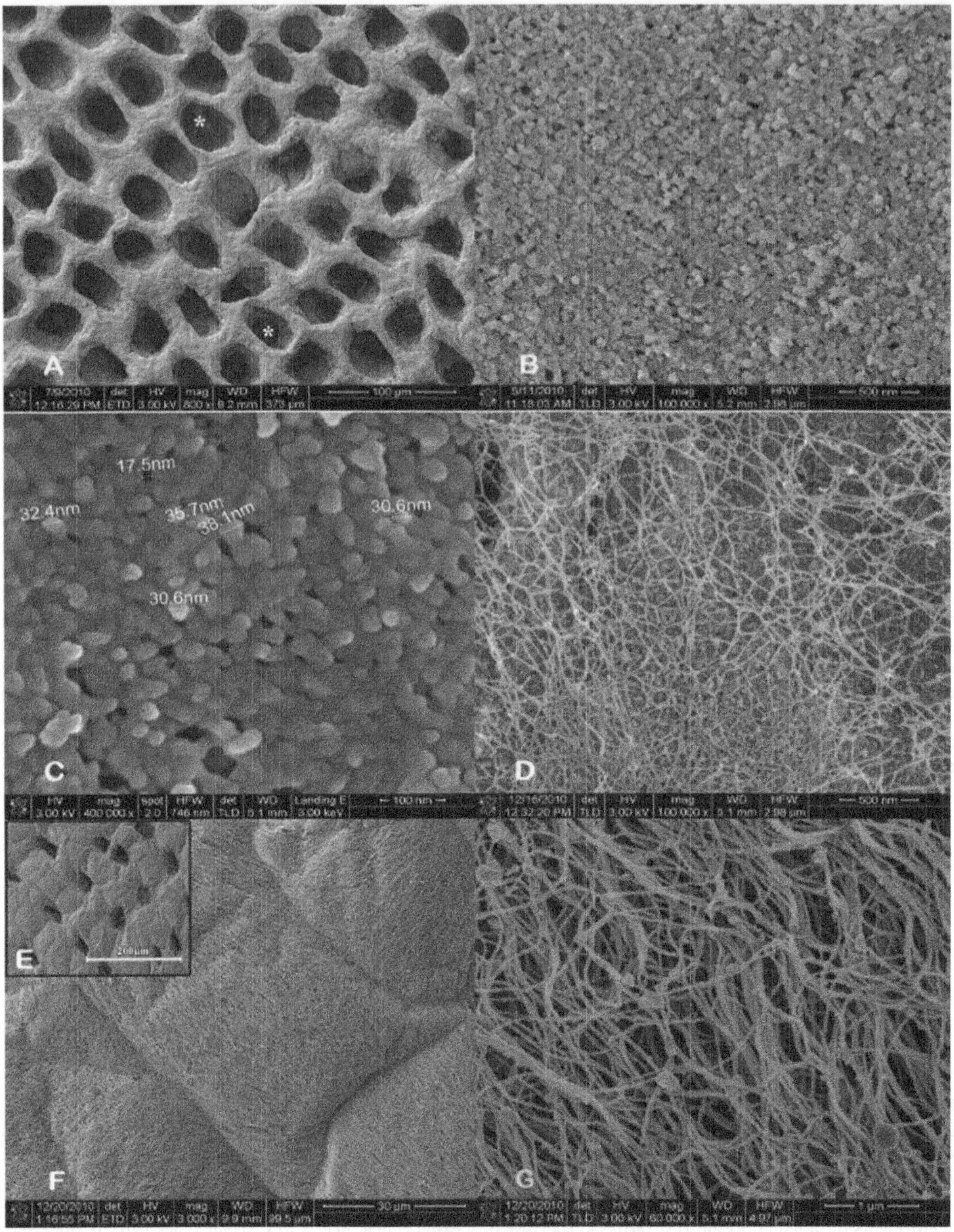

Fig. 8.4 Scanning electron microscopy of the luminal surface of the colonic mucosa after decellularization (reproduced with permission of reference Lopez Gomez et al. 2018). **a** Overview showing the cell-free BM surface and the mouths of the crypts (*). **b** Epithelial side of the BM membrane at higher magnification. Note the presence of globules. **c** Measurements of particle diameter or globules. Some globules rise to the surface, whereas others appear embedded in the BM membrane. **d** Surface of the BM membrane after SDS treatment. Note the presence of a network of fibers, which subsequently disappeared upon the application of the enzyme DNase. **e** Insert showing the surface of the mucosa after trypsin treatment. Note the regular distribution of the crypts (dark mouth). **f** The enzymatic action of trypsin has hydrolyzed the BM membrane rendering the collagen fibers underneath directly visible. **g** Detail of Fig. 8.9f. The collagen fibers do not appear to have been affected by the action of trypsin, indicating that only the BM has been removed

would be lost. To render them conductive, the samples were sputter-coated with platinum (Lopez Gomez et al. 2018) and examined with a Nano-SEM 240 FEG FEI scanning electron microscope. Figure 8.4 shows various SEM aspects of the luminal surface of the colon mucosa after decellularization with detergents and the enzyme trypsin.

For TEM, the samples were treated as described above, but instead of being dried, they were carefully detached from the holder and embedded in epoxy resin (TAAB). Semi-thin Sects. (0.5–1.0 μm thick) were stained Azur II and methylene blue (Richardson et al. 1960) and examined with a light microscope. Ultrathin Sects. (80 nm thick) were stained with uranyl acetate and lead citrate (Reynolds 1963) and examined in a Jeol JEM 1010 transmission electron microscope (for details see Lopez Gomez et al. 2018).

In summary, it can be said that microscopic studies have shown that, with the decellularization protocol used, complete decellularization is achieved with practically no contamination with DNA fragments (Fig. 8.4). The network of the ECM is well preserved as shown by the SEM images (Fig. 8.4). On the other hand, TEM reveals that the luminal basement membrane remains thinned only at the side where the epithelium is attached (Fig. 8.5). On the underside of the mucosa, the relation between collagen fibers and BM appears to have been well preserved. As explained elsewhere (Lopez Gomez et al. 2018) the release of enterocytes ruptures the links between them and the BM so that molecules located in the BM lamina lucida (LL) are lost, and consequently, this layer apparently disappears (Lopez Gomez et al. 2018). On the other hand, as the released enterocytes remain vital, the action of EDTA appears not to be lethal for these cells (Diener et al. 1989).

The ICC and TEM studies show that part of the BM corresponding approximately to the BM lamina densa (LD) retains its surface properties, enabling the renewed attachment of cells when the specimen is recellularized. On the other hand, these BM components provide the scaffold here with special functional properties (Lopez Gomez et al. 2018). The nanotopography of the BM surface after these treatments is characterized by small globules approx. 40 nm in diameter, a size which seems to be their original size and is not significantly different after the two treatments described here (Fig. 8.4) (Mestres et al. 2014; Lopez Gomez et al. 2018).

Not only was the mucosa isolated successfully, but also very good results were achieved in experiments to remove the cells of the isolated muscle layer and the entire colon wall (Fig. 8.6).

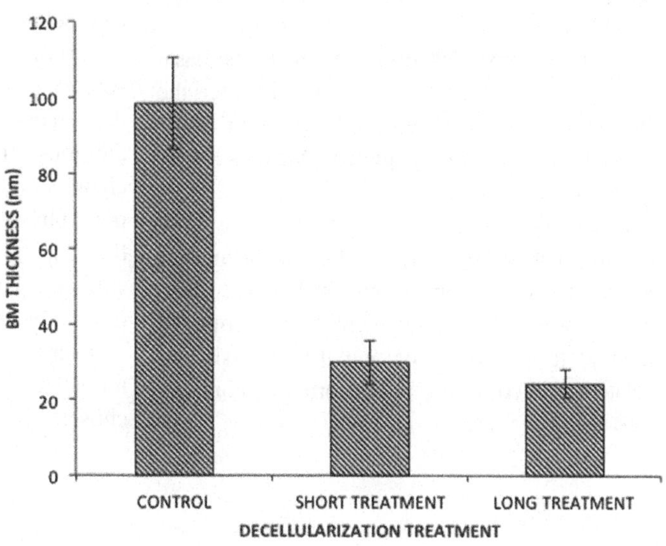

Fig. 8.5 Thickness of luminal colonic BM after decellularization (short treatment). X = Treatments. Y = Thickness of the BM in nm. Mean values with standard deviation: Control: 98.35 ± 12.1; Short treatment: 29.92 ± 5.78; Long treatment: 24.4 ± 3.75

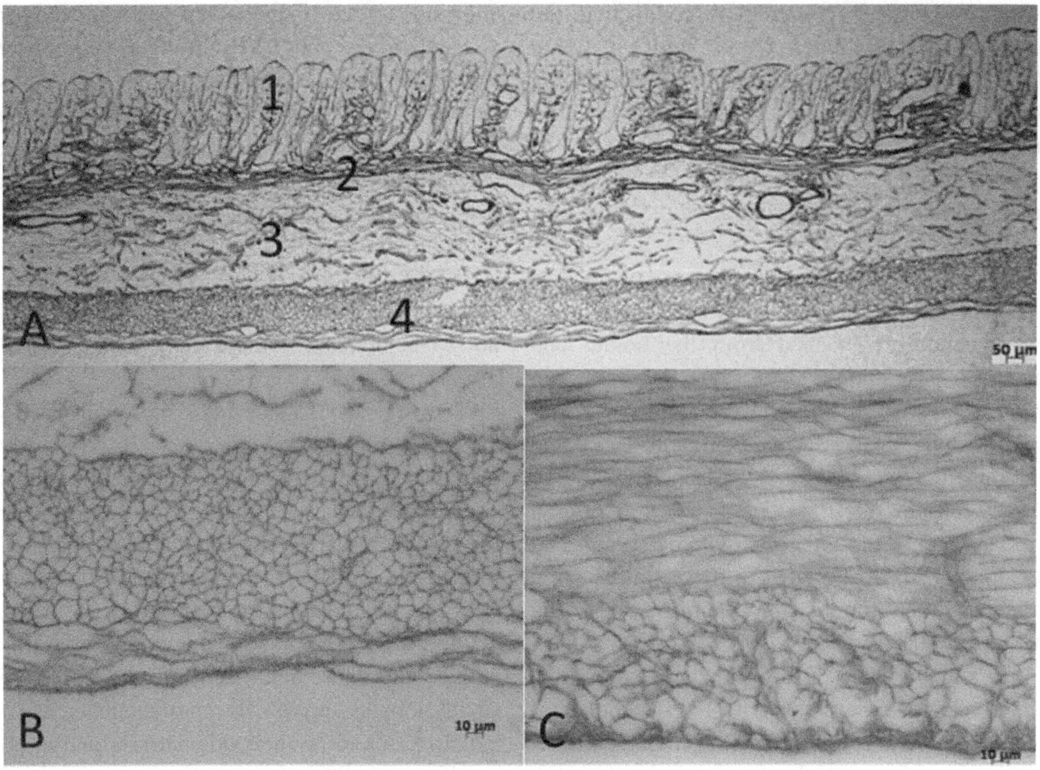

Fig. 8.6 The whole colon wall was decellularized. Paraffin section was stained after the PAS reaction. **a** Overview of the decellularized colon wall in longest Sect. 1: mucosa, 2: muscularis mucosae, 3: submucosa and 4: tunica muscularis. **b** Detail of the tunica muscularis in long section. **c** Detail of the tunica muscularis in cross section

These tests were not part of the routine procedure but merely intended to test the effectiveness of the decellularization procedure and to examine possible further scaffolds.

With the above-described treatment, the basal membrane was preserved and already after the short treatment complete decellularization of the isolated mucosa was achieved (Mestres et al. 2014; Lopez Gomez et al. 2018).

If the BM is removed, for instance with trypsin, then the remaining fibers of the lamina propria can be visualized (Fig. 8.4). The difference between the arrangement of the collagen fibers at the luminal side and the underside of isolated mucosa is noteworthy (compare Figs. 8.4 and 8.9).

8.6 Cell Cultures

Cell cultures on the isolated colon mucosa served as proof of compatibility between cells and substrate. In general, the suitability of a substrate depends not only on its properties but also on the affinities of cells for it. One property is the chemical composition, another is the nanotopography of the surfaces on which the cells settle, and also mechanical properties, such as stiffness. The first two points will be examined more closely.

In the present study, both cells from primary human cultures and permanent cell lines were chosen:

- Primary cultures of human fibroblasts isolated from the umbilical cord, generously provided by Dr. Fernando Serrano Gómez (University Hospital Foundation Alcorcón, Madrid).
- Primary cultures of human endothelial cells isolated from umbilical veins and supplied by Dr. Jose Uranga Ocio (Department of Basic Health Sciences, Faculty for Health Sciences. University Rey Juan Carlos, Alcorcón, Madrid).
- And cell lines commercially available such as 804 G cells (access: CVCL_J122; Izumi et al. 1981), PC3 cells, a line of epithelial nature from a prostate adenocarcinoma (ATCC, Kaighn et al. 1979) and P21 (hPCP) cells, a cell line isolated from stroma of a prostate adenocarcinoma were used particularly in the co-culture experiments (Janssen et al. 2000).

Human fetal fibroblasts. Before seeding, the mucosa, isolated and decellularized as described, was incubated in DMEM culture medium with antibiotics (1% penicillin/streptomycin) with FCS (10%) or without serum at 4 °C and under constant stirring.

For the culture of the cells, an incubation chamber was constructed as shown in Fig. 8.7. The mucosa was attached to a Macrolon™ ring using Histoacryl™ with the luminal side of the mucosa, i.e., with the BM, facing the inside of the ring. The ring has a height of 1 cm and also an internal diameter of 1 cm. This enables the construction of a small container to safely keep the cells in. The Macrolon rings with the attached mucosa were placed in a 12-Transwell plate (Falcon-Corning) (Fig. 8.7). Fetal fibroblasts in a concentration of 6 × 10^4 were placed in such small containers and both the ring and the 12-Transwell were filled with culture medium. The cultures were maintained in an incubator at 37 ° C, with humidity control, an atmosphere with 5% CO_2 and frequent changes of medium.

As the mucosa offers very poor optical properties, which do not permit direct microscopically monitoring of culture growth, cells were parallel cultivated on Thermanox™ coverslips.

For fine structural studies, the samples were processed for electron microscopy as described above and semi-thin and ultrathin sections were examined with light and electron microscopes respectively.

Human fetal fibroblasts adhere well to the surface of the mucosa, adopting a very flat form and thus become adapted to the topography of

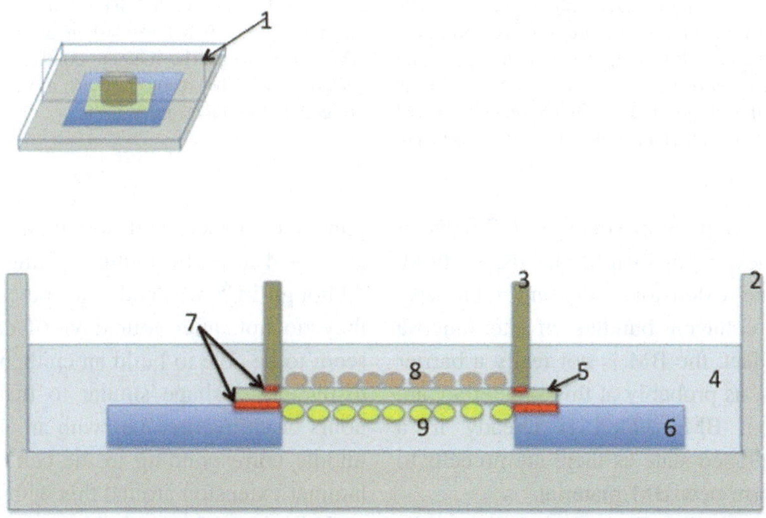

Fig. 8.7 Scheme of the chamber for in vitro experiments. (1): 3D view where the cutting plane represented in the second diagram is marked. (2): Transwell; (3): Makrolon ring; (4): medium; (5): isolated and decellularized colon mucosa; (6) plastic support on which the Makrolon ring is located; (7) places where Histoacryl has been used; (8) cells on the BM and (9) cells on the downside of the scaffold in the co-cultures

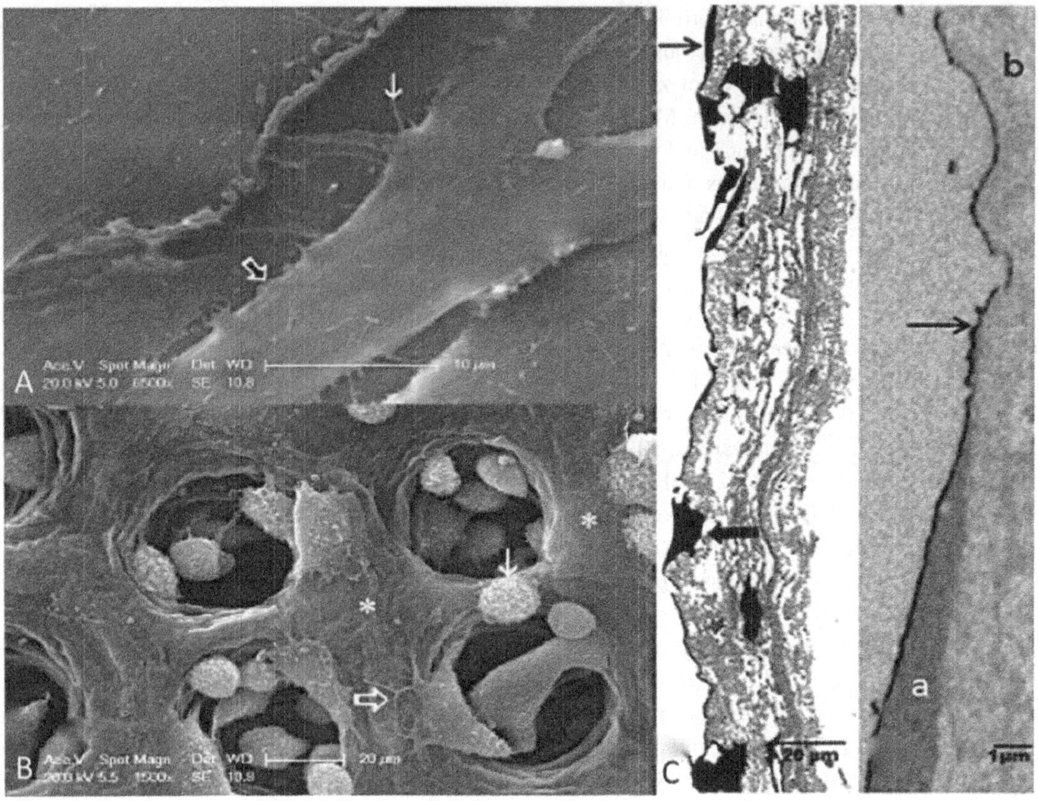

Fig. 8.8 SEM of fetal human fibroblasts in vitro. (Reproduced with permission of reference Lopez Gomez et al. 2018). **a** Fibroblasts in vitro are already morphologically developed. The cells show fine ruffled borders and thin filopodia, making contact with neighboring cells and with the substrate, in this case, the BM. **b** Fibroblasts in an early phase of cultivation. About four crypts and BM between the crypt openings can be seen. Several cells have settled in the crypts. **c** 1 µm thick section stained with Azure II and methylene blue. The fibroblasts are stained dark. At the surface, strongly flattened cells can be observed (thin arrow). Other cells display a more stellate shape with thin processes penetrating into the neighboring ECM, indicating migration into deeper layers (thick arrow). **d** Correlative SEM/TEM image where the cytoplasmic border of a fibroblast in close contact with the BM can be seen. The metallic conductive coating of the specimen can be recognized as a very dark line that protects the surface (arrow)

the substrate. After approximately 4–7 days of cultivation, they begin to infiltrate the scaffold, presenting fine extensions winding their way between the collagen bundles of the mucosa (Fig. 8.8). In fact, the BM is not really a barrier for these cells, as probably at this stage, they are encountering a BM, which is already in a somewhat degraded state as there are no cells to renew it or form new BM material.

Endothelial cells. The endothelial cells became attached very early after seeding. They display a more or less round shape with only punctual contacts and still maintain this shape after 3–4 days of culture (Jaffe et al. 1973). Although FCS was added to the culture medium they do not, after four days of cultivation, yet seem to be able to build an epithelial carpet, still displaying a shape similar to that of the early hours of cultivation, i.e., with an elevation in the middle corresponding to the cell nucleus and a laminar extension around this with thin processes at the borders. It is possible that they lack some factors which they need to be able to grow and differentiate, and that this could be due to the

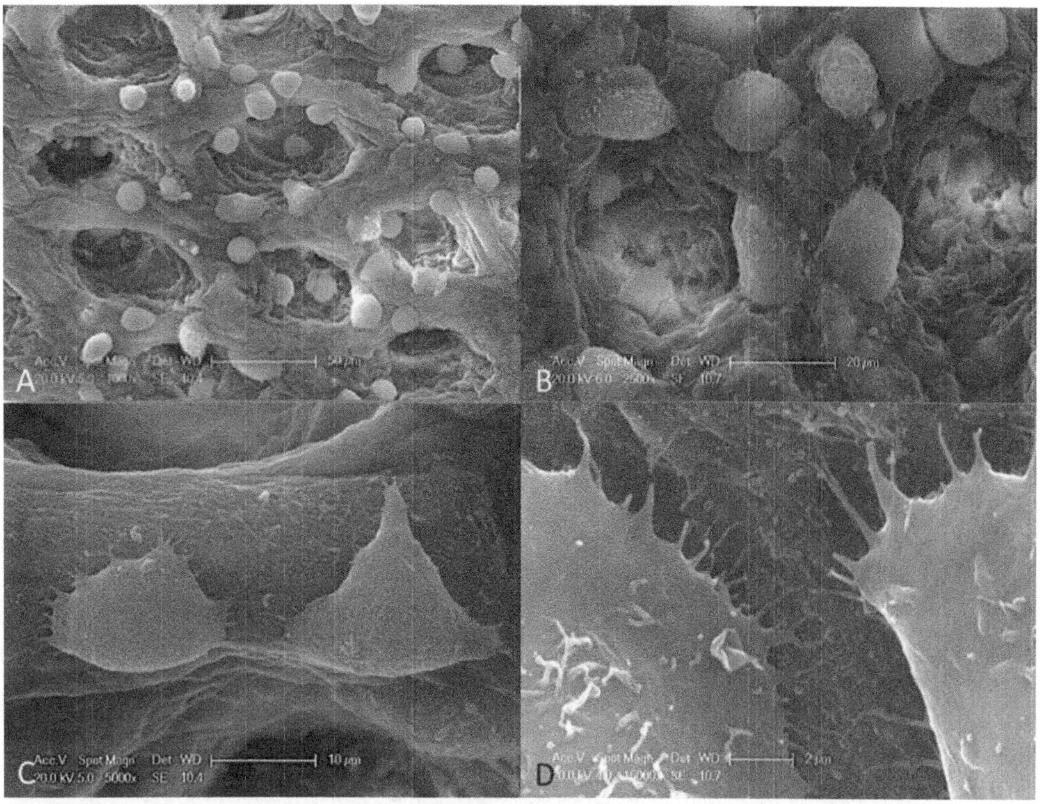

Fig. 8.9 SEM of 804G cells growing on the isolated colonic mucosa. **a** Cells recently attached and in contact with others; **b** cells attached to the BM and openings of two adjacent crypts can be distinguished; **c** Two cells with fine filopodia on the edge and in close contact by means thin processes; **d** detail of the filopodia and cell surface

extraction of materials from the ECM during the process of decellularization (not shown). Attempts to confirm or refute this hypothesis are in preparation.

804G cells—The 804G cells were cultivated for 3, 4, and 7 days. From the very beginning, they adhered satisfactorily, showing normal growth patterns (Fig. 8.9) (Langhofer et al. 1993).

The next point deals with the experiments carried out by Pötschke (2012) in which two cell lines then separated through the mucosa scaffold were simultaneously cultivated. The chamber described above was used in these experiments (Fig. 8.7). In cultures with only the PC3 cells, they were seeded on the side of the scaffold with a basement membrane. These cells attached very well, developing contacts and building a cellular carpet of polarized epithelial cells (Fig. 8.10).

In the co-cultures, the P21 cells (hPCP) derived from the prostatic stroma were placed on the underside of the scaffold where there is no BM the cells established direct contact with the ECM (Fig. 8.10). In fact, this arrangement of the cells represents a further development of our initial experimental design, giving information about the suitability of decellularized mucosa for studies in vitro.

Both cell lines require RPMI 1640 medium, supplemented with 10% FCS, as well as 2.5% HEPES and must be kept in the incubator at 37 ° C with 5% CO_2.

This arrangement was tested not only to see how both cell lines behave on this substrate but also to detect possible interactions which could modify, for example, the growth rates of one or both cell types involved.

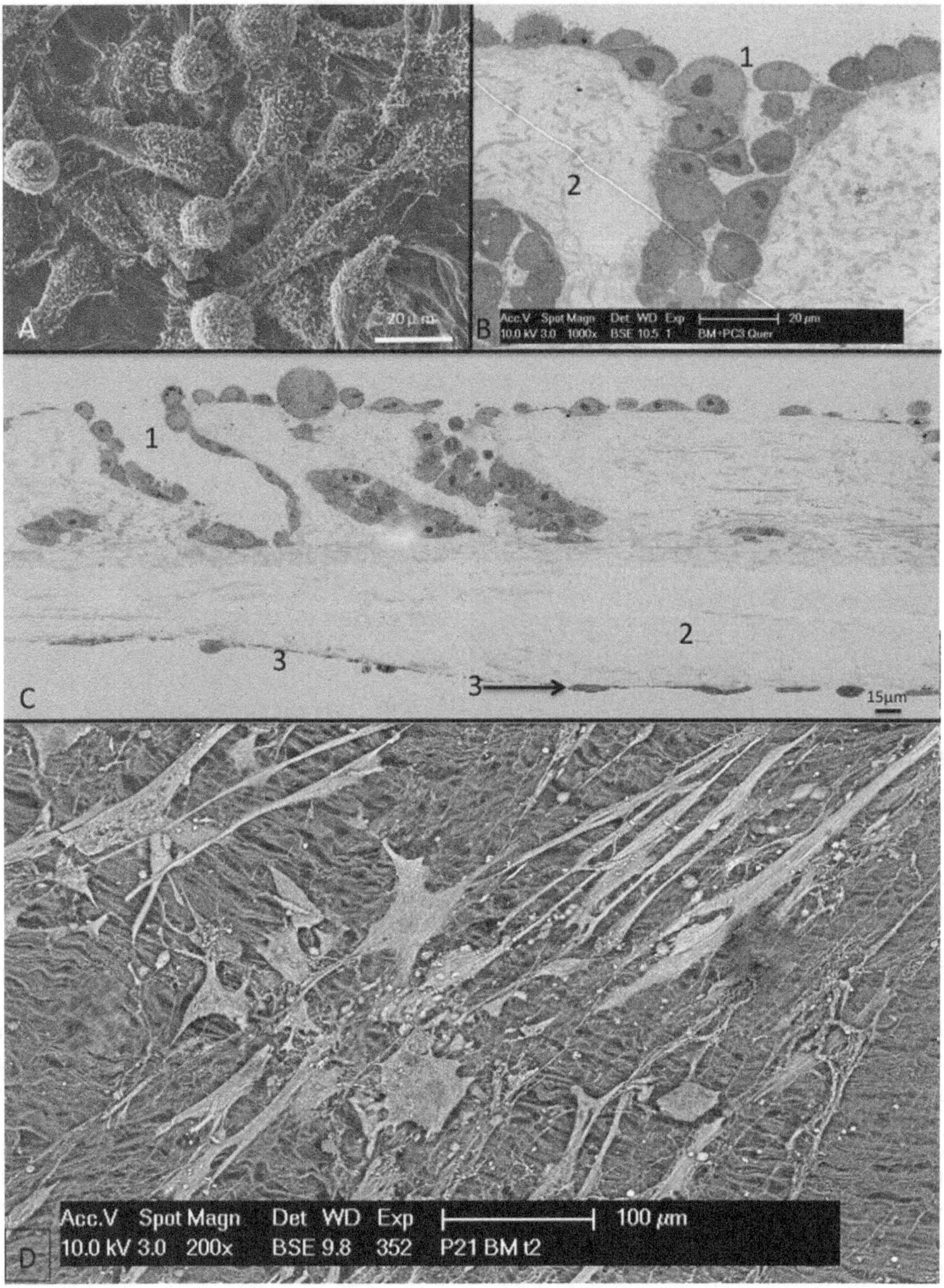

Fig. 8.10 Electron microscopy of co-cultures of PC3 and P21 (hPCP) cells. **a** Cell surface of PC3 cells after 4 days' culture. **b** Block-face SEM image. Crypts and surface of scaffold occupied by PC3 cells arranged in an epithelial manner. 1: opening of a crypt with cells; notice the nucleus with prominent nucleoli. **c** Decellularized colonic mucosa with cells growing on both sides. The PC3 cells lie directly on top (BM), while the P21 cells are found directly on the extracellular matrix to the underside of the scaffold. (1): Crypts and free surface occupied by PC 3 cells; (2) scaffold; (3) P21 cells forming a flat layer. **d** Underside of the scaffold. P21 cells appear bright in the picture and show a rather elongated form with processes in contact with each other. In between, there are fields where the fibrous structure of the scaffold ECM at this side can be clearly seen

The cells and their relationship with the substrate were investigated with scanning electron microscopy and the so-called block-face SEM technique, a method in which the tip of the resin block, already prepared for ultra-microtomy, is examined in the SEM, mainly using the backscattered electrons (BSE) signal to obtain the images (Laue et al. 2005; Nuñez-López et al. 2018).

In single and co-cultures both PC3 and P21 cell lines appear well adapted to the culture conditions. The PC3 cells form a monolayer, which extends across the free surface and also occupies the crypts as well (Fig. 8.10). The stromal P21 cells adhere to the lower face of the mucosa scaffold, where collagen fibers of the ECM are found, and developed stellate shapes with long extensions, forming close relations to each other (Fig. 8.10). There are indications that, under the described conditions, the epithelial PC3 cells could positively influence the proliferation of stromal cells (Pötschke 2012).

8.7 Discussion

In the present study, it has been shown that the mucosa layer of the rat colon can be isolated and decellularized, producing a scaffold of ECM, which can serve as a substrate for tissue engineering. Other studies deal with the colon mucosa of the pig (Keane et al. 2015) and the entire colon wall of the sheep (Kajbafzadeh et al. 2014) and have also prepared scaffolds, but this has—so far—not been the case yet with rat mucosa.

Our aim was to decellularize the mucosa while, at the same time, preserving the BM on the luminal side of the colon wall. In this way, the scaffold would have a top side with BM and an underside with decellularized muscularis mucosae, which is partially covered by submucosa tissue, the latter resulting from the separation of the mucosa from the rest of the colonic wall.

The decellularization procedure foresees the use of a chelator, EDTA, and two detergents, an ionic one, Triton X-100, and a non-ionic one, SDS. This is a procedure that has been used successfully in the past in our laboratory (Mestres et al. 2014; Lopez Gomez et al. 2018). In principle, the decellularization methods are all rather similar, using detergents and other substances such as proteolytic enzymes and physical factors like hypo and hyperosmotic shocks (Liao et al. 2007; Du and Wu 2011; Soto-Gutierrez et al. 2011; Keane et al. 2013; Friedrich et al. 2014; Ji et al. 2018). The combination of the above-mentioned agents results in different recipes, thus permitting the adaptation of decellularization to the various organs, from which ECM scaffolds can be obtained.

To determine the extent of decellularization, histological procedures such as staining with H and E and fluorescence microscopy were used (Badylak et al. 1998; Bhrany et al. 2006; Narayanan et al. 2009; Brown and Badylak 2013; Oliveira et al. 2013; Keane et al. 2017; Lopez Gomez et al. 2018). The presence of cell nuclei or residues thereof was monitored in serial sections, where nucleic acids were stained with propidium iodide (Crapo et al. 2011; Lopez Gomez et al. 2018). After treatment with detergents, our protocol has an added step with DNAse to eliminate the DNA residues that decorate the surface of the specimens in the form of filaments and complex networks (Inaga et al. 1991) and are also present inside the sample (Narayanan et al. 2009; Keane et al. 2013). The degradation and rinsing off of such cell debris from the sample is of crucial importance as otherwise the suitability of the scaffold might be affected. This would in turn influence the cells which could be cultured on it and could even cause further problems if the sample is implanted in vivo, where significant inflammatory reactions could be triggered which might prevent the process of tissue reconstruction (Zhang et al. 2010; Brown and Badylak 2013; Keane et al. 2015).

The BM is a distinct entity of the extracellular matrix, located between parenchymal cells and the connective tissues (Brown et al. 2006; Vllasaliu et al. 2014). Our experiments have shown that with the removal of enterocytes, the basement membrane becomes significantly thinner, with a loss of material at the ad-epithelial side of the BM, the LL. The LD of the BM, located at the ab-epithelial side, seems to be more stable and, after decellularization, still maintains the fine structure and the relationship with the underlying ECM of the lamina propria (Lopez Gomez et al. 2018). Such changes in BM thickness had already been observed after EDTA treatment, but not quantified (Mestres et al. 1991; Lopez Gomez et al. 2018). Although this is a phenomenon which presumably also occurs in many other decellularization procedures, there are as yet no studies with the electron microscope (TEM) to examine this question and compare the various procedures (Lopez Gomez et al. 2018).

On the other hand, our various examinations of the BM marker proteins using ICC show that collagen IV, laminin, and GAG perlecan remain in the BM; molecules which are very important for adhesion, growth, and differentiation of cells (Massie et al. 2017; Ozeki et al. 2006; Brown et al. 2006; Song and Ott 2011; Lopez Gomez et al. 2018).

In general, cells respond to topographical features of the substrate such as ridges, sulci, protrusions of different sizes, etc. with evident changes in metabolism, in orientation and arrangement of the cells, in their motility, adhesion, and cell shape (Folkman and Moscona 1978; Curtis and Wilkinson 1997; Kim et al., 2012).

Scanning electron microscopy has shown that the nanotopography of the BM surface is not completely smooth and is characterized by very small globules of a relatively constant diameter, regardless of the decellularization method used. These globules could be considered as a type of basic construction unit of BM (Carlson and Carlson 1991; Mestres et al. 2014) and may offer clues relating to cell attachment and other functional cyto-parameters.

The surface of the BM with the described topographical and chemical properties has proven to be a useful substrate for cell cultures and the morphology and dynamics of the cells is in first approximation an important indicator for the assessment of the biocompatibility of a scaffold. In this context, it should also be remembered that not only is the composition of the scaffold important, but also that of the other substances used for decellularization, which have to be washed out to avoid negative effects on the cell cultures (Crapo et al. 2011; Friedrich et al. 2014). However, thorough washing of the sample means that several necessary factors such as growth factors attached to the ECM (sponge effect) will also be removed and must subsequently be replaced (Reing et al. 2010). We have observed the dramatic effects of such deficits on our cultures, especially when working without serum and especially on endothelial cells.

Fibroblasts growing on the rat scaffold develop their own characteristic cell form and arrangement (Chou et al. 1995).

The endothelial cells also adhere, but more slowly than the fibroblasts. This behavior could be due to the absence or reduced amount of factors in the medium which the cells themselves are not able to produce. They thus remain vital but arrested during cultivation (Jaffe et al. 1973).

The culture chamber developed in our laboratory allows the cultivation of one or even two cell types, in this case, separated from each other by the substrate, i.e., the isolated colonic mucosa. This is a special situation in which possible interactions between the two cell types in terms of adhesion, growth, and differentiation can be investigated. Studies have reported that chondrocytes were potentiated by co-culture with mesenchymal stem cells, being the first cells to move in a growth phase and differentiate prior to implantation in an in vivo model (Meretoja et al. 2012). The studies performed by Pötschke (2012) have shown that both cell lines used to adhere to the mucosa scaffold and cell counts suggest an influence of PC3 cells on the growth rates of P21 (hPCP) cells. This is remarkable as, in our setting, such an influence could only take place through cell release of substances, which then diffuse through the scaffold.

The stiffness and mechanical properties of the scaffold can also affect the behavior of the cells. The isolated mucosa of the colon is rather thin, approximately 200–250 μm thick, depending on the extent of colon dilatation, but once isolated, it can be stretched, thus automatically becoming thinner, but not significantly stiffer (Wiesner et al. 2002). In our preparations, we have tried to compensate these weaknesses by clamping the isolated mucosa over the Macrolon ring. This measure also seems to be adequate for other cultures, as the various cell types display normal growth and typical cell morphology (Emonard et al. 1987; Kleinman and Martin 2005). Another way of stabilizing the mucosa is the application of cross-linkers such as genipin or even the assembly of several layers of the substrate (Koch et al. 2012). Some such mono- and multilayered scaffolds are commercially available (Badylak 2007; Badylak et al. 2009).

With regard to future experiments with the implantation of mucosa scaffolds, a comparison with the small intestine submucosa (SIS) might prove interesting. While the thickness and structure of the connective fibers appear to be very similar, the difference lies in the presence of the BM in the mucosa scaffold. Such comparisons might lead to the extrapolation of similar applications for such a scaffold despite the size difference.

One problem that can arise with implantation is the presence in the scaffold of factors such as the alpha-gal epitopes, which can trigger a negative reaction in the recipient (McPherson et al. 2000). However, the fact that SIS fails to activate the complement in vitro leads us to suspect that the density of these epitopes is very low, especially after decellularization treatment; a situation that could also be possible in the mucosa of the colon of the rat (McPherson et al. 2000; Naso et al. 2011). And finally, SIS reabsorbs relatively quickly (Badylak et al. 1998). This could be due to the architecture of the connective tissue fibers, which again is similar in the colonic mucosa.

8.8 Conclusions

The present studies show that an ECM scaffold can be prepared from the colon of the rat and that it is comparable to other ECM scaffolds derived from other sections of the gastrointestinal tract.

However, this scaffold has characteristics that distinguish it from others. The scaffold of the colonic mucosa remains continuous on one side with basement membrane material, while on the other side remnants of submucosa and the muscle layer of mucosa tissue remain. With these surface differences, this scaffold gains a certain singularity. The method we have followed here foresees the use of a chelator, two detergents, and no proteolytic enzymes. The application schedule is flexible as far as it is possible to change the duration of the different steps. In such cases, it is advisable to reduce the temperature more the longer the treatment is applied. We have achieved complete decellularization, even in the case of the entire rat colon wall.

The chamber for cell cultures was designed especially for experiments with rat material, but it can, in principle, be modified and adapted to suit other objectives or questions. The strategy of co-cultures described here could be of interest for questions related to tissue reconstruction involving, for instance, stem cells.

As both primary cultures and cell lines grew well on the scaffold of the colon mucosa, the biocompatibility of this ECM scaffold can be considered proven. The slightly different behavior of the endothelia indicates that they require additives in the medium and seems to be not so much a problem of the scaffold itself.

With regard to possible applications, a comparison of this scaffold to the SIS is justified in view of the thickness and the microscopic and 3D structure of the ECM. These similarities suggest that the ECM scaffold of the colonic mucosa might be rapidly resorbed in implantation trials—similar to SIS. It is also important to note that SIS only triggers moderate inflammatory responses. This is mostly due to the fact that certain epitopes are reduced in quantity after decellularization. By analogy, similar responses could also be expected for ECM scaffolds of the colonic mucosa. Future studies will be devoted to clarifying these last points in order to propose potential applications for this new scaffold, be it on animal models or for clinical issues.

Acknowldgements For the linguistic revision of the manuscript, we thank Mrs. Ann Soether.This research has received generous support from Saarland University (to P. M.) and from the Multimate-Challenge Project funded by the Madrid Region under program S2013/MIT-2862 (to G. del R.).

References

Andres H, Bock R, Bridges RJ, Rummel W, Schreiner J (1985) Submucosal plexus and electrolyte transport across rat colonic mucosa. J Physiol 364:301–312

Badylak SF (2007) The extracellular matrix as a biological scaffold material. Biomaterials 28(25):3587–3593. https://dio.org/10.1016/j.biomaterials.2007.04.043

Badylak SF, Kropp B, McPherson T, Liang H, Snyder PW (1998) Small intestine submucosa: a rapid resorbed bioscaffold for augmentation cystoplasty. Tissue Eng 4(4):379–387

Badylak SF, Freytes DO, Gilbert TW (2009) Extracellular matrix as a biological scaffold material: structure and function. Acta Biomater 5:1–13

Barnes CA, Brison J, Michel R, Brown BN, Castner DG, Badylak SF, Ratner BD (2011) The surface molecular functionality of decellularized extracellular matrices. Biomaterials 32(1):137–143. https://doi.org/10.1016/j.biomaterials.2010.09.007

Bhrany AD, Beckstead BL, Lang TC, Farwell DG, Giachelli CM, Ratner BD (2006) Development of an esophagus acellular matrix tissue scaffold. Tissue Eng 12(2):319–330. https://doi.org/10.1089/ten.2006.12.319

Böhme M, Diener M, Mestres P, Rummel W (1992) Direct and indirect actions of $HgCl_2$ and methyl mercury chloride on permeability and chloride secretion across the rat colonic mucosa. Toxicol Appl Pharmacol 114(2):285–294

Bridges RJ, Rack M, Rummel W, Schreiner J (1986) Mucosal plexus and electrolyte transport across the rat colonic mucosa. J Physiol 376:531–542

Brown BN, Badylak SF (2013) Expanded applications, shifting paradigms and an improved understanding of host-biomaterial interactions. Acta Biomater 9 (2):4948–4955. https://doi.org/10.1016/j.actbio.2012.10.025

Brown B, Lindberg K, Reing J, Stolz DB, Badylak SF (2006) The basement membrane component of biologic scaffolds derived from extracellular matrix. Tissue Eng 12(3):519–526

Carlson EC, Carlson BM (1991) A method for preparing skeletal-muscle fiber basal laminae. Anat Rec 230 (3):325–331

Chou L, James D, Firth JD, Veli-Jukka Uitto VJ, Donald M, Brunette DM (1995) Substratum surface topography alters cell shape and regulates fibronectin mRNA level, mRNA stability, secretion and assembly in human fibroblasts. J Cell Sci 108(Pt4):1563–1573

Crapo PM, Gilbert TW, Badylak SF (2011) An overview of tissue and whole organ decellularization processes. Biomaterials 32(12):3233–3243

Curtis A, Wilkinson Ch (1997) Topographical control of cells (review). Biomaterials 18:1573–1583

Denost Q, Adam JP, Pontallier A, Montembault A, Bareille R, Siadous R, Delmond S, Rullier E, David L, Bordenave L (2015) Colorectal tissue engineering: a comparative study between porcine small intestinal submucosa (SIS) and chitosan hydrogel patches. Surgery 158(6):1714–1723. https://doi.org/10.1016/j.surg.2015.06.040

Diener M, Rummel W, Mestres P, Lindemann B (1989) Single chloride channels in colon mucosa and isolated colonic enterocytes of the rat. J Membr Biol 108:21–30

Du L, Wu X (2011) Development and characterization of a full-thickness acellular porcine cornea matrix for tissue engineering. Artif Organs 35(7):691–705. https://doi.org/10.1111/j.1525-1594.2010.01174.x

Emonard H, Calle A, Grimaud JA, Peyrol S, Castronovo V, Noel A, Lapière CM, Klein-man HK, Foidart JM (1987) Interactions between fibroblasts and a reconstituted basement membrane matrix. J Invest Dermatol 89:156–163

Folkman J, Moscona A (1978) Role of cell shape in growth control. Nature 273:345–349

Friedrich LH, Jungebluth P, Sjöqvist S, Lundin V, Haag JC, Lemon G, Gustafsson Y, Ajalloueian F, Sotnichenko A, Kielstein H, Burguillos MA, Joseph B, Teixeira AI, Lim ML, Macchiarini P (2014) Preservation of aortic root architecture and properties using a detergent-enzymatic perfusion protocol. Biomaterials 35(6):1907–1913

Inaga S, Osatake H, Tanaka K (1991) SEM images of DNA double helix and nucleosomes observed by ultrahigh-resolution scanning electron microscopy. J Electron Microsc (Tokyo) 40(3):181–186

Izumi W, Hirao Y, Hopp L, Oyasu R (1981) In vitro induction of ornithine decarboxylase in urinary bladder carcinoma cells. Cancer Res 41:405–409

Jaffe EA, Nachman RL, Becker CG, Minick CR (1973) Culture of human endothelial cells derived from umbilical veins. Identification by morphologic and immunologic criteria. J Clin Invest 52(11):2745–2756

Janssen M, Albrecht M, Möschler O, Renneberg H, Fritz B, Aumüller G, Konrad L (2000) Cell lineage characteristics of human prostatic stromal cells cultured in vitro. Prostate 43(1):20–30

Ji Y, Zhou J, Sun T, Tang K, Xiong Z, Ren Z, Yao S, Chen K, Yang F, Zhu F, Guo X (2018) Diverse preparation methods for small intestinal submucosa (SIS): decellularization, components, and structure. J Biomed Mater Res A. https://doi.org/10.1002/jbm.a.36582

Kaighn ME, Narayan KS, Ohnuki Y, Lechner JF, Jones LW (1979) Establishment and characterization of a human prostatic carcinoma cell (PC-3). Invest Urol 17(1):16–23

Kajbafzadeh AM, Masoumi A, Hosseini M, Borjian MA, Akbarzadeh A, Mohseni MJ (2014) Sheep colon acellular matrix: immunohistologic, biomechanical, scanning electron microscopic evaluation and collagen quantification. J Biosci Bioeng 117(2):236–241. https://doi.org/10.1016/j.jbiosc.2013.07.006

Keane TJ, Londono R, Carey RM, Carruthers CA, Reing JE, Dearth CL, D'Amore A, Medber-Ry CJ, Badylak SF (2013) Preparation and charac-terization of a biologic scaffold from esophageal mucosa. Biomaterials 34(28):6729–6737

Keane TJ, Swinehart IT, Badylak SF (2015) rization used for preparation of biological scaffolds and in vivo relevance. Methods 84:25–34

Keane TJ, Dziki J, Castelton A, Faulk DM, Messer-schmidt V, Londono R, Reing JE, Velankar SS, Badylak SF (2017) Preparation and characterization of a biologic scaffold and hydrogel derived from colonic mucosa. J Biomed Mater Res Part B Appl Mater 105 (2):291–306. https://doi.org/10.1002/jbm.b.33556

Kim YH, Kwon HJ, Kim DS (2012) Matrix metalloproteinase 9 (MMP-9)-dependent processing of βig-h3 protein regulates cell migration, invasion, and adhesion. J Biol Chem 287(46):38957–38969. https://doi.org/10.1074/jbc.M112.357863

Kleinman HK, Martin GR (2005) Matrigel: basement membrane matrix with biological activity. Sem Cancer Biol 15(5):378–386

Koch H, Graneist C, Emmrich F, Till H, Metzger R, Aupperle H, Schierle K, Sack U, Boldt A (2012) Xenogenic esophagus scaffolds fixed with several agents: comparative in vivo study of rejection and inflammation. J Biomed Biotechnol 948320. https://doi.org/10.1155/2012/948320

Langhofer M, Hopkinson SB, Jones JC (1993) The matrix secreted by 804G cells contains laminin-related components that participate in hemi-desmosome assembly in vitro. J Cell Sci 105(Pt3):753–764

Laue M, Kiefer G, Leis B, Pütz N, Mestres P (2005) Use of environmental scanning electron microscopy to study a resin block face. Microsc Anal 19(5):17–19

Liao J, Joyce EM, Sacks MS (2007) Effects of decellularization on the mechanical and structural properties of the porcine aortic valve leaflet. Biomaterials 29 (8):1065–1074

Lopez Gomez L, Nuñez Lopez T, del Rosario Hernandez G, Mestres-Ventura P (2018) The isolated colonic mucosa: a new substrate for tissue reconstruction. Isolation, decellularization and characterization by immunohistochemistry and electron microscopy. Eur J Anat 22(1):1–15

Lopez-Gomez L (2015) Ph. Thesis. Descripción y evaluación de un método para descelularizar la mucosa aislada del colon de la rata y la posible aplicación de ésta en ingeniería de tejidos (in Spanish). https://eciencia.urjc.es/handle/10115/13181

Maghsoudlou P, Totonelli G, Loukogeorgakis SP, Eaton S, De Coppi P (2013) A decellularization methodology for the production of a natural acellular intestinal matrix. J Vis Exp 80:1–6, e50658. https://doi.org/10.3791/50658

Massie I, Spaniol K, Barbian A, Poschmann G, Stühler K, Geerling G, Metzger M, Mertsch S, Schrader S (2017) Evaluation of decellularized porcine jejunum as a matrix for lacrimal gland reconstruction in vitro for treatment of dry eye syndrome. Invest Ophthalmol Vis Sci 58:5564–5574. https://doi.org/10.1167/iovs.16-20759

McPherson TB, Liang H, Record RD, Badylak SF (2000) Gala (1, 3) Gal epitope in porcine small intestinal submucosa. Tissue Eng 6(3):233–239

Meretoja VV, Dahlin RL, Kasper FK, Mikos AG (2012) Enhanced chondrogenesis in co-cultures with articular chondrocytes and mesenchymal stem cells. Biomaterials 33:6362–6369. https://doi.org/10.1016/j.biomaterials.2012.05.042

Mestres P, Diener M, Mai H, Rummel W (1991) The epithelial basal lamina of the isolated colonic mucosa —scanning and transmission electron-microscopy. Acta Anat 141(1):74–81

Mestres P, Diener M, Rummel W (1992a) Histo- and immunocytochemical characterization of the neurons of the mucosal plexus in the rat colon. Acta Anat 143:268–274

Mestres P, Diener M, Rummel W (1992b) Electron microscopy of the mucosal plexus of the rat colon. Acta Anat 143:275–284

Mestres P, Lopez Gomez l, Nuñez Lopez T, del Rosario Hernandez G, Lukas SW, Hartmann U (2014) The basement membrane of the isolated rat colonic mucosa. A light, electron and atomic force microscopy study. Ann Anat 196(2–3):108–118. https://doi.org/10.1016/j.aanat.2014.01.001

Narayanan K, Leck KJ, Gao S, Wan ACA (2009) Three-dimensional reconstituted extracellular matrix scaffolds for tissue engineering. Biomaterials 30:4309–4317

Naso F, Gandaglia A, Iop L, Spina M, Gerosa G (2011) First quantitative assay of alpha-Gal in soft tissues: presence and distribution of the epitope before and after cell removal from xenogeneic heart valves. Acta Biomater 7:1728–1734

Nietzer S, Baur F, Sieber S, Hansmann J, Schwarz T, Stoffer C, Häfner H, Gasser M, Waaga-Gasser AM, Walles H, Dandekar G (2016) Mimicking metastases including tumor stroma: a new technique to generate a three-dimensional colorectal cancer model based on a biological decellularized intestinal scaffold. Tissue Eng Part C Methods 22(7):621–635. https://doi.org/10.1089/ten.TEC.2015.0557

Núñez-López T, del Rosario-Hernández G, López-Gómez L, Mestres-Ventura P (2018) Specimen preparation for "block face" scanning electron microscopy (BFSEM). An energy-dispersive X-ray spectroscopy study. Eur J Anat 22(6):471–481

Oliveira AC, Garzón I, Ionescu AM, Carriel V, Cardona JL, González-Andrades M, Pérez MM, Alaminos M, Campos A (2013) Evaluation of small intestine grafts decellularization methods for corneal tissue engineering. PLOS One (6):e66538. https://doi.org/10.1371/journal.pone.0066538

Ozeki M, Narita Y, Kagami H, Ohmiya N, Itoh A, Hirooka Y, Niwa Y, Ueda M, Goto H (2006) Evaluation of decellularized esophagus as a scaffold for cultured esophageal epithelial cells. J Biomed Mater Res A 79(4):771–778

Parsons DS, Paterson CR (1965) Fluid and solute transport across rat colonic mucosa. Exp Physiol 50 (2):220–231

Pötschke N (2012) Diplomarbeit. Epitheliale und stromale Interaktionen von humanen Prostatazelllinien unter dem Einfluss von Ratten-Basalmembran (in German). Zentrum für Human- und Molekularbiologie (ZHMB), Naturwissenschaftliche Technische Fakultät und Medizinische Fakultät der Universität des Saarlandes, Saarbrücken, Germany

Reing JE, Brown BN, Daly KA, Freund JM, Gilbert TW, Hsiong SX, Huber A, Kullas KE, Tottey S, Wolf MT, Badylak SF (2010) The effects of processing methods upon mechanical and biologic properties of porcine dermal extracellular matrix scaffolds. Biomaterials 31 (33):8626–8633. https://doi.org/10.1016/j.biomaterials.2010.07.083

Reynolds ES (1963) The use of lead citrate at high pH as an electron-opaque stain in electron microscopy. J Cell Biol 17(1):208–212. https://doi.org/10.1083/jcb.17.1.208.

Riccardi C, Nicoletti I (2006) Analysis of apoptosis by propidium iodide staining and flow cytometry. Nat Protoc 1(3):1458–1461

Richardson KC, Jarett L, Finke EH (1960) Em-bedding in epoxy resins for ultrathin sectioning in electron microscopy. Stain Technol 35:313–325

Song JJ, Ott HC (2011) Organ engineering based on decellularized matrix scaffolds. Trends Mol Med 17 (8):424–432. https://doi.org/10.1016/j.molmed.2011.03.005

Soto-Gutierrez A, Zhang L, Medberry Ch, Fukumitsu K, Faulk D, Jiang H, Reing J, Gramignoli R, Komori J, Ross M, Nagaya M, Lagasse E, Stolz D, Strom SC, Fox IJ, Badylak SF (2011) A whole-organ regenerative medicine approach for liver replacement. Tissue Eng Part C Methods 17(6):677–686. https://doi.org/10.1089/ten.tec.2010.0698

Vllasaliu D, Falcone FH, Stolnik S, Garnett M (2014) Basement membrane influences intestinal epithelial cell growth and presents a barrier to the movement of macromolecules. Exp Cell Res 323(1):218–231. https://doi.org/10.1016/j.yexcr.2014.02.022

Wiesner W, Mortelé KJ, Ji H, Ros PR (2002) Normal colonic wall thickness at CT and its relation to colonic distension. J Comput Assist Tomogr 26(1):102–106

Zhang Q, Raoof M, Chen Y, Sumi Y, Sursal T, Junger W, Brohi K, Itagaki K, Hauser CJ (2010) Circulating mitochondrial DAMPs cause inflammatory responses to injury. Nature 464(7285):104–107. https://doi.org/10.1038/nature08780

The Renal Extracellular Matrix as a Supportive Scaffold for Kidney Tissue Engineering: Progress and Future Considerations

9

Shabnam Sabetkish
and Abdol-Mohammad Kajbafzadeh

Abstract

During the past decades, diverse methods have been used toward renal tissue engineering in order to replace renal function. The goals of all these techniques included the recapitulation of renal filtration, re-absorptive, and secretary functions, and replacement of endocrine/metabolic activities. It is also imperative to develop a reliable, up scalable, and timely manufacturing process. Decellularization of the kidney with intact ECM is crucial for in-vivo compatibility and targeted clinical application. Contemporarily there is an increasing interest and research in the field of regenerative medicine including stem cell therapy and tissue bioengineering in search for new and reproducible sources of kidneys. In this chapter, we sought to determine the most effective method of renal decellularization and recellularization with emphasis on biologic composition and support of stem cell growth. Current barriers and limitations of bioengineered strategies will be also discussed, and strategies to overcome these are suggested.

Keywords

Decellularization · Recellularization · Kidney matrices · Scaffold · Renal stem cells

9.1 History

End-stage renal disease (ESRD) is a disturbing problem which may culminate in the progressive worsening of renal function with donor organ transplantation (Sullivan et al. 2012). The ways by which kidney function can be restored in humans consist of dialysis and renal allotransplantation. Furthermore, Medicare expenses of these patients exceed $29 billion, yearly (Saran et al. 2017). Due to decreased organ donation, poorly tolerated dialytic therapies, and the morbidity related to immunosuppression, regenerative medicine methods have been considered as proficient curative alternatives (Song et al. 2013). In 2014, a report into organ transplantation in the USA showed that kidneys are the most recurrently transplanted organ with 15,978 procedures (Alachkar et al. 2011). Dialysis and allogenic transplantation are the widespread types of pyelonephritis treatment, although their noteworthy complications and high cost, accessibility, and long-term dependency are still main

S. Sabetkish · A.-M. Kajbafzadeh (✉)
Pediatric Urology and Regenerative Medicine Research Center, Section of Tissue Engineering and Stem Cells Therapy, Children's Medical Center, Pediatric Center of Excellence, Tehran University of Medical Sciences, No. 62, Dr. Gharib's Street, Keshavarz Boulevard, 1419433151 Tehran, Iran
e-mail: kajbafzd@sina.tums.ac.ir

A. Kajbafzadeh (ed.), *Decellularization Methods of Tissue and Whole Organ in Tissue Engineering*, Advances in Experimental Medicine and Biology 1345,
https://doi.org/10.1007/978-3-030-82735-9_9

concerns of these modalities (Kajbafzadeh et al. 2010).

The function of the mature kidney with an outstandingly complex structure is dependent on the growth and differentiation of its precursor cells within the intermediate mesoderm into a mature organ with numerous diverse cell types (Hammerman 2003). Definition of twenty six terminally differentiated nephron cell types, as mentioned by Awqati and Oliver, takes into account cell morphology, location, and function (Al-Awqati and Oliver 2002). The nephrons should be incorporated in three dimensions with a collecting system and with one another and, the origin of which is the ureteric bud, for the purpose of glomerular filtration, reabsorption, and secretion of fluid and electrolytes (Horster et al. 1999; Sariola et al. 1983; Hyink et al. 1996; Rogers and Hammerman 2001). The whole kidney regeneration may be intricate; however functional recovery of about 10% of kidney filtration function will withdraw the dialysis in patients with ESRD, resulting in significantly increasing quality of life (Locatelli et al. 2005). The regeneration of a functional kidney signifies a gigantic convenient challenge. However, with the progress in the field of regeneration, we are optimistic about overcoming ESKD through kidney regeneration.

State-of-the-art tissue engineering techniques try to construct decellularized scaffolds that could avoid allograft rejection and prevent the need for immunosuppressive therapy (Perin et al. 2008; Crapo et al. 2011). The novel therapeutic potential of tissue engineering and stem cell therapy with anti-inflammatory properties and immune modulation has been the focus of recent studies. In an ideal tissue-engineered kidney, the function of erythropoietin, renin synthesis, 1-α hydroxylation of 25(OH)D38, and 5′ deiodination of thyroid hormone should be recapitulated (Yasuoka et al. 2020; Martini and Danser 2017; Bland et al. 2000; Benedetti et al. 2019). Natural extra cellular matrices (ECM) and synthetic scaffolds are the two major categories of 3D scaffolds. However, ECMs have demonstrated superior performance compared to synthetic scaffolds. Preservation of the ultrastructure, integrity, and bioactivity of the ECM can be optimized by using an effective decellularization methodology (Kawecki et al. 2018). As the kidney organ has complexity with more than thirty different cell types, using tissue engineering and regenerative medicine techniques with the unique microenvironment and preserved ECM may pave a road for the bioengineering of a whole transplantable kidney (Figliuzzi et al. 2017). Extracorporeal bioartificial kidney (BAK) support systems were first developed in the 1980s when synthetic scaffolds were merged with cellular constituents (Aebischer et al. 1987; Ip and Aebischer 1989). In these studies, human proximal renal tubular cells were cultured on hollow fiber scaffolds and put in sequences with a hemofiltration path. In phase I/II clinical trials, bioartificial kidney systems were successful in filtering urine, develop metabolic parameters, diminish pro-inflammatory cytokine levels, and enhance cardiovascular stability (Humes et al. 2003, 2004). Nevertheless, a considerable impact on survival was hard to determine due to the fact that the Phase II clinical trial achieved in 2004–2005 was underpowered (Tumlin et al. 2008; Chertow and Waikar 2008). Presently, the usage of BAKs is limited by the survival of tubular cells and the cost-effective developing of the device (Humes et al. 2014; Sanechika et al. 2011).

Our group has extensive experience with regard to the structure of the kidney, which makes us to our decision to continue with the preparation of kidney, with the aim of developing a new ECM scaffold. Herein, we aim to review the state-of-art and tissue engineering and regenerative medicine technology highlighting the major achievements toward the production of bioengineered renal obtained decellularization and recellularization techniques.

This chapter will delineate the progress to date recorded for approaches of converse the kidney as a subject for tissue engineering paradigm, discuss about the recent developments made in the fields of renal tissue engineering and stem cells, as well as emphasize the challenges which we may confront with when applying such constructs in a clinical setting.

9.2 Materials and Methods

Native kidney ECM, which plays a crucial role in kidney development and repair, has been demonstrated to provide a scaffold for cell seeding and a niche for stem cells to differentiate into a whole renal organ (Ohata and Ott 2020; Oliver 1953). Findings of previous studies in whole organ decellularization (Sabetkish et al. 2015) encouraged scientists to reconstruct entire kidneys by a combination of different decellularization kidneys. Moreover, in the case of whole organ decellularization, it is of utmost importance to maintain an intact vascular network. Organogenesis and repair of the renal organ are influenced by molecules of the ECM and their receptors. These scaffolds provide spatial organization of cells with secreting and storing growth factors and cytokines and regulating signal transduction (Yi et al. 2017). Biomaterials such as polymer scaffolds have been examined for kidney tissue engineering with a limited capacity. It has been also demonstrated that combination of synthetic polymers, such as PLGA with natural ECM may enhance biocompatibility and decrease hydrophobicity of scaffolds compared to PLGA alone (Lih et al. 2016). Collagen, elastin, glycosaminoglycans (GAGs), growth factors, fibronectin, and laminin also participate in the development and preservation of vascular structures which is remarkable when the ECM is used as a scaffold for whole organ reconstruction. The preservation of these components is of great importance in the decellularization method (Badylak 2004). BFGF and VEGF are among the most important growth factors in the decellularized kidney matrix. BFGF is recognized in the developing kidney and early renal organization, while VEGF is critical for angiogenesis and renal podocyte development (Wang et al. 2018; Little and McMahon 2012). Using high concentration of SDS (1%) may completely eliminate the bFGF or VEGF within the decellularized tissues, as reported by Nakayama et al. and Soto-Gutierrez et al.

The meticulous structural components of the kidney ECM such as glomerular basement membrane, are decisive for suitable tissue-specific function, and integrin-binding ligands of the basement membrane and bound growth factors as extracellular signaling cues, are imperative for driving renal cell growth (Uzarski et al. 2014). Detergents, enzymes, or other cell-lysing solutions are perfused antegrade via the renal vasculature to eliminate the antigenic parenchyma from the whole renal ECM.

ECM scaffolds from animal and whole human-cadaveric organs can be produced by detergent-based decellularization (Orlando et al. 2011). The choice of the detergents in decellularization method is significantly essential, as they can harm the microstructure and composition of the obtained scaffold and may indirectly affect the mechanical properties of the final product. The widespread use of detergents, such as the nonionic Triton X-100 (Song et al. 2013; Ross et al. 2009; Ross et al. 2012; Kajbafzadeh et al. 2019) and the anionic sodium dodecyl sulfate (SDS) (Kajbafzadeh et al. 2019; Burgkart et al. 2014; Bonandrini et al. 2014; Nakayama et al. 2013) to solubilize and wash out cellular components, has been reported in a number of kidney decellularization protocols. In our center, no enzymatic methods are used owing to the possibility of adverse effects of enzymes on sensitive molecules in the ECM. Additionally, no snap freezing method is performed as extremely low temperatures can alter the 3D architecture of the scaffolds. Guan et al. used both 1% SDS and 1% Triton X-100 to decellularize porcine kidneys, with preserved ECM components and the microarchitecture of the kidneys (Guan et al. 2015a). The choice of sterilization method has also an impact on the quality of the scaffold. Pressurized steam and dry heat are among the commonly used sterilization methods which have the potential to cause protein denaturation (Song and Ott 2011). Several studies have applied decellularization and recellularization techniques for kidney regeneration. Many animals such as rats (Ross et al. 2009), rhesus monkeys

(Nakayama et al. 2010), and pigs (Sullivan et al. 2012) have been used for decellularization studies. Zebrafish are also a brilliant animal model utilized in studying kidney regeneration. To date, zebrafish have been applied to progress our awareness of renal development as well as characteristics of numerous kidney disease phenotypes (Swanhart et al. 2011; Lieschke and Currie 2007).

Perfusion-based and static-based decellularization methods have been considered as effective methods for preserving the 3D architecture and vascular tree of the native ECM scaffolds which can guide organ development and reconstruction (Song and Ott 2011). Bioreactor systems have been also designed to enable the perfusion of decellularized whole kidneys through their native vasculature. This process assists standardized delivery of nutrient-rich culture media to cells seeded throughout the scaffold (Ross et al. 2012; Bijonowski et al. 2013; Fraser and Endres 2014; Kajbafzadeh et al. 2017). Enhancing the effectiveness of cell seeding process by the application of novel bioreactors and creative cell delivery approaches is fundamental to scale-up recellularization techniques to human-sized kidney scaffolds with a significantly more number of cells.

Regardless of the decellularization technique used, the obtained renal ECM must be meticulously assessed to confirm sufficient cell removal and preserved ECM. DNA quantification, DAPI staining, histological loss of nuclei or MHC antigens, maintenance of basement membrane proteins and growth factors through immunohistochemical (IHC) staining, ELISA, proteomic analysis, scanning electron microscopy (SEM), and histological staining are among the applied examinations (Crapo et al. 2011). Measurement of hydroxyproline content, sirius red staining, movat's pentachrome staining, biophysical properties (tensile test), and computerized tomography (CT) angiography are all among the other valuable examinations in renal regeneration. The competence of the obtained scaffold should be also verified to support cellular adhesion, growth, and stem cell differentiation and authenticate the bioactivity of the renal matrix (Uzarski et al. 2014). The required criteria for efficient decellularization including <50 ng dsDNA per mg ECM dry weight, <200 bp DNA fragment length, and lack of visible nuclear material in tissue sections stained with DAPI or H&E (Crapo et al. 2011).

To date, few cases have been reported regarding the transplantation of decellularized whole kidney ECM scaffolds, due to the inherent thrombogenicity of the non-endothelialized vasculature. However, in our recent study, we evaluated the efficacy of two different whole organ decellularization protocols, vasculature integrity, and in-vivo transplantation of sheep kidneys (Kajbafzadeh et al. 2019). The results verified the efficacy of well-preserved decellularized scaffold and vasculature network in post renal transplant outcome in a sheep model.

We also introduced simultaneous transplantation of bilateral decellularized kidneys in a rat model as a more feasible and practical approach (Kajbafzadeh et al. 2018). Various microvascular techniques of arterial and venous anastomosis in renal transplant in rat models have been previously described (Soma et al. 2009; Asfar et al. 1988). In one study in 2012, the renal artery and vein of decellularized porcine kidneys were anastomosed to the aorta and vena cava, respectively, in pig model. Although sufficient blood flow without bleeding was observed during one hour of intra-operative monitoring, extensive thrombosis was noted throughout the kidney scaffolds 2 weeks after surgery (Orlando et al. 2012).

The on the rise field of tissue engineering is also to produce structures and devices to restore lost tissue or organ functions (MacKay et al. 1992; Humes et al. 1999). The combination of regenerative medicine and bioengineering proposes promise for the regeneration of the whole renal organ (Salvatori et al. 2014). A renal tubule assist device (RAD) with living renal proximal tubule cells demonstrated differentiated absorptive, metabolic, and endocrine functions similar to host kidneys in animal experiments (Song and Humes 2009). Application of 3D bioprinting techniques was also flourishing in reconstructing the complex structures of proximal tubules and vasculatures in vitro (Mota et al. 2020), although the physiological functions reproduced by these technologies reflect only a small part of organs. An overview of the kidney decellularization literature has been provided in Table 9.1 (Sullivan et al. 2012; Song et al. 2013; Bonandrini et al. 2014; Guan et al. 2015a, b; Nakayama et al. 2010,2013, 2011; Ross et al. 2012, 2009; Vishwakarma et al. 2014; Orlando et al. 2013; Peloso et al. 2015; Rafighdoust et al. 2015; Sambi et al. 2017; Baptista et al. 2009; Park and Woo 2012; Choi et al. 2015; Poornejad et al. 2015, 2016a, b; Willenberg et al. 2015; He et al. 2017; Liu et al. 2015; Yu et al. 2014; Burgkart et al. 2014; Caralt 2015; Uzarski et al. 2015).

ECM, extracellular matrix; ddH_2O, double-distilled water; PBS, phosphate-buffered saline; SDS, sodium dodecyl sulfate; NaDOC, sodium deoxycholate; DNase, deoxyribonuclease; DMEM, Dulbecco's modified Eagle's medium; EGTA, ethylene glycol tetraacetic acid; FGF, fibroblast growth factor; hESCs, human embryonic stem cells; dH2O, distilled water; NaCl, sodium chloride; KCl, potassium chloride; HEPES, 4-(2-hydroxyethyl)-1-piperazineethanesulfonic acid; PAA, peracetic acid; iPSC, induced pluripotent stem cell.

9.3 Cell Seeding

Different cell types are considered to be proficient to contribute directly to renal reconstruction after injury or considerably improve renal damage without directly contributing to the renal epithelium. For the treatment of patients with early stage renal disease, cell-based therapies have been considered as a crucial step. However, these approaches may not be effective in ESRD cases with severe fibrosis (Bandeiras et al. 2019). Hence, whole organ regeneration using seeded cells would be a fascinating conception. Generating precise renal progenitor cells is indispensable for the development of a whole kidney de novo. How well the decellularized ECM preserves the significant functions is likely to be influenced by the decellularization technique (Caralt et al. 2015) the age and health of the donor tissue (Nakayama et al. 2010, 2011). In addition, different routes of cell administration such as renal artery vs. ureter perfusion; or medullary vs. cortical injection have been used for reseeding the renal scaffold. Each of these techniques may influence migration, differentiation, and functional results (Uzarski et al. 2014).

A variety of stem cell sources have been verified to be capable to form kidney cell types such as human fetal kidney stem cells, cells from the amniotic fluid, or the directed differentiation of human PSCs to kidney endpoints.

Pluripotent stem cells (PSCs) which include both embryonic stem cells (ESCs) and iPSCs, signify an essential cell type for bioengineering techniques aimed at kidney regeneration. Embryonic stem (ES) cells have been considered as a useful starting material for some clinical purposes of tissue engineering (Hielscher et al. 2018). As iPSCs can be produced effortlessly from any patient, they are considered as the promise of disease modeling and stem cell therapies without any immunologic rejection. In order to entirely recapitulate kidney improvement, ureteric bud progenitor, metanephric mesenchyme progenitor, and angioblasts are required as three different types of progenitors, all of which are successfully generated from hPSCs.

The existence, location, and contribution of renal progenitor cells (RPCs) to epithelial repair have been investigated in many studies and the application of these cells has been depicted to be successful in both acute and chronic animal models (Bussolati et al. 2005; Angelotti et al. 2012; Romagnani and Remuzzi 2013). Additionally, the isolation and culture of RPCs from human urine may pave the road for personalized disease modeling (Lazzeri et al. 2015).

The kidney is originated from intermediate mesoderm (IM) and arises from the ureteric bud and metanephric mesenchyme (MM) cells, following interactions between several signals (Blake et al. 2014). Recent advance in stem cell

Table 9.1 Previous studies regarding advances within the field of in-vitro and in-vivo kidney tissue engineering

Source of kidney	Method of decellularization	Overall time	Results	References
Goat	Perfusion-based 0.1% SDS; gradient of SDS (0.5%, 1.0%); 0.1%Triton X-100; 5 mM calcium chloride and magnesium sulfate	5–6 days	Preserved the structure and composition of renal ECM and vascular structures within the scaffold without residual cellular components	Vishwakarma et al. (2014)
Human	Perfusion-based 0.5% SDS for 48 h, PBS 5 days at a flow rate of 6 mL/min	7 days	SDS-based decellularization protocol removed cellular compartment, while the innate ECM framework maintained its architecture and biochemical properties	Orlando et al. (2013)
Human	Perfusion-based 0.5% SDS, DNase for 6 h at a flow rate of 6 mL/h and then with PBS at the same flow rate for 5 days	7–8 days	A well-preserved structure and function of the vasculature, as well as growth factors	Peloso et al. (2015)
Mouse	Perfusion-based Nitrogen for 2 min, PBS at 37 °C	1–2 days	Complete removal of cells and nuclei	Rafighdoust et al. (2015)
Mouse	Perfusion-based 0.1% SDS 0.1% Triton X100 for 24–72 h 0.4% Sodium deoxycholine for 24–72 h ± 90 U/mL benzonase for 2 h	1–2 days	Scaffolds provided regionalized factors, resulting in organized kidney structures within the acellular kidney	Sambi et al. (2017)
Porcine	Perfusion-based 1% of the detergent Triton X-100 and 0.1% ammonium hydroxide in ddH_2O, 10–60 mL/h for 24 h or until translucent. Perfused with ddH_2O prior to sterilization (gamma irradiation)	2 days	Preserved vascular network	Baptista et al. (2009)
Porcine	Perfusion-based 0.5% SDS in 1 × PBS, 0.25% SDS in 1 × PBS, or 1% Triton X-100/0.1% Ammonium Hydroxide in 1 × PBS were perfused through the kidneys for a total of 36 h	6–7 days	Satisfactory decellularization of kidneys of a clinically relevant size	Sullivan et al. (2012)
Porcine	Perfusion-based 1% (v/v) SDS in dH_2O	2–3 days	Preservation of major architecture and vasculature	Park and Woo (2012)
Porcine	Perfusion-based 1% Triton X-100 or 1% SDS	Not reported	Verified that 1% Triton X-100 is a more suitable decellularizing agent than SDS regarding structural, biochemical integrity, and biocompatibility of the scaffold	Choi et al. (2015)
Porcine	Perfusion-based 0.5% SDS	2–3 days	Freeze porcine kidneys before decellularization. The decellularized organs can be preserved for months without	Poornejad et al. (2015)

(continued)

Table 9.1 (continued)

Source of kidney	Method of decellularization	Overall time	Results	References
			cryoprotectants and thawed just prior to recellularization	
Porcine	Perfusion-based dH_2O; 1% SDS; 1% Triton X-100	3–4 days	Scaffolds maintain their basic components and show intact vasculature system	Guan et al. (2015a)
Porcine	Perfusion-based 1.0% Triton X-100; PBS; 0.25% or 0.75% SDS; PBS; DNase; 1% antibiotics/antimycotics; sterilized with 1% MIN-NCARE® (4.5% PAA and 22.0% hydrogen peroxide), or by irradiation with 12–16 kGy over 24–30 h	4 days	Preservation of diverse vascular and collecting system sections	Willenberg et al. (2015)
Porcine	Solutions of 0.1 N NaOH (pH 11.8–12), 1% (w/v) PAA (pH 2.6), 3% (v/v) Triton X-100 (pH 7.2), 1% (w/v) SDS (pH 8.1), and 0.05% Trypsin/ethylenediaminetetraacetic acid (EDTA)	24 h	The NaOH solution caused efficient cell removal with the highest amount of cell viability and proliferation after recellularization. The most significant damage to collagenous fiber networks The SDS solution led to less severe damage to the ECM structure but with lower metabolic activity and less proliferation PAA and Triton X-100 caused minimum disruption of ECMs and the most preserved glycosaminoglycans (GAGs) and FGF. These last two agents were not as efficient in removing cellular materials as NaOH and SDS	Poornejad et al. (2016a)
Porcine	Hypertonic solution (0.5 M NaCl in H_2O) was pumped into the kidneys for 30 min. 0.5% w/w SDS solution for 30 min, followed by DI water (hypotonic solution) for 30 min	2 days	Preservation of the microstructure and complete cell removal	Poornejad et al. (2016b)
Rat	SDS at differing concentrations and durations (1.0, 0.125, 0.25, and 0.5%), PBS for 1 h	4 hat 24 h	Preservation of both structural and functional components of the whole kidney ECM bioscaffold	He et al. (2017)
Rat	1% SDS for 12 h, ddH_2O for 12 min, 1% Triton X-100 for 30 min, PBS for 48 h, and antibiotic containing PBS	3 days	No cell residue was found in the scaffolds	Liu et al. (2015)
Rat	NaDOC as the ionic detergent: Triton X-100 at 0.5, 3, 6, and 10% solutions; ddH_2O; DNase;4% NaDOC; use of SDS:	5 days	The detergent-based perfusion protocols produced decellularized kidneys, yet	Ross et al. (2009)

(continued)

Table 9.1 (continued)

Source of kidney	Method of decellularization	Overall time	Results	References
	3% Triton X-100, DNase, repeat 3% Triton X-100, and then the 4% SDS. 0.05% sodium azide		retained the web-like appearance of the basement membrane	
Rat	Multiple sequential solutions that included Triton X-100 and SDS detergents, DNase, and deionized water rinses	Over 5 days	New evidence for matrix-to-cell signaling in Decellularized whole organ scaffolds that induces differentiation of pluripotent precursor cells to endothelial lineage	Ross et al. (2012)
Rat	12 h of 1% SDS in ddH_2O, 15 min of ddH_2O, and 30 min of 1% Triton X-100 in ddH_2O, Washed the kidney scaffolds with PBS containing 10,000 U/mL penicillin G, 10 mg/mL streptomycin, and 25 µg/mL amphotericin-B at 1.5 mL/min constant arterial perfusion for 96 h	4 days	Yield decellularized scaffolds with vascular, cortical, and medullary architecture, collecting system, and ureters	Song et al. (2013)
Rat	1% SDS in PBS for 17 h at a flow rate of 0.4 mL/ min	17 h	Rat kidneys were efficiently decellularized to produce renal ECM scaffolds in a relatively short time and rapid recellularization of vascular structures and glomeruli	Bonandrini et al. (2014)
Rat	0.1% Triton X-100 for 3 h, ddH_2O for 30 min, 0.8% (v/v) SDS for 3 h, and ddH_2O containing 100 U/mL penicillin and 100 mg/mL streptomycin for 24 h. Kidney scaffolds were kept in 50 mL of ddH_2O containing the penicillin and streptomycin at 4 °C for less than 7 days	8 days	Decellularized kidney scaffolds could be able to promote renal recovery in the treatment of chronic kidney disease	Yu et al. (2014)
Rat	ddH_2O for 10 min. 1st: SDS concentrations of 0.25, 0.5, 0.66, and 1% combined with a perfusion time of 0.5, 1, 2, and 4 h. 2nd: concentration of SDS was always 0.66% and the perfusion time was 1 h. After the first 30 min of perfusion with SDS, the kidneys were washed for 10 min with dH2O and then the organs were perfused for another 30 min with the SDS solution	5 h	Novel standardized, time-efficient, and reproducible protocol for the decellularization of solid tissues to derive a ready-to-use biomatrix within only 5 h	Burgkart et al. (2014)
Rat	0.5% SDS	1–2 days	Preserved ECM 3D architecture, an intact vascular tree, and biochemical components	Guan et al. (2015b)

(continued)

Table 9.1 (continued)

Source of kidney	Method of decellularization	Overall time	Results	References
Rat	Protocol 1: 1% Triton X-100; Protocol 2: 1% Triton X-100; 0,1% SDS; Protocol 3: 0.02% Trypsin-0.05% EGTA; 1% Triton X-100	1–2 days	Triton and Triton/ SDS preserved renal microarchitecture and retained matrix-bound basic FGF and vascular endothelial growth factor Trypsin caused structural deterioration and growth factor loss. Triton/ SDS-scaffolds with 3 h of leak-free blood flow in a rodent transplantation model and supported repopulation with human iPSC-derived endothelial cells and tubular epithelial cells ex vivo	Caralt (2015)
Rat	Demonstrated non-invasive monitoring capabilities for tracking dynamic changes within scaffolds as the native cellular component is removed during decellularization and model human cells are introduced into the scaffold during recellularization and proliferate in maintenance culture	7 days	1% Triton X-100, 1% Triton X-100/0.1% SDS and 0.02% Trypsin-0.05% EGTA/1% Triton X-100	Uzarski et al. (2015)
Rhesus monkeys	1% SDS	10–14 days	Scaffolds with intrinsic spatial ability to influence hESC differentiation by physically shaping cells into tissue-appropriate structures and phenotypes	Nakayama et al. (2013)
Rhesus monkeys	1% (v/v) SDS or 1% (v/v) Triton X-100 diluted in dH2O at either 48 °C or 37 °C. Decellularization solution was changed 8 h after the initial tissue harvest and then every 48 h until the tissues were transparent	7–10 days	SDS was the most effective for decellularization of kidney sections. Triton X-100 was unable to completely decellularize the tissues and caused greater disruption of the basement membrane and connective tissue ECM	Nakayama et al. (2010)
Rhesus monkeys	1% (v/v) SDS diluted in dH2O at 4 °C. The solution was changed 8 h after initial tissue harvest and then every 48 h until the tissue was transparent, washed with PBS; 10% (v/v) penicillin/streptomycin (Gibco, Invitrogen) in PBS at 4 °C until use	7–10 days	Removal of cellular components while preserving the structural and functional properties of the native ECM	Nakayama et al. (2011)

therapy research has presented human nephron progenitor cells, including intermediate mesoderm (IM) and metanephric mesenchyme (MM) cells (Lam et al. 2014; Mae et al. 2013; Takasato et al. 2014).

Mesenchymal stem cells (MSCs) have attracted extreme attention for their probable therapeutic applications in the treatment of kidney disease, as a regenerative approach already in translation (Humphreys 2014). MSCs have been considered to apply anti-inflammatory, prorepair, immunomodulatory effects by secretion of soluble factors including growth factors and angiogenic cytokines, in a paracrine fashion (Humphreys and Bonventre 2008; Bruno et al. 2009). Although some studies have investigated those MSCs remain in kidney only transiently after injection (Togel et al. 2005; Taylor et al. 2019), numerous investigations also verified the satisfactory results of these cells to improve a wide variety of kidney diseases (Wang et al. 2013). Despite the convincing development shown by the application of MSCs, unanswered questions about how best to apply MSCs to kidney disease remain concerning.

Adipose-derived mesenchymal stem cells (ADMSC) are a flexible and proficient stem cell source with practicable and therapeutic potentials that can be used for the regeneration of damaged tissues. ADMSCs can be regained in high numbers from adipose tissue fragments and can resourcefully be expanded in vitro (Bruno et al. 2009). We hypothesized that the source of ADMSCs may be fundamental for cell therapy reasons and better outcomes may be obtained if the cells are harvested from adjacent adipose tissue of an organ for which cell therapy is intended to be performed. In our previous study, we compared the effect of autologous and heterologous ADMSC injections in rabbit models of pyelonephritis. The efficacy of two different sources of ADMSCs (neck subcutaneous adipose tissue and perirenal adipose tissue) was also evaluated. The results indicated that the usage of perirenal autologous ADMSCs is a superior alternative in the amelioration of renal scarring in the rabbit model of pyelonephritis compared to autologous ADMSCs derived from nape. Moreover, autologous cells were shown to be a better modality for cell therapy in comparison with non-autologous cells (Sabetkish et al. 2018).

Nephron progenitors, isolated from human amniotic fluid collected between 12 and 18 weeks of gestation, are another approach to sourcing cells capable of acting as nephron progenitors (Villani et al. 2018). These cells, with a capacity for long-term expansion in culture, have been depicted to be competent to self-renew in a chromosomally stable fashion for >350 doublings (Spitzhorn et al. 2017).

Whole kidney ECM seeding has been performed by perfusion of cells through the renal artery (Ross et al. 2012; Burgkart et al. 2014) or ureter (Song et al. 2013; Ross et al. 2009). Retrograde injection of cells through the ureter culminated in an irregular cellular allocation, with cells failing to reach glomeruli (Maruyama et al. 2002). In the novel strategy introduced by Song et al., a negative pressure gradient was applied to the bioreactor chamber while injecting neonatal rat renal cells into decellularized rat kidney scaffolds through the ureter. The results showed improvement in the distribution and retention of cells within the collecting system (Song et al. 2013). In a recent study, an isolated rat kidney was decellularized with a detergent-based method and seeded with renal progenitor cells, the results of which represented a remarkable technical feat. The vasculature, the tubules, and glomeruli were seeded with cultured endothelial cells and rat kidney progenitor cells isolated from embryonic rat kidneys, respectively. The outcomes demonstrated that pre-seeded kidneys produced urine and filtrate suggesting active absorptive and secretory function (Song et al. 2013). In another study, rat kidneys were used as native, decellularized, or recellularized with human umbilical vein endothelial cells and rat neonatal kidney cells and were implanted into immunodeficient rats. The outcomes did not show any specific duration of in-vivo perfusion however, no thrombosis or bleeding was reported over the implantation period (Song et al. 2013).

9.4 Clinical Applications

Latest technological progressions in clinical tissue engineering and regenerative medicine can pave the road for regenerating compound spatial arrangement of kidneys with a vascular network and urinary excretion pathway. There are currently no published clinical reports regarding the regeneration of functional organs for the treatment of terminal organ failure. Currently, human MSCs have been used to more than one hundred patients by autologous or allogenic transplantation (Rabelink and Little 2013; Salem and Thiemermann 2010; Leuning et al. 2017). In the study of Tan et al. (2012) patients showed reduced transplant rejection and improvement in renal function at 1 year of given autologous infusion of MSCs. The bioengineered artificial kidney is presently being assessed in patients with only short-term use in the setting of severe acute renal failure (Fissell et al. 2001; Humes et al. 2002). To be clear, it is still uncertain whether stem cells have long-term efficacy or adverse side effects. Therefore, we need a specific guideline for isolation, expansion, and characterization protocols in order to extend the use of stem cells in different trials. It is also imperative that an understanding of the innate inconsistency or strength of any iPSC differentiation practice is achieved before its use for patient disease modeling purposes. Cost-effective techniques to raise cell yield from differentiation protocols will be necessitated to permit the usage of these methods in clinical medicine.

9.5 Limitations

The practicability of incorporating new nephrons into a damaged kidney using current approaches is most likely pretty low. The diseased host kidney would limit the utility of this technique, even if a way was established to connect the new nephron units to the host's collecting system. In addition, massive thrombi, despite strong anticoagulation prophylaxis, have been considered as the problem of decellularized cadaveric scaffolds. The use of nephron progenitor populations originated from human fetal kidney tissue, has been considered as one of the regenerative approaches that has been investigated. However, in some countries, it still presents ethical and technical challenges as there is yet no good understanding of how to preserve a human nephron progenitor population outside the stem cell niche (Brown et al. 2015). Despite the possible application of hPSC-derived kidney tissue for modeling genetic renal disease, present differentiation procedures will remain limited in their capability to model the mass of adult onset kidney disease using patient-derived iPSC. Kidney regeneration is particularly complicated due to the anatomical complexity of the organ, and reconstruction of the kidney's cubic organization is hard. To date, regeneration of all the component cells of the kidney has not been attained. However, we will hopefully resolve these issues and open the door to new therapeutic strategies for kidney regeneration by advances in stem cell research and cellular engineering. Another major challenge and limitation is regarding the selection of a way for cell delivery. The tail vein is frequently used as a place for intravenous injection of cells in murine injury models. However, injection of cells into the tail vein will mean the instant delivery of the injected cells into the microvasculature of the lungs and possible elimination from circulation within the spleen. As a result, only a few injected cells will reach the kidney. Injection of cells via the vasculature is another challenge regarding entering the cells in the renal parenchyma. The risk of extreme bleeding and the long-term risk of interstitial fibrosis have been mentioned as complications after direct renal parenchymal injection. Briefly, scale, structure, and functional maturation are the key challenges include that represent noteworthy obstacles to renal bioengineering.

9.6 Conclusion

The development of tissue engineering technologies for the regaining of renal function will take time. Regarding the advancement in renal tissue engineering, we should consider that we are only one or two decades into renal tissue engineering. There will be more renal tissue engineering techniques to arrive. Further investigations to examine the unique ECM configuration associated with the diverse sections of the renal organ are of great importance to inform further tissue engineering strategies for renal tissue. Although all the approaches have offered very promising results, many challenges such as size, sufficient vascularization, immunological issues, and proper connection to the host vascular and draining system must be overcome before engineered kidneys become clinically useful. This method should ensure structural integrity of the ECM and vascular tree. The selection of cell type, and an understanding of each cell type's differentiation and proliferation, as well as the cultural environment required to support a kidney scaffold for the production of whole kidney organ capable of replicate native kidney function, requires considerable further research. In conclusion, suitable, systematic, and reproducible protocols must be further optimized for decellularization and recellularization for future clinical application.

References

Aebischer P, Ip TK, Panol G, Galletti PM (1987) The bioartificial kidney: progress towards an ultrafiltration device with renal epithelial cells processing. Life Support Syst: J Eur Soc Artif Organs 5(2):159–168

Alachkar N, Rabb H, Jaar BG (2011) Urinary biomarkers in acute kidney transplant dysfunction. Nephron Clin Pract 118(2):c173–c181

Al-Awqati Q, Oliver JA (2002) Stem cells in the kidney. Kidney Int 61(2):387–395

Angelotti ML, Ronconi E, Ballerini L, Peired A, Mazzinghi B, Sagrinati C et al (2012) Characterization of renal progenitors committed toward tubular lineage and their regenerative potential in renal tubular injury. Stem Cells 30(8):1714–1725

Asfar SK, Catto GR, Engeset J (1988) The rat renal transplant model. Details of microsurgical technique and complications. J R Coll Surg Edinb 33(6):314–317

Badylak SF (2004) Xenogeneic extracellular matrix as a scaffold for tissue reconstruction. Transpl Immunol 12(3–4):367–377. https://doi.org/10.1016/j.trim.2003.12.016

Bandeiras C, Cabral JM, Gabbay RA, Finkelstein SN, Ferreira FC (2019) Bringing stem cell-based therapies for type 1 diabetes to the clinic: early insights from bioprocess economics and cost-effectiveness analysis. Biotechnol J 14(8):1800563

Baptista PM, Orlando G, Mirmalek-Sani SH, Siddiqui M, Atala A, Soker S (2009) Whole organ decellularization—a tool for bioscaffold fabrication and organ bioengineering. Conference proceedings: annual international conference of the IEEE Engineering in Medicine and Biology Society IEEE Engineering in Medicine and Biology Society annual conference. 2009:6526–6529. https://doi.org/10.1109/iembs.2009.5333145

Benedetti V, Lavecchia AM, Locatelli M, Brizi V, Corna D, Todeschini M et al (2019) Alteration of thyroid hormone signaling triggers the diabetes-induced pathological growth, remodeling, and dedifferentiation of podocytes. JCI Insight. 4(18)

Bijonowski BM, Miller WM, Wertheim JA (2013) Bioreactor design for perfusion-based, highly-vascularized organ regeneration. Curr Opin Chem Eng 2(1):32–40. https://doi.org/10.1016/j.coche.2012.12.001

Blake J, Rosenblum ND (eds) (2014) Renal branching morphogenesis: morphogenetic and signaling mechanisms. Seminars in cell & developmental biology. Elsevier

Bland R, Zehnder D, Hewison M (2000) Expression of 25-hydroxyvitamin D3-1α-hydroxylase along the nephron: new insights into renal vitamin D metabolism. Curr Opin Nephrol Hypertens 9(1):17–22

Bonandrini B, Figliuzzi M, Papadimou E, Morigi M, Perico N, Casiraghi F et al (2014) Recellularization of well-preserved acellular kidney scaffold using embryonic stem cells. Tissue Eng Part A 20(9–10):1486–1498. https://doi.org/10.1089/ten.TEA.2013.0269

Brown AC, Muthukrishnan SD, Oxburgh L (2015) A synthetic niche for nephron progenitor cells. Dev Cell 34(2):229–241

Bruno S, Grange C, Deregibus MC, Calogero RA, Saviozzi S, Collino F et al (2009) Mesenchymal stem cell-derived microvesicles protect against acute tubular injury. J Am Soc Nephrol 20(5):1053–1067. https://doi.org/10.1681/asn.2008070798

Burgkart R, Tron A, Prodinger P, Culmes M, Tuebel J, van Griensven M et al (2014) Decellularized kidney matrix for perfused bone engineering. Tissue Eng Part C Methods 20(7):553–561. https://doi.org/10.1089/ten.TEC.2013.0270

Bussolati B, Bruno S, Grange C, Buttiglieri S, Deregibus MC, Cantino D et al (2005) Isolation of renal progenitor cells from adult human kidney. Am J Pathol 166(2):545–555

Caralt M (2015) Present and future of regenerative medicine: liver transplantation. Transpl Proc 47 (8):2377–2379. https://doi.org/10.1016/j.transproceed.2015.08.029

Caralt M, Uzarski JS, Iacob S, Obergfell KP, Berg N, Bijonowski BM et al (2015) Optimization and critical evaluation of decellularization strategies to develop renal extracellular matrix scaffolds as biological templates for organ engineering and transplantation. Am J Transplant 15(1):64–75

Chertow GM, Waikar SS (2008) Toward the promise of renal replacement therapy. J Am Soc Nephrol 19 (5):839–840. https://doi.org/10.1681/asn.2008030291

Choi SH, Chun SY, Chae SY, Kim JR, Oh SH, Chung SK et al (2015) Development of a porcine renal extracellular matrix scaffold as a platform for kidney regeneration. J Biomed Mater Res, Part A 103(4):1391–1403. https://doi.org/10.1002/jbm.a.35274

Crapo PM, Gilbert TW, Badylak SF (2011) An overview of tissue and whole organ decellularization processes. Biomaterials 32(12):3233–3243

Figliuzzi M, Remuzzi G, Remuzzi A (2017) Recellularization of kidney scaffold with stem cells. In: Kidney transplantation, bioengineering and regeneration. Elsevier, p 877–86

Fissell WH, Kimball J, Mackay SM, Funke A, Humes HD (2001) The role of a bioengineered artificial kidney in renal failure

Fraser SJ, Endres C (2014) Quorus bioreactor: a new perfusion-based technology for microbial cultivation. Adv Biochem Eng Biotechnol 138:149–177. https://doi.org/10.1007/10_2013_238

Guan Y, Liu S, Liu Y, Sun C, Cheng G, Luan Y et al (2015a) Porcine kidneys as a source of ECM scaffold for kidney regeneration. Mater Sci Eng, C Mater Biol Appl 56:451–456. https://doi.org/10.1016/j.msec.2015.07.007

Guan Y, Liu S, Sun C, Cheng G, Kong F, Luan Y et al (2015b) The effective bioengineering method of implantation decellularized renal extracellular matrix scaffolds. Oncotarget 6(34):36126–36138. https://doi.org/10.18632/oncotarget.5304

Hammerman MR (2003) Tissue engineering the kidney. Kidney Int 63(4):1195–1204

He M, Callanan A, Lagaras K, Steele JAM, Stevens MM (2017) Optimization of SDS exposure on preservation of ECM characteristics in whole organ decellularization of rat kidneys. J Biomed Mater Res B Appl Biomater 105(6):1352–1360. https://doi.org/10.1002/jbm.b.33668

Hielscher D, Kaebisch C, Braun BJV, Gray K, Tobiasch E (2018) Stem cell sources and graft material for vascular tissue engineering. Stem Cell Rev Rep 14 (5):642–667

Horster MF, Braun GS, Huber SM (1999) Embryonic renal epithelia: induction, nephrogenesis, and cell differentiation. Physiol Rev 79(4):1157–1191

Humes HD, Mackay SM, Funke AJ, Buffington DA (1999) Tissue engineering of a bioartificial renal tubule assist device: in vitro transport and metabolic characteristics. Kidney Int 55(6):2502–2514

Humes HD, Fissell WH, Weitzel WF, Buffington DA, Westover AJ, MacKay SM et al (2002) Metabolic replacement of kidney function in uremic animals with a bioartificial kidney containing human cells. Am J Kidney Dis 39(5):1078–1087

Humes HD, Weitzel WF, Bartlett RH, Swaniker FC, Paganini EP (2003) Renal cell therapy is associated with dynamic and individualized responses in patients with acute renal failure. Blood Purif 21(1):64–71. https://doi.org/10.1159/000067864

Humes HD, Weitzel WF, Bartlett RH, Swaniker FC, Paganini EP, Luderer JR et al (2004) Initial clinical results of the bioartificial kidney containing human cells in ICU patients with acute renal failure. Kidney Int 66(4):1578–1588. https://doi.org/10.1111/j.1523-1755.2004.00923.x

Humes HD, Buffington D, Westover AJ, Roy S, Fissell WH (2014) The bioartificial kidney: current status and future promise. Pediatr Nephrol 29(3):343–351. https://doi.org/10.1007/s00467-013-2467-y

Humphreys BD (2014) Kidney injury, stem cells and regeneration. Curr Opin Nephrol Hypertens 23(1):25

Humphreys BD, Bonventre JV (2008) Mesenchymal stem cells in acute kidney injury. Annu Rev Med 59:311–325

Hyink DP, Tucker DC, St John P, Leardkamolkarn V, Accavitti M, Abrass C et al (1996) Endogenous origin of glomerular endothelial and mesangial cells in grafts of embryonic kidneys. Am J Physiol-Renal Physiol 270(5):F886–F899

Ip TK, Aebischer P (1989) Renal epithelial-cell-controlled solute transport across permeable membranes as the foundation for a bioartificial kidney. Artif Organs 13 (1):58–65. https://doi.org/10.1111/j.1525-1594.1989.tb02833.x

Kajbafzadeh AM, Elmi A, Talab SS, Sadeghi Z, Emami H, Sotoudeh M (2010) Autografting of renal progenitor cells ameliorates kidney damage in experimental model of pyelonephritis. Cell Med 1(3):115–122. https://doi.org/10.3727/215517910x551044

Kajbafzadeh AM, Abbasioun R, Sabetkish N, Sabetkish S, Habibi AA, Tavakkolitabassi K (2017) In vivo human corpus cavernosum regeneration: fabrication of tissue-engineered corpus cavernosum in rat using the body as a natural bioreactor. Int Urol Nephrol 49(7):1193–1199. https://doi.org/10.1007/s11255-017-1582-2

Kajbafzadeh AM, Khorramirouz R, Kameli SM, Nabavizadeh B (2018) Microsurgical anastomosis of renal vasculature in rats: a practical platform for acellular kidney transplantation. J Pediatr Urol 14(2):194–195. https://doi.org/10.1016/j.jpurol.2018.01.008

Kajbafzadeh A-M, Khorramirouz R, Nabavizadeh B, Seyedian S-SL, Akbarzadeh A, Heidari R et al (2019) Whole organ sheep kidney tissue engineering and in vivo transplantation: effects of perfusion-based decellularization on vascular integrity. Mater Sci Eng C 98:392–400

Kawecki M, Łabuś W, Klama-Baryla A, Kitala D, Kraut M, Glik J et al (2018) A review of decellurization methods caused by an urgent need for quality control of cell-free extracellular matrix'scaffolds and their role in regenerative medicine. J Biomed Mater Res B Appl Biomater 106(2):909–923

Lam AQ, Freedman BS, Morizane R, Lerou PH, Valerius MT, Bonventre JV (2014) Rapid and efficient differentiation of human pluripotent stem cells into intermediate mesoderm that forms tubules expressing kidney proximal tubular markers. J Am Soc Nephrol 25(6):1211–1225

Lazzeri E, Ronconi E, Angelotti ML, Peired A, Mazzinghi B, Becherucci F et al (2015) Human urine-derived renal progenitors for personalized modeling of genetic kidney disorders. J Am Soc Nephrol 26 (8):1961–1974

Leuning DG, Lievers E, Reinders ME, van Kooten C, Engelse MA, Rabelink TJ (2017) A novel clinical grade isolation method for human kidney perivascular stromal cells. J Visualized Exp 126:e55841

Lieschke GJ, Currie PD (2007) Animal models of human disease: zebrafish swim into view. Nat Rev Genet 8 (5):353–367. https://doi.org/10.1038/nrg2091

Lih E, Park KW, Chun SY, Kim H, Kwon TG, Joung YK et al (2016) Biomimetic porous PLGA scaffolds incorporating decellularized extracellular matrix for kidney tissue regeneration. ACS Appl Mater Interfaces 8(33):21145–21154. https://doi.org/10.1021/acsami.6b03771

Little MH, McMahon AP (2012) Mammalian kidney development: principles, progress, and projections. Cold Spring Harb Perspect Biol 4(5). https://doi.org/10.1101/cshperspect.a008300

Liu RF, Gao JS, Yang YF, Zeng WX (2015) Preparation of rat whole-kidney acellular matrix via peristaltic pump. Urol J 12(6):2457–2461

Locatelli F, Buoncristiani U, Canaud B, Köhler H, Petitclerc T, Zucchelli P (2005) Dialysis dose and frequency. Nephrol Dial Transplant 20(2):285–296

MacKay SM, Funke AJ, Buffington DA, Humes HD (1998) Tissue engineering of a bioartificial renal tubule. ASAIO J (American Society for Artificial Internal Organs: 1992) 44(3):179–183

Mae S-I, Shono A, Shiota F, Yasuno T, Kajiwara M, Gotoda-Nishimura N et al (2013) Monitoring and robust induction of nephrogenic intermediate mesoderm from human pluripotent stem cells. Nat Commun 4(1):1–11

Martini AG, Danser AJ (2017) Juxtaglomerular cell phenotypic plasticity. High Blood Pressure Cardiovasc Prev 24(3):231–242

Maruyama H, Higuchi N, Nishikawa Y, Hirahara H, Iino N, Kameda S et al (2002) Kidney-targeted naked DNA transfer by retrograde renal vein injection in rats. Hum Gene Ther 13(3):455–468. https://doi.org/10.1089/10430340252792585

Mota C, Camarero-Espinosa S, Baker MB, Wieringa P, Moroni L (2020) Bioprinting: from tissue and organ development to in vitro models. Chem Rev. https://doi.org/10.1021/acs.chemrev.9b00789

Nakayama KH, Batchelder CA, Lee CI, Tarantal AF (2010) Decellularized rhesus monkey kidney as a three-dimensional scaffold for renal tissue engineering. Tissue Eng Part A 16(7):2207–2216. https://doi.org/10.1089/ten.tea.2009.0602

Nakayama KH, Batchelder CA, Lee CI, Tarantal AF (2011) Renal tissue engineering with decellularized rhesus monkey kidneys: age-related differences. Tissue Eng Part A 17(23–24):2891–2901. https://doi.org/10.1089/ten.TEA.2010.0714

Nakayama KH, Lee CCI, Batchelder CA, Tarantal AF (2013) Tissue specificity of decellularized rhesus monkey kidney and lung scaffolds. PloS one 8(5): e64134

Nakayama KH, Lee CC, Batchelder CA, Tarantal AF (2013) Tissue specificity of decellularized rhesus monkey kidney and lung scaffolds. PLoS ONE 8(5): e64134. https://doi.org/10.1371/journal.pone.0064134

Ohata K, Ott HC (2020) Human-scale lung regeneration based on decellularized matrix scaffolds as a biologic platform. Surg Today 1–11

Oliver J (1953) Correlations of structure and function and mechanisms of recovery in acute tubular necrosis. Am J Med 15(4):535–557

Orlando G, Baptista P, Birchall M, De Coppi P, Farney A, Guimaraes-Souza NK et al (2011) Regenerative medicine as applied to solid organ transplantation: current status and future challenges. Transpl Int 24 (3):223–232

Orlando G, Farney AC, Iskandar SS, Mirmalek-Sani SH, Sullivan DC, Moran E et al (2012) Production and implantation of renal extracellular matrix scaffolds from porcine kidneys as a platform for renal bioengineering investigations. Ann Surg 256(2):363–370. https://doi.org/10.1097/SLA.0b013e31825a02ab

Orlando G, Booth C, Wang Z, Totonelli G, Ross CL, Moran E et al (2013) Discarded human kidneys as a source of ECM scaffold for kidney regeneration technologies. Biomaterials 34(24):5915–5925. https://doi.org/10.1016/j.biomaterials.2013.04.033

Park KM, Woo HM (2012) Porcine bioengineered scaffolds as new frontiers in regenerative medicine. Transpl Proc 44(4):1146–1150. https://doi.org/10.1016/j.transproceed.2012.03.043

Peloso A, Petrosyan A, Da Sacco S, Booth C, Zambon JP, O'Brien T et al (2015) Renal extracellular matrix scaffolds from discarded kidneys maintain glomerular morphometry and vascular resilience and retains critical growth factors. Transplantation 99(9):1807–1816. https://doi.org/10.1097/tp.0000000000000811

Perin L, Giuliani S, Sedrakyan S, Da Sacco S, De Filippo RE (2008) Stem cell and regenerative science applications in the development of bioengineering of renal tissue. Pediatr Res 63(5):467–471. https://doi.org/10.1203/PDR.0b013e3181660653

Poornejad N, Frost TS, Scott DR, Elton BB, Reynolds PR, Roeder BL et al (2015) Freezing/thawing without cryoprotectant damages native but not decellularized porcine renal tissue. Organogenesis 11(1):30–45. https://doi.org/10.1080/15476278.2015.1022009

Poornejad N, Schaumann LB, Buckmiller EM, Momtahan N, Gassman JR, Ma HH et al (2016a) The impact of decellularization agents on renal tissue extracellular matrix. J Biomater Appl 31(4):521–533. https://doi.org/10.1177/0885328216656099

Poornejad N, Momtahan N, Salehi AS, Scott DR, Fronk CA, Roeder BL et al (2016b) Efficient decellularization of whole porcine kidneys improves reseeded cell behavior. Biomedical Materials (bristol, England). 11(2):025003. https://doi.org/10.1088/1748-6041/11/2/025003

Rabelink TJ, Little MH (2013) Stromal cells in tissue homeostasis: balancing regeneration and fibrosis. Nat Rev Nephrol 9(12):747–753

Rafighdoust A, Shahri NM, Baharara J (2015) Decellularized kidney in the presence of chondroitin sulfate as a natural 3D scaffold for stem cells. Iran J Basic Med Sci 18(8):788–798

Rogers SA, Hammerman MR (2001) Transplantation of rat metanephroi into mice. Am J Physiol-Regul Integr Comp Physiol 280(6):R1865–R1869

Romagnani P, Remuzzi G (2013) Renal progenitors in non-diabetic and diabetic nephropathies. Trends Endocrinol Metab 24(1):13–20

Ross EA, Williams MJ, Hamazaki T, Terada N, Clapp WL, Adin C et al (2009) Embryonic stem cells proliferate and differentiate when seeded into kidney scaffolds. J Am Soc Nephrol 20(11):2338–2347

Ross EA, Abrahamson DR, St. John P, Clapp WL, Williams MJ, Terada N et al (2012) Mouse stem cells seeded into decellularized rat kidney scaffolds endothelialize and remodel basement membranes. Organogenesis 8(2):49–55

Ross EA, Abrahamson DR, St John P, Clapp WL, Williams MJ, Terada N et al (2012) Mouse stem cells seeded into decellularized rat kidney scaffolds endothelialize and remodel basement membranes. Organogenesis 8(2):49–55. https://doi.org/10.4161/org.20209

Sabetkish S, Kajbafzadeh AM, Sabetkish N, Khorramirouz R, Akbarzadeh A, Seyedian SL et al (2015) Whole-organ tissue engineering: decellularization and recellularization of three-dimensional matrix liver scaffolds. J Biomed Mater Res, Part A 103 (4):1498–1508. https://doi.org/10.1002/jbm.a.35291

Sabetkish S, Sabetkish N, Talebi MA, Halimi S, Kajbafzadeh AM (2018) The role of nonautologous and autologous adipose-derived mesenchymal stem cell in acute pyelonephritis. Cell Tissue Banking 19 (3):301–309. https://doi.org/10.1007/s10561-017-9674-x

Salem HK, Thiemermann C (2010) Mesenchymal stromal cells: current understanding and clinical status. Stem Cells 28(3):585–596

Salvatori M, Peloso A, Katari R, Orlando G (2014) Regeneration and bioengineering of the kidney: current status and future challenges. Curr Urol Rep 15 (1):379

Sambi M, Chow T, Whiteley J, Li M, Chua S, Raileanu V et al (2017) Acellular mouse kidney ECM can be used as a three-dimensional substrate to test the differentiation potential of embryonic stem cell derived renal progenitors. Stem Cell Rev Rep 13(4):513–531. https://doi.org/10.1007/s12015-016-9712-2

Sanechika N, Sawada K, Usui Y, Hanai K, Kakuta T, Suzuki H et al (2011) Development of bioartificial renal tubule devices with lifespan-extended human renal proximal tubular epithelial cells. Nephrology, dialysis, transplantation: official publication of the European Dialysis and Transplant Association—European Renal Association 26(9):2761–2769. https://doi.org/10.1093/ndt/gfr066

Saran R, Robinson B, Abbott KC, Agodoa LY, Albertus P, Ayanian J et al (2017) US renal data system 2016 annual data report: epidemiology of kidney disease in the United States. Am J Kidney Dis 69(3): A7–A8

Sariola H, Ekblom P, Lehtonen E, Saxén L (1983) Differentiation and vascularization of the metanephric kidney grafted on the chorioallantoic membrane. Dev Biol 96(2):427–435

Soma T, Lerut E, Billiau A, Waer M, Goebels J, Koshiba T et al (2009) An easy and reproducible model of kidney transplantation in rats. Transpl Proc 41(8):3422–3424. https://doi.org/10.1016/j.transproceed.2009.09.027

Song JH, Humes HD (2009) The bioartificial kidney in the treatment of acute kidney injury. Curr Drug Targets 10(12):1227–1234

Song JJ, Ott HC (2011) Organ engineering based on decellularized matrix scaffolds. Trends Mol Med 17 (8):424–432. https://doi.org/10.1016/j.molmed.2011.03.005

Song JJ, Guyette JP, Gilpin SE, Gonzalez G, Vacanti JP, Ott HC (2013) Regeneration and experimental orthotopic transplantation of a bioengineered kidney. Nature medicine. 19(5):646–651. https://doi.org/10.1038/nm.3154

Spitzhorn L-S, Rahman MS, Schwindt L, Ho H-T, Wruck W, Bohndorf M et al. (2017) Isolation and molecular characterization of amniotic fluid-derived mesenchymal stem cells obtained from caesarean sections. Stem Cells Int 2017

Sullivan DC, Mirmalek-Sani S-H, Deegan DB, Baptista PM, Aboushwareb T, Atala A et al (2012) Decellularization methods of porcine kidneys for whole organ engineering using a high-throughput system. Biomaterials 33(31):7756–7764

Swanhart LM, Cosentino CC, Diep CQ, Davidson AJ, de Caestecker M, Hukriede NA (2011) Zebrafish kidney development: basic science to translational research. Birth Defects Res C Embryo Today 93(2):141–156. https://doi.org/10.1002/bdrc.20209

Takasato M, Er P, Becroft M, Vanslambrouck JM, Stanley E, Elefanty AG et al (2014) Directing human embryonic stem cell differentiation towards a renal lineage generates a self-organizing kidney. Nat Cell Biol 16(1):118–126

Tan J, Wu W, Xu X, Liao L, Zheng F, Messinger S et al (2012) Induction therapy with autologous mesenchymal stem cells in living-related kidney transplants: a randomized controlled trial. JAMA 307(11):1169–1177

Taylor A, Sharkey J, Harwood R, Scarfe L, Barrow M, Rosseinsky MJ et al. (2019) Multimodal imaging techniques show differences in homing capacity between mesenchymal stromal cells and macrophages in mouse renal injury models. Mol Imag Biol 1–10

Togel F, Hu Z, Weiss K, Isaac J, Lange C, Westenfelder C (2005) Administered mesenchymal stem cells protect against ischemic acute renal failure through differentiation-independent mechanisms. Am J Physiol-Ren Physiol 289(1):F31–F42

Tumlin J, Wali R, Williams W, Murray P, Tolwani AJ, Vinnikova AK et al (2008) Efficacy and safety of renal tubule cell therapy for acute renal failure. J Am Soc Nephrol 19(5):1034–1040. https://doi.org/10.1681/asn.2007080895

Uzarski JS, Xia Y, Belmonte JC, Wertheim JA (2014) New strategies in kidney regeneration and tissue engineering. Curr Opin Nephrol Hypertens 23(4):399–405

Uzarski JS, Bijonowski BM, Wang B, Ward HH, Wandinger-Ness A, Miller WM et al (2015) Dual-purpose bioreactors to monitor noninvasive physical and biochemical markers of kidney and liver scaffold recellularization. Tissue Eng Part C Methods 21 (10):1032–1043. https://doi.org/10.1089/ten.TEC.2014.0665

Villani V, Petrosyan A, De Filippo RE, Da Sacco S (2018) Amniotic fluid stem cells for kidney regeneration. Perinatal stem cells. Elsevier, p 85–95

Vishwakarma SK, Bhavani PG, Bardia A, Abkari A, Murthy GS, Venkateshwarulu J et al (2014) Preparation of natural three-dimensional goat kidney scaffold for the development of bioartificial organ. Indian J Nephrol 24(6):372–375. https://doi.org/10.4103/0971-4065.133008

Wang Y, He J, Pei X, Zhao W (2013) Systematic review and meta-analysis of mesenchymal stem/stromal cells therapy for impaired renal function in small animal models. Nephrology 18(3):201–208

Wang P, Chen Y, Yong J, Cui Y, Wang R, Wen L et al (2018) Dissecting the global dynamic molecular profiles of human fetal kidney development by single-cell RNA sequencing. Cell Rep 24(13):3554–67 e3. https://doi.org/10.1016/j.celrep.2018.08.056

Willenberg BJ, Oca-Cossio J, Cai Y, Brown AR, Clapp WL, Abrahamson DR et al (2015) Repurposed biological scaffolds: kidney to pancreas. Organogenesis 11(2):47–57. https://doi.org/10.1080/15476278.2015.1067354

Yasuoka Y, Fukuyama T, Izumi Y, Nakayama Y, Inoue H, Yanagita K et al (2020) Erythropoietin production by the kidney and the liver in response to severe hypoxia evaluated by Western blotting with deglycosylation. Physiol Rep 8(12):e14485

Yi S, Ding F, Gong L, Gu X (2017) Extracellular matrix scaffolds for tissue engineering and regenerative medicine. Curr Stem Cell Res Ther 12(3):233–246

Yu YL, Shao YK, Ding YQ, Lin KZ, Chen B, Zhang HZ et al (2014) Decellularized kidney scaffold-mediated renal regeneration. Biomaterials 35(25):6822–6828. https://doi.org/10.1016/j.biomaterials.2014.04.074

10 Production and Preparation of Porcine Urinary Bladder Matrix (UBM) for Urinary Bladder Tissue-Engineering Purposes

N. F. Davis and A. Callanan

Abstract

Surgical repair for the end stage bladder disease utilises vascularised, autogenous and mucus-secreting gastrointestinal tissue to replace the diseased organ or to augment inadequate bladder tissue. Post-operatively, the compliance of the bowel is often enough to restore the basic shape, structure and function of the urinary bladder; however, lifelong post-operative complications are common. Comorbidities that result from interposition of intestinal tissue are metabolic and/or neuromechanical, and their incidence approaches 100%. The debilitating comorbidities and complications associated with such urological procedures may be mitigated by the availability of alternative, tissue-engineered, animal-derived extracellular matrix (ECM) scaffolds such as porcine urinary bladder matrix (UBM). Porcine UBM is a decellularized biocompatible, biodegradable biomaterial derived from the porcine urinary bladder. This chapter aims to describe the production and preparation techniques for porcine UBM for urinary bladder regenerative purposes.

Keywords

Tissue engineering · Regenerative medicine · Urology · Urinary bladder · Urinary bladder matrix · UBM · Extracellular matrix

10.1 History

The bladder's function is compromised by malignant and benign conditions such as bladder cancer, bladder exstrophy, spinal cord injury, myelomeningocele, multiple sclerosis and interstitial cystitis (Davis et al. 2014). For some patients, conservative treatments such as pelvic floor physical therapy, pharmacological therapy (e.g. treatment with antimuscarinic drugs) and non-therapeutic interventions (e.g. self-intermittent catheterisation) are enough to improve bladder stability (Davis et al. 2017). However, there is a subset of patients for whom malignancy, intractable incontinence or pain is incompatible with an acceptable quality of life, and this can necessitate surgical intervention. Surgical repair for the end stage bladder disease utilises vascularised, autogenous and mucus-secreting gastrointestinal tissue to either *replace* the diseased organ or to *augment* inadequate bladder tissue (Davis et al. 2018a).

N. F. Davis (✉)
Department of Surgery, Royal College of Surgeons in Ireland (RCSI), Dublin, Ireland
e-mail: nialldavis@rcsi.ie

A. Callanan
School of Engineering, Institute for Bioengineering, University of Edinburgh, Edinburgh, UK

A. Kajbafzadeh (ed.), *Decellularization Methods of Tissue and Whole Organ in Tissue Engineering*, Advances in Experimental Medicine and Biology 1345,
https://doi.org/10.1007/978-3-030-82735-9_10

Augmentation cystoplasty involves the addition of viscoelastic ileal tissue to defective bladder tissue to improve functional bladder capacity. Post-operatively, the compliant smooth muscle of the bowel is often sufficient to restore the basic shape, structure and function of the urinary bladder, however, lifelong post-operative complications are common (Davis et al. 2010). Morbidities that result from interposition of intestinal tissue are usually sub-classified into three broad areas: metabolic, neuromechanical and technical–surgical (Davis et al. 2010).

Metabolic complications are the result of altered solute reabsorption by the intestine of the urine that it contains. Neuromechanical aspects involve the configuration of the bowel, which affects the storage volume and contraction of the intestine that may lead to difficulties in storage (Davis et al. 2018b). Technical–surgical complications involve perioperative aspects of the procedure that result in surgical morbidity. Post-operative complications associated with these urological procedures are largely attributable to the mucus-producing absorptive bowel epithelium. Therefore, it seems logical that the removal of gastrointestinal epithelium inner prior to bladder reconstruction should negate such complications. However, experimental animal models have shown that the augmentation with de-epithelialised bowel segments results in significant fibrosis and shrinkage (Aktuğ et al. 2001; Flood et al. 1995).

Limitations associated with gastrointestinal tissue may be moderated by the availability of alternative, readily available and animal-derived (i.e. xenogenic) tissue sources (Davis et al. 2010). Ideally, an engineered xenograft should restore or preserve the normal function of the organ it is augmenting or replacing. Furthermore, it should be biologically inert to minimise rejection or foreign body inflammatory reactions. Throughout the twentieth century, several synthetic scaffold materials were investigated for potential use in reconstructive urology. In the 1950s, non-biodegradable synthetic materials like polytetrafluoroethylene (PTFE), silicone, rubber, polyvinyl and polypropylene were investigated but rapidly encrusted with prolonged urinary contact. Additionally, synthetic biomaterials were susceptible to bacterial colonisation and foreign body reactions (Davis et al. 2010). Therefore, contemporary research has focused on the use of biodegradable materials derived from the extracellular matrix (ECM) layer. ECMs such as porcine small intestine submucosa (SIS) and porcine urinary bladder matrix (UBM) are advantageous as they are minimally immunogenic. Both biomaterials are porcine in their origin and are prepared by mechanical, chemical and enzymatic treatments to yield tissue that is minimally immunogenic but retains its basic structural elements, including collagen, glycosaminoglycans, fibronectin, laminins and other intrinsic growth factors to facilitate the attachment, growth and differentiation of host cells (Davis et al. 2010). There are two approaches for manipulating the biodegradable scaffolds after they are prepared: **seeded** and **unseeded** techniques. The seeded method requires the in vitro culture and expansion of various cell types onto a scaffold to create composite 'cellular-scaffold' template for grafting in vivo. The unseeded method involves the use of a bare scaffolding material in vivo to provide a framework for the ingrowth of native tissue. The aim of this chapter is to provide an overview on the production and preparation techniques for ECMS, namely porcine UBM for urinary bladder tissue-engineering purposes.

10.2 Materials and Methods

10.2.1 ECM Tissue Harvest and Preparation

ECM sheets are manufactured by harvesting porcine urinary bladders from market weight pigs (110–130 kg) from an abattoir immediately after euthanasia. Tissues are then immediately transferred to the laboratory on ice for further processing. After draining the urine, bladders are

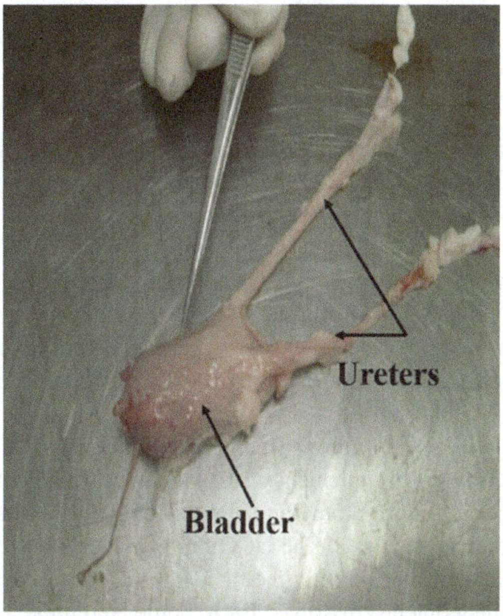

Fig. 10.1 Harvested porcine lower urinary tract prior to preparation of UBM

cleaned in the laboratory sink by rinsing under a cold tap (Fig. 10.1). Excess fatty or connective tissue is resected from the edges and outer surface of the bladder. Bladders are then placed in type 1 water (e.g. Barnstead™ Filtered water). Each bladder is then sliced open lengthwise to form a sheet like structure with the abluminal surface of the bladder facing outwards.

Muscular layers are then manually stretched by applying a bevelled acrylic across the muscular surface of the bladder. This process is performed by applying the scraper to the centre of the sheet and working towards the peripheral muscular layers, thereby loosening the underlying muscle. Occasionally, gauze is necessary to achieve optimal tissue grip. A scissors is used to create a horizontal incision across the middle of the bladder perpendicular to the original incision from the apex to the bladder neck. Each horizontal incision involves cutting through the muscularis externa and submucosal layers of the bladder tissue. Muscle layers are manually delaminated away with forceps, and the final product contained only tunica propria and an underlying basement membrane (Fig. 10.2).

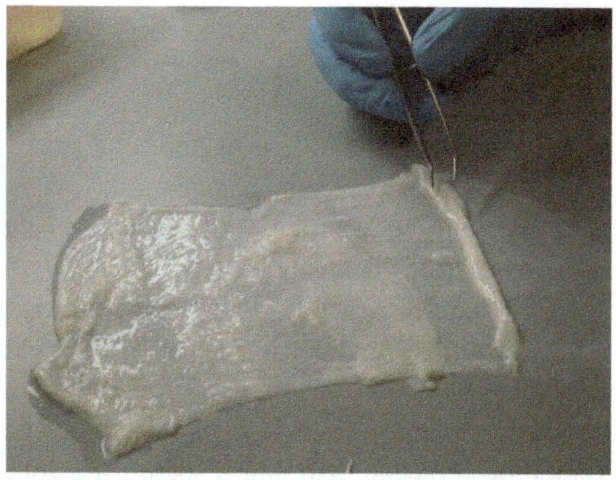

Fig. 10.2 Manual delamination of bisected porcine bladder

10.2.2 UBM Sheet Manufacture

UBM is stored in type 1 water at 4 °C. A second stage of muscle removal may be required if any muscle content remains on the scaffold. Decellularisation is continued by exposing the scaffold to peracetic acid (ml) solutions (Fig. 10.3).

The amount of peracetic acid solution required is calculated by multiplying the weight of the ECM material (grams) by a factor of 20, thereby creating a ratio of 20:1 (volume:weight). Peracetic acid solution consists of appropriately pure water (96%) and 100% EtOH alcohol (4%). Alternatively, the volume of peracetic acid required can be calculated by the following equation:

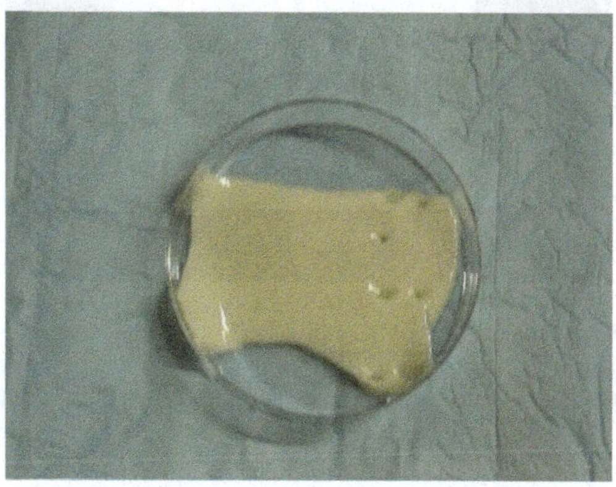

Fig. 10.3 Decellularised UBM after exposure to ethylene peracetic acid

$$a \times y = b \times x$$

where

a = percent concentration of peracetic acid

y = amount of peracetic acid required

b = percent concentration of final peracetic acid solution for disinfection

x = volume of peracetic acid stock solution.

Typically, the final concentration of peracetic acid solution is 0.1%. An appropriately sized calibrated pipette is used to aspirate the correct volume of peracetic acid and dispense the solution into a disinfection chamber. UBM is then added to the peracetic acid solution and placed on a shaker for 2 h. After this timeframe, the peracetic acid solution is disposed into a waste disposal container.

Contamination is avoided by refilling the container with Dulbecco's phosphate-buffered saline (DPBS, pH 7.4) and placing the container on the shaker for a further 15 min. The DPBS solution is drained off, and the container is refilled with the same volume of filtered water and placed on the shaker for a further 15 min. This step is repeated twice. Finally, terminal sterilisation is achieved by exposure to γ-irradiation (Fig. 10.4).

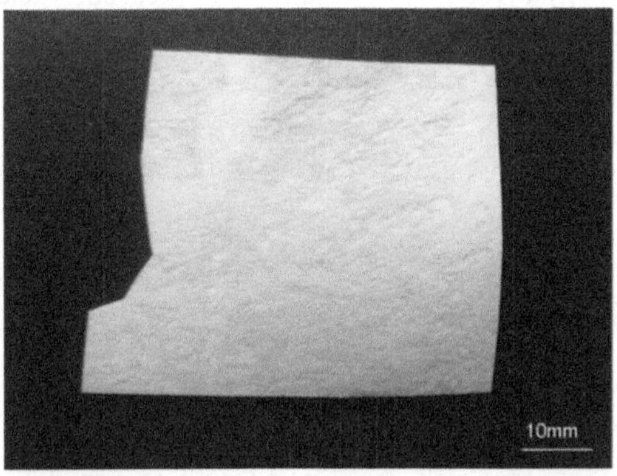

Fig. 10.4 Sheet of one-ply UBM scaffold that was prepared according to the standardised manufacturing protocol. (scale bar represents 10 mm)

10.3 Cell Seeding

10.3.1 Cell Culture

Human urothelial cell (HUC) lines are initially obtained from a bladder biopsy and frozen in liquid nitrogen. Prior to cell culture, vials are then stored at 37 °C in a water bath and rotated gently until the contents are completely thawed. Vials are then rinsed with 70% ethanol and wiped to remove excess. The cap is removed with care ensuring that the interior thread remains sterile.

Five millilitres (mls) of Urothelial Cell Growth Supplement UCGS and 5mls of penicillin/streptomycin solution P/S solution are thawed at 37 °C. The UCGS tube is gently tilted several times during the thawing process to help the contents dissolve. Each bottle and tube is rinsed with 70% ethanol and wiped to remove excess. UCGS and P/S solution are then added into the basal medium in a sterile field and mixed until the reconstituted medium was ready for use.

Using a 1 ml pipette, the contents of the vial are gently dispensed into equilibrated, poly-L-lysine-coated vessels (Sarstedt Ltd®, Wexford, Ireland). The cover is replaced, and the vessel slowly rocked to distribute the cells evenly. Culture vessels are returned to the incubator 16 h after initiation. Growth medium is removed the following day to remove residual DMSO and any unattached cells, then every second day thereafter. A healthy urothelial culture displayed polygonal, cobblestone-shaped sheets of contiguous cells (Fig. 10.5).

10.3.2 Maintenance of Culture

Cell medium is changed to fresh supplemented medium after 16 h when a culture from cryopreserved cells was established. For subsequent subcultures, medium is changed every 48 h after establishing the initial subculture until the culture

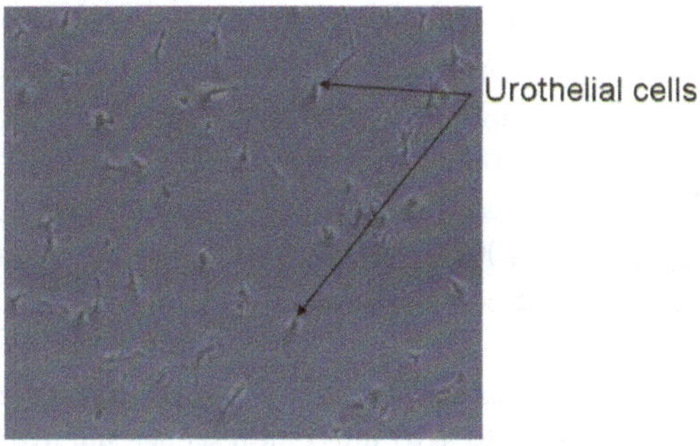

Fig. 10.5 Urothelial cells cultured and expanded prior to seeding onto UBM

is approximately 50% confluent. Once the culture reaches 50%, confluence medium is changed every 24 h until the culture is approximately 80% confluent.

10.3.3 Subcultures

Cells are subcultured/passaged at 80% confluency. Poly-L-lysine-coated cell culture flasks are prepared by warming media, trypsin/EDTA solution, trypsin neutralisation solution (TNS) and Dulbecco's phosphate-buffered saline (DPBS) to room temperature.

Cell medium is removed from the culture vessel by aspiration and the monolayer washed with DPBS to remove all traces of serum. DPBS solution is removed by aspiration. Subsequently, cells are incubated with 3 ml of trypsin/EDTA solution and placed in an incubator for approximately 2 min or until 80% of the cells are rounded up (microscope monitored). Three ml of trypsin neutralisation solution is added to the digestion, and the culture vessel is gently rocked. Released cells are harvested and transferred into a 15 ml centrifuge tube, and the vessel is rinsed with another 3 ml of growth medium to collect any residual cells. Vessels are examined under a microscope to ensure cell harvesting is successful. Ideally, there should be less than 5% of the cell population remaining in the flask. Harvested cell suspensions are centrifuged at 1000 rpm for 5 min and re-suspended in growth medium. Cells are counted and plated in a new poly-L-lysine-coated flask with a selected cell density.

10.3.4 Cell-Seeding Techniques

Specimens of UBM are cut into circles of 2 cm diameter, transferred into tissue culture plates and weighted with stainless steel rings also 2 cm in diameter to inhibit their lifting (Fig. 10.6).

Each stainless steel ring is autoclaved prior to insertion to ensure sterility. The biomaterials are then seeded with urothelial cells ($2.5–5.0 \times 10^4$ cells/cm^2 per well).

10.3.5 AlamarBlue™ Experimental Protocol to Assess Cell Viability on UBM

Viability and proliferative activity are assessed with the AlamarBlue™ cell viability reagent (Fig. 10.7).

In the presence of viable cells, the dye reduces, turns red and becomes highly fluorescent. One-tenth of the volume of AlamarBlue™ reagent is added directly into culture medium based on the protocol summarised below (Table 10.1).

After adding the reagent, the solution is incubated for 4 h at 37 °C in a cell culture incubator that is protected from direct light.

Fig. 10.6 Cell-seeded UBM weighted with stainless steel rings

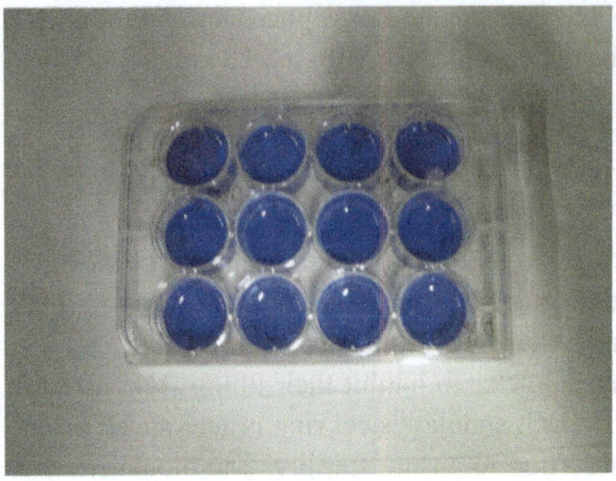

Fig. 10.7 AlamarBlue™ cell viability reagent cultured with urothelial cells in 12-well plate

Table 10.1 AlamarBlue™ protocol for assessing viable cell numbers

Format	Volume of cells and medium	Volume of 10 × AlamarBlue® to add (μL)
Cuvette	1 mL	100
96-well plate	100 μL	10
284-well plate	40 μL	4

Sensitivity for detecting cell viability generally increases with longer duration times. Therefore, samples with fewer cells should use incubation times up to 24 h. For greater sensitivity, the extent of AlamarBlue™ dye reduction is quantified by fluorescence spectrophotometry. Proliferation rate is measured as a fold increase in the number of viable cells per day.

10.3.6 Fluorescence

Fluorescence is read using a fluorescence excitation wavelength of 540–570 nm (peak excitation is 570 nm). Read fluorescence emission is read at 580–610 nm (peak emission is approximately 585 nm). Fluorescence methods are more sensitive than absorbance methods. Assay plates can be wrapped in foil, stored at 4 °C and read within 1–3 days without affecting the fluorescence or absorbance values.

To plot a calibration, curve cells are plated in 100 μL medium into 96-well tissue culture plates after performing cell number titration in the range of 40–10,000 for adherent cells and 2000–500,000 for suspension cells. A background control of 100 μL of medium without cells should be used. Subsequently, 10 μL of AlamarBlue™ is added into the medium, and cells are incubated at 37 °C overnight. Fluorescence is measured at 580–610 nm. Fluorescence intensity versus concentration of the test compound is then plotted (Fig. 10.8).

10.4 Clinical Applications

Reconstructing genitourinary tract tissue with ECM scaffolds has been extensively described (Table 10.2). An important urological advantage of ECMs is their multifunctional potential after implantation due to their versatility. Acellular ECM scaffolds are useful for providing intraoperative haemostasis through compression and for repairing small tissue defects by releasing bio-inductive growth factors. In addition, cell-seeded and composite ECM scaffolds may be capable of repairing larger and perhaps more complicated genitourinary tract defects and organs (Atala et al. 2006, 1999; Gilbert et al. 2006). Initially, most of these studies were performed on animal models. However, reports of successful human trials with encouraging short- and long-term follow-up results are now becoming more frequent.

Bladder augmentation with a tissue-engineering approach has been described in 22 patients to date in three different clinical trials (Table 10.3) (Atala et al. 2006; Joseph et al. 2014; Caione et al. 2012). Cell-seeded UBM/bladder submucosal matrix (BSM) scaffolds cultured in conjunction with biodegradable synthetic materials did show early promise for augmenting or replacing the urinary bladder as exemplified in a phase 2 clinical trial study by Atala et al. in 2006 (2006). Native urothelial and smooth muscle cells were cultured onto UBM/BSM scaffolds or a composite scaffold composed of collagen and polyglycolic acid

Fig. 10.8 Calibration curve with polynomial fit to quantify the number of viable seeded cells with spectrophotometry

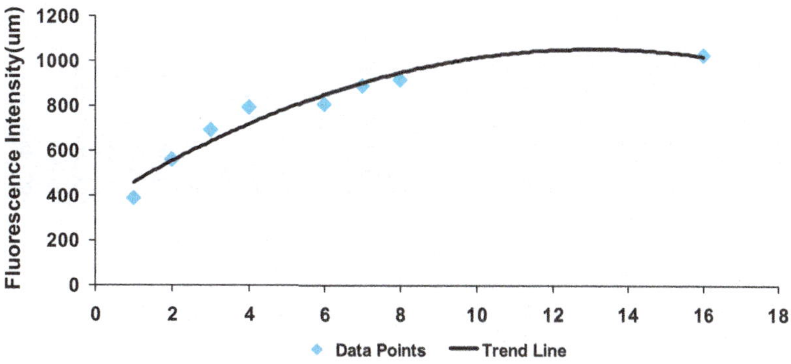

Table 10.2 Current applications of ECMs for genitourinary procedures

Organ/tissue	Pathology	ECM	Urological application
Kidney	Renal tumour	SIS	*Haemostasis in nephron sparing surgery for renal tumours (Simon et al. 2008)
Ureter	Ureteral stricture	SIS	*Ureteral replacement post excision of stricture or low-grade TCC (Liatsikos et al. 2001)
Bladder	Myelomeningocele	UBM	Urinary bladder reconstruction (Atala et al. 2006)
	Neuropathic urinary incontinence	SIS	Bladder neck slings (Misseri et al. 2005)
	Stress urinary incontinence	SIS	Commercially available as a pubovaginal sling (Giri et al. 2006)
Urethra	Urethral stricture	UBM	*'Onlay Graft' for defects <0.5 cm post-urethroplasty (Dorin et al. 2008)
	Hypospadias	Tubularised cell-seeded UBM	Repair of defects >0.5 cm post-urethroplasty (Fu and Deng 2006)

UBM urinary bladder matrix, *SIS* small intestinal submucosa
*Procedure carried out on animal models

(PGA) in paediatric patients requiring augmentation cystoplasty for myelomeningocele (n = 7). The tissue-engineered scaffolds were implanted with or without an omental wrap, and no post-operative complications were noted after 46 months. Furthermore, post-operative cystograms and urodynamic studies demonstrated an increase in bladder capacity and compliance values that were 1.58-fold to 2.79-fold improved compared to baseline values. Mean bladder leak point pressure at capacity decreased post-operatively by 56% (67–37.5 cm H_2O) (Atala et al. 2006).

Notably, one other recent phase 2 clinical trial by Joseph et al. was unable to replicate these encouraging results when an autologous cell-seeded polyglycolide/polylactide (PGA/PLA) composite scaffold was utilised for augmentation cystoplasty in patients with spina bifida (n=10) (Joseph et al. 2014). There was no improvement in bladder capacity on urodynamics after 1 year or 3 years, and serious adverse events occurred in four patients with five patients requiring re-operation in the form of a conventional ileo-cystoplasty (Joseph et al. 2014). Such findings demonstrate that further prospective studies are needed to demonstrate the clinical effectiveness of tissue engineering for reconstructing the urinary bladder.

Most recently, urologists have developed a PGA urinary conduit scaffold as an alternative to a conventional ileal conduit for urine drainage after cystectomy. The 'neo-conduit' was seeded with autologous smooth muscle cells (SMCs),

Table 10.3 Summary and clinical outcomes of studies that describe tissue-engineering methods for reconstructing the urinary bladder in human patients

Author	Year	Patient No	Age (Years)	Underlying Pathology	Scaffold source	Operative procedure	Mean length of follow-up	Clinical outcome
Atala et al. (2006)	2006	7	4–19	Myelomeningocele	UCs + SMCs + PGA + collagen + omentum	Augmentation cystoplasty	46 months	Improved compliance up to 56% No complications
Caione et al. (2012)	2012	5	10	Previous bladder exstrophy repair	SIS	Augmentation cystoplasty	18 months	30% increase in bladder capacity and compliance
Joseph et al. (2014)	2014	10	3–21	Spina bifida	Adipose stem cells + PGA + PLGA	Augmentation cystoplasty	36 months	No significant improvement in overall compliance.4/10 experienced a major complication

UCs urothelial cells, *SMCs* smooth muscle cells, *PGA* polyglycolic acid, *SIS* small intestinal submucosa, *PLGA* poly-DL-lactide-co-glycolide

grown from adipose-derived mesenchymal stem cells, for patients undergoing radical cystectomy for bladder cancer (Sopko et al. 2015; Kates et al. 2015). Eight patients have been enrolled in this phase 2 clinical trial to date, and early findings have demonstrated the regeneration of urothelium, smooth muscle and neuronal tissue on histopathology (Kates et al. 2015). Long-term functional results are currently awaited.

10.5 Limitations

At present, cell-seeded UBM is not being implanted into urological patients on a regular basis as ethical considerations, cost, regulation, manufacturing and reimbursement need to be fully clarified and transparent. The concept of unregulated application of these exploratory therapies into patients without evidence-based clinical trials is concerning to researchers. Furthermore, the increasing popularity of tissue-engineered interventions in non-scientific media should be cautiously perceived by scientists and clinicians. The effect of a patient's microenvironment on porcine UBM should be fully considered prior to implantation for lower urinary tract reconstruction. Surrounding mechanical forces, pH, cytotoxic agents, signalling agents and oxygen levels need to be carefully considered. Most importantly, international collaboration with recognised consensus guidelines is required to facilitate standards that allow safe use of these therapies in human patients with therapeutic benefit.

Specific urological factors that have inhibited the progression of UBM include the cytotoxic effects of urine and the presence of uropathogens in the upper and lower urinary tracts. One in vitro study demonstrated that 'off-the-shelf' UBM is limited by their inability to induce urothelial proliferation in the host's natural urine environment (Davis et al. 2011). Cytotoxicity is attributable to cationic substances and low molecular weight products that are normally found in urine (Davis et al. 2011). It appears that a pre-established impermeable urothelial layer prior to in vivo implantation is a prerequisite. Another limiting factor is the absence of a sophisticated vascular network when implanted in vivo. Although angiogenesis can develop in vivo, the process is not sufficient for clinical urological applicability (Bertassoni et al. 2014). Prefabrication of biomaterials and stimulation with pro-angiogenic bioactive factors such as VEGF and bFGF are limited due to their inability to develop a vascular network over a short period of time (Baptista et al. 2011).

10.6 Conclusion

Significant experimental and clinical progress has been made with tissue-engineered xenografts in urological surgery. Importantly, concerns over poor regenerative in vivo capability need to be addressed to maintain the established clinical applicability of unseeded and cell-seeded UBM. If future studies can address and resolve this issue through the development of effective, standardised and repeatable UBM preparation techniques, tissue-engineered scaffolds may continue to contribute to reconstructive urology.

References

Aktuğ T, Özdemir T, Ağartan C, Özer E, Olguner M, Akgür FM (2001) Experimentally prefabricated bladder. J Urol 65(6 I)

Atala A, Guzman L, Retik AB (1999) A novel inert collagen matrix for hypospadias repair. J Urol 1148–1151

Atala A, Bauer SB, Soker S, Yoo JJ, Retik AB (2006) Tissue-engineered autologous bladders for patients needing cystoplasty. Lancet 367(9518):1241–1246

Baptista PM, Siddiqui MM, Lozier G, Rodriguez SR, Atala A, Soker S (2011) The use of whole organ decellularization for the generation of a vascularized liver organoid. Hepatology 53(2):604–617

Bertassoni LE, Cecconi M, Manoharan V, Nikkhah M, Hjortnaes J, Cristino AL et al (2014) Hydrogel bioprinted microchannel networks for vascularization of tissue engineering constructs. Lab Chip [Internet] 14(13):2202–2211. http://xlink.rsc.org/?DOI=C4LC00030G

Caione P, Boldrinic R, Salerno A, Nappo SG (2012) Bladder augmentation using acellular collagen biomatrix: A pilot experience in exstrophic patients. Pediatr Surg Int 28(4):421–428

Davis NF, McGuire BB, Callanan A, Flood HD, McGloughlin TM (2010) Xenogenic extracellular matrices as potential biomaterials for interposition grafting in urological surgery. J Urol 184(6)

Davis NF, Callanan A, McGuire BB, Flood HD, McGloughlin TM (2011) Evaluation of viability and proliferative activity of human urothelial cells cultured onto xenogenic tissue-engineered extracellular matrices. Urology 77(4)

Davis NF, Brady CM, Creagh T (2014) Interstitial cystitis/painful bladder syndrome: epidemiology, pathophysiology and evidence-based treatment options. Eur J Obstet Gynecol Reprod Biol 175(1)

Davis NF, Cunnane EM, O'Brien FJ, Mulvihill JJ, Walsh MT (2017) Tissue engineered extracellular matrices (ECMs) in urology: evolution and future directions. Surgeon

Davis N, Cunnane E, Mulvihill J, Quinlan M, Bolton D, Walsh M et al (2018a) The role of stem cells for reconstructing the lower urinary tracts. Curr Stem Cell Res Ther

Davis NF, Mulvihill JJE, Mulay S, Cunnane EM, Bolton DM, Walsh MT (2018b) Urinary bladder vs gastrointestinal tissue: a comparative study of their biomechanical properties for urinary tract reconstruction. Urology 113

Dorin RP, Pohl HG, De Filippo RE, Yoo JJ, Atala A (2008) Tubularized urethral replacement with unseeded matrices: what is the maximum distance for normal tissue regeneration? World J Urol 26 (4):323–326

Flood HD, Malhotra SJ, O'Connell HE, Ritchey MJ, Bloom DA, McGuire EJ (1995) Long-term results and complications using augmentation cystoplasty in reconstructive urology. Neurourol Urodyn 14 (4):297–309

Fu Q, Deng CL (2006) Ten-year experience with composite bladder mucosa-skin grafts in hypospadias repair. Urology

Gilbert TW, Sellaro TL, Badylak SF (2006) Decellularization of tissues and organs. Biomaterials 3675–3683

Giri SK, Hickey JP, Sil D, Mabadeje O, Shaikh FM, Narasimhulu G et al (2006) The long-term results of pubovaginal sling surgery using acellular cross-linked porcine dermis in the treatment of urodynamic stress incontinence. J Urol

Joseph DB, Borer JG, De Filippo RE, Hodges SJ, McLorie GA (2014) Autologous cell seeded biodegradable scaffold for augmentation cystoplasty: phase II study in children and adolescents with spina bifida. J Urol [Internet]. Elsevier Ltd 191(5):1389–94. http://dx.doi.org/10.1016/j.juro.2013.10.103

Kates M, Singh A, Matsui H, Steinberg GD, Smith ND, Schoenberg MP et al (2015) Tissue-engineered urinary conduits. Curr Urol Rep

Liatsikos EN, Dinlenc CZ, Kapoor R, Alexianu M, Yohannes P, Anderson AE et al (2001) Laparoscopic ureteral reconstruction with small intestinal submucosa. J Endourol

Misseri R, Cain MP, Casale AJ, Kaefer M, Meldrum KK, Rink RC et al (2005) Small intestinal submucosa bladder neck slings for incontinence associated with neuropathic bladder. J Urol

Simon J, De Petriconi R, Meilinger M, Hautmann RE, Bartsch G (2008) Optimized haemostasis in nephron-sparing surgery using small-intestine submucosa. BMC Urol

Sopko NA, Kates M, Bivalacqua TJ (2015) Use of regenerative tissue for urinary diversion. Curr Opin Urol [Internet] 25(6):578–85. https://www.scopus.com/inward/record.uri?eid=2-s2.0-84943142918&partnerID=40&md5=7efda37d933ddcb46391eb193fc620e1

11 Decellularization Methods of Ovary in Tissue Engineering

Hannes Campo, Sara López-Martínez, and Irene Cervelló

Abstract

The ovaries or female gonads are situated in the ovarian fossa of the abdominal cavity. These are paired, almond-shaped organs measuring about 3.5 cm long and 1.5 cm thick and exist out of a central medullary zone and a peripheral cortex that are enclosed in a fibrous capsule called the tunica albuginea. The ovaries serve 2 main functions, the first one being the production of female gametes called oocytes (oogenesis). Interestingly, the number of primary oocytes that reside in the ovary is determined at birth. About 400 oocyte-containing follicles successfully go through all the developmental stages from this limited pool during folliculogenesis throughout the female reproductive life. In this process, primordial follicles grow and advance until forming a mature or Graafian follicle; during ovulation, secondary oocytes are released and the remaining follicular wall collapses and forms the highly vascularized corpus luteum or luteal gland. This ovarian cycle is regulated by several hormones secreted from the adenohypophysis and lasts about 28 days. During this cycle, the ovaries also serve as endocrine glands and produce female sex hormones such as estrogens and progesterone (steroidogenesis), influencing the growth and development of tissues sensitive to these hormones such as the endometrium. Hence, the endometrial cycle goes synchronized with the ovarian cycle.

Keywords

Ovary · Decellularization · Regeneration · Fertility

11.1 History

While the ovary is considered as a non-vital organ, the option to store and transplant ovarian tissue is considered as a major possible improvement on the quality of life for many women, and this can be based from the preservation of fertility to the improvement of health (Andersen and Kristensen 2015; Shea et al. 2014). Investigation behind the transplantation and cryopreservation of the ovary started as early as the 1950s and found its way into clinical practice half a century later (Deanesly 1954; Donnez et al. 2004). In the pursuit of replacing or regenerate human cells, tissues, or organs in order to restore or establish normal function, a new multidisciplinary field was born, namely regenerative medicine (Mason and Dunnill

H. Campo · S. López-Martínez · I. Cervelló (✉)
Fundación Instituto Valenciano de Infertilidad (FIVI), Instituto de Investigación Sanitaria La Fe, Valencia, Spain
e-mail: Irene.Cervello@ivirma.com

A. Kajbafzadeh (ed.), *Decellularization Methods of Tissue and Whole Organ in Tissue Engineering*, Advances in Experimental Medicine and Biology 1345,
https://doi.org/10.1007/978-3-030-82735-9_11

2007). By using cells and designing materials mimicking their native reproductive tissues and organs, the subfield "reproductive tissue engineering" (REPROTEN) has been coined recently (Amorim 2017). In this chapter, we intend to give an overview of the different methods used in a novel technique used within this field, namely the decellularization (DC) and recellularization (RC) of the ovary.

11.2 Materials and Methods

Decellularization intends to obtain the tissue-specific and complex extracellular matrix (ECM) and utilize it as a bioscaffold for tissue engineering. By removing the cells and cellular antigens, detrimental inflammatory reactions are attenuated. There has been much progress reported in virtually all vital organs such as the liver (Baptista et al. 2011) and heart (Kim et al. 2016b) or in the case of reproductive medicine, the uterus (Campo et al. 2017) to name a few. In this section, we will provide an overview of the methodologies and reagents used for the decellularization of the ovary, and a summary of all relevant publications can be found in Table 11.1.

Laronda et al. were the first to publish the successful decellularization of human and bovine ovarian tissues (Laronda et al. 2015). The human samples originated from cancer patients, most of which had a history of treatment (radiation, chemotherapy, immunosuppression) who donated ovarian tissue for fertility preservation and investigation, and the bovine ovaries came from young cows and were processed in 500-micron-thick slices. The presence of cancer cells was also demonstrated in approximately half of the human samples. To decellularize human and bovine tissues, an excess of 0.1% sodium dodecyl sulfate (SDS) solution was used, and samples were left for 24 h in agitation at room temperature. Complete decellularization was demonstrated first by histological analysis, where DAPI and hematoxylin and eosin (HE) staining did not show any cells nor nuclei, which was corroborated by DNA quantification. Furthermore, after DC the SALL4-positive cells in the human samples were also completely removed. In this study, the DC of the whole organ was also attempted by immersing whole bovine ovaries in 0.1% SDS for 3 weeks at 4 °C. From these samples, an analysis of the ultrastructure was performed by scanning electron microscopy, showing the preservation of the major hierarchical organization containing important structural fibers. This was further analyzed by immunohistochemistry. Signals for collagen IV, laminin and fibronectin were present after DC in the bovine samples, with some not presenting in the cortex or medulla layer. Human samples stained positive for collagen I and IV, but very little signal was present for fibronectin and laminin, suggesting that this protocol has a relative harsh effect on the human cortical ECM. This two to four week long protocol was also employed by Jakus et al., confirming the efficacy of the protocol by HE staining and scanning electron microscopy (Jakus et al. 2017). In this study, the acellular ovarian tissue was lyophilized, pulverized and mixed with poly(lactic-co-glycolic acid) (PLGA) polymer (65 vol% and 35 vol%, respectively) to create ECM "inks." These hybrid scaffolds were made into "tissue papers," and a unique tissue-specific surface topology was observed. Furthermore, the ovarian tissue paper (OTP) could absorb up to 300% of its volume, while the elastomeric polymer matrix provided mechanical stability, allowing for easy manipulation (folding, cutting, suturing), resulting in a promising new biomaterial.

Liu et al. on the other hand published a DC protocol for pig ovarian tissue. Here, ovaries were cleaned from connective tissue and fat, were subjected to three freezing and thawing cycles (from −80–37 °C) and were cut in 1.5-mm-thick pieces. The DC protocol included a wash in deionized water (H_2Od), and fragments were agitated for 9 and 3 h at room temperature in 1% Triton X-100 and 0.5% SDS, respectively. After a 12-h wash in distilled water (H_2Od), the residual DNA was removed by 200 U/ml of DNAse I at 37 °C for 12 h and washed for another 4 h in H_2Od. Sterilization was performed with 3% peracetic acid. Correct DC was first assessed visually and was confirmed by histology

Table 11.1 Summary of all relevant publications in the de- and recellularization of the ovary

Decellularization				In vitro experiments/tissue engineered construct			In vivo experiments		Results (in vitro and/or in vivo)	References
Tissue	Species	Method	Chemicals	Cells/tissue/follicles used	Time	Culture conditions	Implanted species	Time		
500-µm-thick sections	Human and Bovine	Immersion	24 h in 0.1% SDS at RT	Murine primary ovarian cells (mainly granulosa cells)	2 days	Static culture in DMEM:F12, 1X insulin–transferrin–selenium, 12.75 mM HEPES 1X Penicillin/Streptomycin, 10% FBS and 40 mg/ml hydrocortisone	Mouse	2 weeks	Protocol decellularizes ovaries from both species, bovine DOM supports mouse ovarian cells, and constructs initiate puberty in ovariectomized mice	Laronda et al.
Chopped ovarian fragments	Bovine	Immersion	24 h in 0.1% SDS at RT	Human BM-MSCs	28 days	Static culture in 1 × low glucose DMEM + 10% FBS, HEPES buffer, l-glutamine, 10 units antibiotic antimyotic	/	/	hMSCs adhered, migrated and proliferated until confluence without degrading/deform OTP	Jakus et al.
				Secondary mouse ovarian follicle	4 days	αMEM Glutamax, 3 mg/ml BSA, 10 mIU/ml rFSH, 1 mg/ml bovine fetuin, 5 µg/ml insulin, 5 µg/ml transferrin and 5 µg/ml selenium	/	/	Attachment, survival and maintaining of spherical cell aggregate for 4 days	
				Human and rhesus macaque ovarian cortical strips (0.5 mm)	56 days	Waymouth's medium with 3 mg/mL HSA, 0.5 mg/mL bovine fetuin, 5 µg/mL insulin, 5 µg/mL transferrin, 5 ng/mL selenium, 64 µg/mL AA and 10 mIU/mL FSH	/	/	Presence of healthy follicles in corical strips	
1.5-mm-thick slices	Porcine	Immersion	Snap freezing–thawing cycles (3), 9 h Triton X-100 1%, 3 h SDS 0.5%. 12 h H_2Od, 12 h 200 U/mL DNase I in PBS, 4 h H_2Od	Rat granulosa cells/explant culture in DOM	9/12 days	Static culture in DMEM:F12 (both in vitro experiments)	Rat	2–4 weeks	DOM ECM improved cell migration in vitro; granulosa cells from explant culture invaded into DOM, forming aggregates. Low density of immune cells in the implanted DOM was reported	Liu et al.

(continued)

Table 11.1 (continued)

Decellularization				In vitro experiments/tissue engineered construct			In vivo experiments		Results (in vitro and/or in vivo)	References
Tissue	Species	Method	Chemicals	Cells/tissue/follicles used	Time	Culture conditions	Implanted species	Time		
2.0-mm-thick cortex tissue slices	Human	Immersion	48 h SLES 1%, 24 h 500 U/mL of deoxyribonuclease at 36 °C	Wharton's jelly mesenchymal stem cells	7 days	Static culture in DMEM/F12, 10% FBS, 2 mM L-glutamine, 100 U/mL penicillin and 100 µg/mL streptomycin	/	/	SLES treatment efficiently removed cellular fraction, and ECM structure and composition were retained. The DOM was cytocompatible and transplantable; recellularized DOMs reconstituted serum progesterone levels in vivo	Hassanpour et al.
				Rat primary ovarian cells	1 day	Static culture in DMEM/F12, 1 × insulin–transferrin–selenium, 12.75 mM HEPES, 1 × penicillin/streptomycin, 10% FBS, 40 mg/mL hydrocortisone	Rat	4 weeks		

AA ascorbic acid, *BM-MSC* bone marrow-derived mesenchymal stem cells, *DOM* decellularized ovarian matrix, *FSH* follicle-stimulating hormone, *HSA* human serum albumin, *OTP* ovarian tissue paper, *RC* recellularized, *rFSH* recombinant follicle-stimulating hormone, *RT* room temperature, *SDS* sodium dodecyl sulfate, *SLES* sodium lauryl ester sulfate

(HE and DAPI) and DNA quantification. The presence of glycosaminoglycans (GAGs) was demonstrated by Alcian blue staining, and quantification did show a significant decrease in GAG content. Total collagen amount was quantified based on hydroxyproline content and showed no significant increase, and collagen type I and III presence was confirmed by immunohistochemistry, as was also the case for fibronectin and laminin. Scanning electron microscopy showed the intact hierarchical composition of the ovaries, showing empty space where follicles, granulosa and theca cells used to reside.

All the protocols mentioned above use the ionic detergent SDS for a certain period of time, which is hard to remove and could have negative impact on cell culture or transplantation if not washed properly (Friedrich et al. 2018). Hassanpour et al. sought an alternative protocol using solely the milder anionic detergent sodium lauryl ester sulfate (SLES) (Hassanpour et al. 2018). It has been reported that SLES improved the preservation of GAGs (mainly glycans on hydrophilic proteins) and growth factors compared to SDS and that after transplantation the SLES decellularized scaffolds showed significantly reduced inflammation and platelet adhesion compared to scaffolds decellularized by SDS (Kawasaki et al. 2015). For DC, 2.0-mm-thick slices of the cortex were agitated in 1% SLES for 48 h at 18–20 °C, and residual DNA was removed by using 500 u/ml DNase I in PBS for 24 h at 36 °C (Hassanpour et al. 2018). Similar to other publications, whole ovarian DC was performed by immersing bisected ovaries for 30–40 days in 1% SLES. Afterward, the ovaries turned white and transparent, and DC was confirmed by histology (HE and Hoechst) and DNA quantification. The composition of the ECM was assessed by histochemical analysis: The presence of GAGs and elastic fibers was confirmed by Alcian blue and Gomori's aldehyde fuchsin, respectively, and Heidenhain's AZAN and Masson Trichrome staining showed the preservation of collagen fibers. This was further corroborated by immunohistochemistry, and collagen I was present throughout the decellularized cortex, while collagen IV was found in the cortex, theca cells compartment and some parts of the medullary stroma. A higher signal for laminin was found near the cortex, while for fibronectin this was the case in the tunica albuginea, theca cells compartment and the medullary stroma.

In conclusion, the studies presented here show various immersion-based DC protocols using both established detergents such as SDS and Triton X-100 and new promising ones such as SLES. While basic (immuno)histological, topological and biochemical analysis has been performed to assess the DC, their effect on the mechanical stability and proteomic profile of the DC ovarian tissues has not been done. The latter would be especially interesting seeing that the matrisome has been recently defined for the ovary (Ouni et al. 2019). Furthermore, while the DC of whole organs was presented, this was not done via perfusion of the vascular system but rather by prolonged immersion (up to 40 days), limiting their future use as vascularized perfusable bioscaffolds for the generation of whole ovarian constructs (Hassanpour et al. 2018; Laronda et al. 2015). Nonetheless, these studies represent an important first step toward the engineering of the ovary and follicle microenvironment.

11.3 Cell Seeding

In vitro cell seeding can be used for many goals, it can be used to make organoids or organotypic 3D culture systems for investigation, to test the cytotoxicity of the bioscaffold or to recellularize whole organs/fragments to create transplantable constructs. The ovary is made out of different cell types that support and interact with follicles of different stages: Some or all of these can be used to recellularize the scaffold. For this, several techniques can be used; sections or fragments can be covered with cells so they adhere, proliferate and penetrate, and they can be injected directly into the scaffold (fragments or whole organs) or delivered via perfusion (via single infusion or by introducing them into the bulk media, only for whole organs).

A natural first step after establishing successful DC is testing the cytotoxicity of the scaffold in vitro. Here, the proliferation and behavior of native primary cultured cells, (stem) cell lines and follicles are tested. Laronda et al. used primary ovarian cells isolated from 3- to 4-week-old mice (Laronda et al. 2015). This cell population exists predominantly out of granulosa cells (theca cells and adherent follicles were also present before trypsinization) and was cultivated for 48 h on decellularized disks of bovine medulla sections. Some penetration of the scaffold was observed, and circular, follicle-like patterns and steroid blebs were observed on the scaffolds by scanning electron microscopy (SEM) as well. To demonstrate the functionality of these recellularized disks, they investigated whether they could restore endocrine function in non-immunodeficient mice prior to puberty. Encouraging results were found in this group: Serum estradiol (E2) levels were restored in most cases, circulating inhibin-α was measured, and vaginal opening (a secondary sex characteristic) was induced. Furthermore, the graft did not induce an immune reaction, and follicle-like structures were found, demonstrating the potential of decellularized scaffolds. Similarly, Jakus et al. confirmed the adhesion, proliferation and infiltration of mesenchymal stem cells (MSCs); after 28 days, live–dead stain showed no cytotoxicity of the OTP (Jakus et al. 2017). The porous tissue-specific structure allowed for the in vitro culture of mouse ovarian follicles. Human and rhesus macaque ovarian cortical strips were also cultured for up to 4 and 56 days, respectively, without observing negative effects or even improving on previously published results (Jakus et al. 2017; Laronda et al. 2014).

After demonstrating no negative effects of culture media incubated with the decellularized ovary, Lui et al. proceeded to demonstrate the scaffolds' biocompatibility in vivo by xenogeneic subcutaneous implantation of decellularized and native ovarian fragments. While only few cells infiltrated the acellular scaffold, immunohistochemistry of CD68, CD86 and CD3 showed a mild immune response compared to the native tissue. In a last in vitro study, rat native tissue was implanted in the decellularized pig ovarian. Here, granulosa cells adhered, migrated and proliferated in the scaffold and E2 secretion functions were sustained (Liu et al. 2017).

Hassanpour et al. seeded Wharton's jelly mesenchymal stem cells on top of decellularized disks to test the biocompatibility. Using a MTT assay, no differential expression with standard 2D culture systems was observed after 3 days; furthermore, after one week of culture the proliferation rate observed was higher than in the controls (Hassanpour et al. 2018). For in vivo studies, rat primary ovarian cells were allowed to adhere for one day in static culture. After this, non-recellularized and recellularized constructs were sutured onto the renal fat pad bilaterally. Neovascularization and invasion of native cells were observed, and furthermore, primordial or primary follicle-like structures were identified in the primary ovarian cell grafts, presenting inhibin-α and steroid receptors. Finally, a restauration of progesterone to physiological levels and an elevation of estradiol levels resulted in the improvement of vaginal patency of the rats.

In these studies, it became clear that decellularized ovarian tissues (and derivatives) are cytocompatible with various cell types such as stem cells (Hassanpour et al. 2018; Jakus et al. 2017), primary culture cells (Laronda et al. 2015) and follicles or cortical strips (Jakus et al. 2017). Furthermore, the biocompatibility was demonstrated by transplanting (non-) recellularized scaffolds in animal models (Hassanpour et al. 2018; Laronda et al. 2015; Liu et al. 2017), paving the way for further improvement of in vitro models and in vivo studies with possible future clinical applications.

11.4 Clinical Applications

Infertility is described by the World Health Organization (WHO) as "a disease of the reproductive system defined by the failure to achieve a clinical pregnancy after 12 months or more of regular unprotected sexual intercourse." Currently, the number of affected couples is around 15% of the world population and a third of infertility problems are because of female causes (Center for Disease Control and Prevention, CDC). There are several diseases that are related to or directly affecting the ovaries and women's fertility, including ovarian cancer, polycystic ovary syndrome (PCOS), endometriosis and premature ovarian failure (POF).

POF, also referred to as primary ovarian insufficiency (POI), is a major cause of female infertility that affects one in every 10.000 women before 20, 1: 1000 before 30 and 1: 100 before 40 years old (Hewlett and Mahalingaiah 2015; Qin et al. 2015). It is characterized by a depleted ovarian reserve that leads to amenorrhea, hypoestrogenism and elevated gonadotropin levels (Shelling 2010). The most prominent potential causes are radio/chemotherapy, genetic alterations, viral infections, metabolic pathologies, immune diseases and environment factors (Kuo et al. 2017; Torrealday et al. 2017). The various possible spontaneous or induced causes of POF are listed in Table 11.2. Women with established POF have a low amount and poor quality of oocytes, for these patients to achieve biological parenting only in vitro fertilization (IVF) with donor gametes remains.

Table 11.2 Causes of premature ovarian failure

Spontaneous
Idiopathic
Genetic
Turner syndrome (45XO) or mosaic Turner (45X/46XX)
Trisomy X (47XXX or mosaic)
Fragile X premutation
Galactosemia (galactose-1-phosphate uridyltransferase deficiency)
Autoimmune polyglandular syndrome (types 1 and 2)
Follicle-stimulating hormone receptor mutations
17α-hydroxylase deficiency
Aromatase deficiency
Blepharophimosis, ptosis, epicanthus inversus syndrome
Bloom syndrome
Ataxia telangiectasia
Fanconi anemia
Autoimmune
Infections
Mumps oophoritis
Tuberculosis, malaria, cytomegalovirus, varicella and shigella
Induced
Bilateral oophorectomy, bilateral ovarian cystectomies
Chemotherapy-primarily, alkylating agents and anthracyclines
Radiation-external beam or intracavitary
Environmental toxins
Pelvic vessel embolization

One group of women that are of special interest in the field of REPROTEN are those receiving cancer treatment; with the rising survival rates, it becomes more pertinent to take the patients' post-treatment quality of life into consideration (Letourneau et al. 2012). Radio/chemotherapy such as abdominal ionizing radiation and alkylating agents (for example, cyclophosphamide) have been described to have devastating effects on the ovarian reserve, inducing POF (Meirow et al. 2010).

Different strategies have been developed to preserve fertility in female patients of reproductive age undergoing radio/chemotherapy, and these include the cryopreservation of oocytes, embryos and ovarian cortical tissue; ovarian shielding by ferto-protective adjuvant therapy and surgical techniques such as ovarian transposition and transplantation (Donnez and Dolmans 2013; Levine et al. 2015; Mahajan 2015). Each of these techniques has their respective advantages and disadvantages, and specific use cases.

Oocyte/embryo cryopreservation, for example, requires controlled ovarian stimulation, which takes 2 weeks starting from the second day of the period, and naturally, this delay is unadvisable for cancer treatment. Furthermore, it is unwise to apply this hormonal treatments in patients with hormone-dependent cancers, and some data even suggest an indirect mitogenic effect of estrogens on hormone receptor-negative cancers (Shea et al. 2014). On the other hand, ovarian transplantation is still an experimental technique that has several limitations such as onset of avascular ischemic period after re-transplantation (Donnez et al. 2013; Lee et al. 2016; Oktay and Oktem 2010).

Cryopreservation and transplantation of ovarian tissue have several advantages and are the only option for prepuberal girls that cannot delay treatment (Jadoul et al. 2010). Firstly, it can effectively preserve large amounts of primordial follicles and can be performed at the time of diagnosis, without the need for hormonal stimulation. Furthermore, transplanting the ovarian tissue not only restores fertility but also endocrine function. While this technique is still considered experimental in nature, it has resulted in at least 60 live births at the time of writing (Donnez and Dolmans 2015). However, cryopreserved ovarian tissue can also contain malignant metastatic cells, and 12.4% of patients died due to recurrence after reimplantation over a 12-year period (Amorim and Shikanov 2016; Imbert et al. 2014).

In other words, current techniques used for treating these disorders are limited by their high risks and suboptimal effectiveness rates, posing a significant threat to the prognoses and quality of life for female patients. Thus, tissue-engineered constructs using ovarian cells and/or follicles could play an important role in the future.

Approaches for in vitro folliculogenesis were proposed along the last years. Three-dimensional culture systems using hydrogels to encapsulate isolated follicles provide physical support that preserves oocyte–somatic cell connections and promotes survival of early stage follicles. Non-tissue-specific hydrogels such as polyethylene glycol (PEG), agarose, alginate, collagen and fibrin are currently tested in vitro supporting successfully the growth and maturation of ovarian follicles in mice, large mammalian species and humans (He 2017; Shea et al. 2014; Skory et al. 2015; Xiao et al. 2015). These systems would be interesting to obtain mature oocytes from primordial follicles, improving on current IVF approaches.

Another major aspect besides fertility preservation is the restauration of the hypothalamus–pituitary–gonadal axis; much like with normal ovarian tissue transplantation, it has also been demonstrated that recellularized DC fragments and other types of constructs are capable to reestablish endocrine functions (Hassanpour et al. 2018; Kim et al. 2016a; Kniazeva et al. 2015; Laronda et al. 2015). This implies that these tissue-engineered constructs could also be used to improve the quality of life of many women not undergoing cancer treatment. An innovative example of this is to postpone menopause by transplanting ovarian tissue that was cryopreserved before menopause (Andersen and Kristensen 2015). REPROTEN approaches could in theory be used in the future to create constructs for menopausal women that did not cryopreserved ovarian tissue.

11.5 Limitations

Even though the decellularization and recellularization of the ovary show promise, relatively few protocols and methods have been published as of yet and more investigation is needed to truly pinpoint the specific limitations of these techniques. While its final use can be up to discussion or example, no perfusion-based whole organ decellularization protocol has been published. The in vivo cell seeding protocols were used to test biocompatibility of the decellularized scaffold; however, no attempts were made to maintain a perfusable construct in a bioreactor. As with all organs, the main limitation in the DC/RC is the recellularization efficiency, the ovary consists of a heterogeneous population of follicles in different phases of development and it is not known whether this dynamic tissue can be completely recapitulated in an in vitro setup. Furthermore, follicles are relatively well isolated from the ovarian stroma, and it is unclear whether pluripotent stem cells can recreate the successes presented in other organs (Ott et al. 2008).

Lastly, it is complicated to obtain whole organ donors, and in some studies, they come from xenogenous sources (Jakus et al. 2017; Liu et al. 2017). Due to the morphological similarities, the best xenogeneic whole organ donor would be the sheep ovary. If the donor tissue would be human, then they could come from females undergoing sexual reassignment surgery (Hassanpour et al. 2018), as small fragments as part of fertility preservation (Laronda et al. 2015) or from deceased donors.

11.6 Conclusion

This chapter describes innovative perspectives and ideas of tissue and whole organ engineering related to the ovaries, including the different methods for whole/partial decellularization, characterization of the extracellular matrix, techniques and functionality tests of the recellularization process and the future clinical applications. The regenerative medicine based on the ovary setting is essential to preserve and to reestablish female fertility as we have explained along these lines.

The decellularization and recellularization of ovarian tissue represent in the therapeutic context an interesting solution; while creating bioscaffolds that are specifically compatible with ovarian cells and follicles, it also effectively removes cancer cells and as such the risk of reintroducing them. Furthermore, the compatibility with the ovarian tissue fragments makes them an interesting biomaterial to improve current ovarian tissue transplantation techniques.

The recent advancements in the bioengineering applied to the ovary, recapitulate the characteristic microenvironment and promote the specific ovarian cellular development as a therapeutic strategy. The advantages of these emerging techniques are encouraging due to the preservation or improvement of female fertility.

References

Amorim CA (2017) Special issue devoted to a new field of regenerative medicine: reproductive tissue engineering. Ann Biomed Eng 45:1589–1591. https://doi.org/10.1007/s10439-017-1862-0

Amorim CA, Shikanov A (2016) The artificial ovary: current status and future perspectives. Future Oncol 12:2323–2332. https://doi.org/10.2217/fon-2016-0202

Andersen CY, Kristensen SG (2015) Novel use of the ovarian follicular pool to postpone menopause and delay osteoporosis. Reprod Biomed Online 31:128–131. https://doi.org/10.1016/j.rbmo.2015.05.002

Baptista PM, Siddiqui MM, Lozier G, Rodriguez SR, Atala A, Soker S (2011) The use of whole organ decellularization for the generation of a vascularized liver organoid. Hepatology 53:604–617. https://doi.org/10.1002/hep.24067

Campo H, Baptista PM, López-Pérez N, Faus A, Cervelló I, Simón C (2017) De- and recellularization of the pig uterus: a bioengineering pilot study. Biol Reprod 96:34–45. https://doi.org/10.1095/biolreprod.116.143396

Deanesly R (1954) Immature rat ovaries grafted after freezing and thawing. J Endocrinol 11:197–200

Donnez J, Dolmans M-M (2013) Fertility preservation in women. Nat Rev Endocrinol 9:735–749. https://doi.org/10.1038/nrendo.2013.205

Donnez J, Dolmans M-M (2015) Ovarian cortex transplantation: 60 reported live births brings the success and worldwide expansion of the technique towards routine clinical practice. J Assist Reprod Genet 32:1167–1170. https://doi.org/10.1007/s10815-015-0544-9

Donnez J, Dolmans MM, Demylle D, Jadoul P, Pirard C, Squifflet J, Martinez-Madrid B, van Langendonckt A (2004) Livebirth after orthotopic transplantation of cryopreserved ovarian tissue. Lancet 364:1405–1410. https://doi.org/10.1016/S0140-6736(04)17222-X

Donnez J, Dolmans M-M, Pellicer A, Diaz-Garcia C, Sanchez Serrano M, Schmidt KT, Ernst E, Luyckx V, Andersen CY (2013) Restoration of ovarian activity and pregnancy after transplantation of cryopreserved ovarian tissue: a review of 60 cases of reimplantation. Fertil Steril 99:1503–1513. https://doi.org/10.1016/j.fertnstert.2013.03.030

Friedrich EE, Lanier ST, Niknam-Bienia S, Arenas GA, Rajendran D, Wertheim JA, Galiano RD (2018) Residual sodium dodecyl sulfate in decellularized muscle matrices leads to fibroblast activation in vitro and foreign body response in vivo. J Tissue Eng Regen Med 12:e1704–e1715. https://doi.org/10.1002/term.2604

Hassanpour A, Talaei-Khozani T, Kargar-Abarghouei E, Razban V, Vojdani Z (2018) Decellularized human ovarian scaffold based on a sodium lauryl ester sulfate (SLES)-treated protocol, as a natural three-dimensional scaffold for construction of bioengineered ovaries. Stem Cell Res Ther 9:252. https://doi.org/10.1186/s13287-018-0971-5

He X (2017) Microfluidic encapsulation of ovarian follicles for 3D culture. Ann Biomed Eng 45:1676–1684. https://doi.org/10.1007/s10439-017-1823-7

Hewlett M, Mahalingaiah S (2015) Update on primary ovarian insufficiency. Curr Opin Endocrinol Diabetes Obes 22:483–489. https://doi.org/10.1097/MED.0000000000000206

Imbert R, Moffa F, Tsepelidis S, Simon P, Delbaere A, Devreker F, Dechene J, Ferster A, Veys I, Fastrez M, Englert Y, Demeestere I (2014) Safety and usefulness of cryopreservation of ovarian tissue to preserve fertility: a 12-year retrospective analysis. Hum Reprod 29:1931–1940. https://doi.org/10.1093/humrep/deu158

Jadoul P, Dolmans M-M, Donnez J (2010) Fertility preservation in girls during childhood: is it feasible, efficient and safe and to whom should it be proposed? Hum Reprod Update 16:617–630. https://doi.org/10.1093/humupd/dmq010

Jakus AE, Laronda MM, Rashedi AS, Robinson CM, Lee C, Jordan SW, Orwig KE, Woodruff TK, Shah RN (2017) "Tissue papers" from organ-specific decellularized extracellular matrices. Adv Funct Mater 27. https://doi.org/10.1002/adfm.201700992

Kawasaki T, Kirita Y, Kami D, Kitani T, Ozaki C, Itakura Y, Toyoda M, Gojo S (2015) Novel detergent for whole organ tissue engineering. J Biomed Mater Res, Part A 103:3364–3373. https://doi.org/10.1002/jbm.a.35474

Kim J, Perez AS, Claflin J, David A, Zhou H, Shikanov A (2016) Synthetic hydrogel supports the function and regeneration of artificial ovarian tissue in mice. npj Regenerative Med 1:16010. https://doi.org/10.1038/npjregenmed.2016.10

Kim JJ, Hou L, Huang NF (2016) Vascularization of three-dimensional engineered tissues for regenerative medicine applications. Acta Biomater 41:17–26. https://doi.org/10.1016/j.actbio.2016.06.001

Kniazeva E, Hardy AN, Boukaidi SA, Woodruff TK, Jeruss JS, Shea LD (2015) Primordial follicle transplantation within designer biomaterial grafts produce live births in a mouse infertility model. Sci Rep 5. https://doi.org/10.1038/srep17709

Kuo C-Y, Baker H, Fries MH, Yoo JJ, Kim PCW, Fisher JP (2017) Bioengineering strategies to treat female infertility. Tissue Eng Part B Rev 23:294–306. https://doi.org/10.1089/ten.TEB.2016.0385

Laronda MM, Duncan FE, Hornick JE, Xu M, Pahnke JE, Whelan KA, Shea LD, Woodruff TK (2014) Alginate encapsulation supports the growth and differentiation of human primordial follicles within ovarian cortical tissue. J Assist Reprod Genet 31:1013–1028. https://doi.org/10.1007/s10815-014-0252-x

Laronda MM, Jakus AE, Whelan KA, Wertheim JA, Shah RN, Woodruff TK (2015) Initiation of puberty in mice following decellularized ovary transplant. Biomaterials 50:20–29. https://doi.org/10.1016/j.biomaterials.2015.01.051

Lee J, Kong HS, Kim EJ, Youm HW, Lee JR, Suh CS, Kim SH (2016) Ovarian injury during cryopreservation and transplantation in mice: a comparative study between cryoinjury and ischemic injury. Hum Reprod 31:1827–1837. https://doi.org/10.1093/humrep/dew144

Letourneau JM, Ebbel EE, Katz PP, Katz A, Ai WZ, Chien AJ, Melisko ME, Cedars MI, Rosen MP (2012) Pretreatment fertility counseling and fertility preservation improve quality of life in reproductive age women with cancer. Cancer 118:1710–1717. https://doi.org/10.1002/cncr.26459

Levine JM, Kelvin JF, Quinn GP, Gracia CR (2015) Infertility in reproductive-age female cancer survivors. Cancer 121:1532–1539. https://doi.org/10.1002/cncr.29181

Liu W-Y, Lin S-G, Zhuo R-Y, Xie Y-Y, Pan W, Lin X-F, Shen F-X (2017) Xenogeneic decellularized scaffold: a novel platform for ovary regeneration. Tissue Eng Part C Methods 23:61–71. https://doi.org/10.1089/ten.tec.2016.0410

Mahajan N (2015) Fertility preservation in female cancer patients: an overview. J Hum Reprod Sci 8:3. https://doi.org/10.4103/0974-1208.153119

Mason C, Dunnill P (2007) A brief definition of regenerative medicine. Regen Med 3:1–5. https://doi.org/10.2217/17460751.3.1.1

Meirow D, Biederman H, Anderson RA, Wallace WHB (2010) Toxicity of chemotherapy and radiation on female reproduction. Clin Obstet Gynecol 53:727–739. https://doi.org/10.1097/GRF.0b013e3181f96b54

Oktay K, Oktem O (2010) Ovarian cryopreservation and transplantation for fertility preservation for medical indications: report of an ongoing experience. Fertil Steril 93:762–768. https://doi.org/10.1016/j.fertnstert.2008.10.006

Ott HC, Matthiesen TS, Goh S-K, Black LD, Kren SM, Netoff TI, Taylor DA (2008) Perfusion-decellularized matrix: using nature's platform to engineer a bioartificial heart. Nat Med 14:213–221. https://doi.org/10.1038/nm1684

Ouni E, Vertommen D, Chiti MC, Dolmans M-M, Amorim CA (2019) A draft map of the human ovarian proteome for tissue engineering and clinical applications. Mol Cell Proteomics 18:S159–S173. https://doi.org/10.1074/mcp.RA117.000469

Qin Y, Jiao X, Simpson JL, Chen Z-J (2015) Genetics of primary ovarian insufficiency: new developments and opportunities. Hum Reprod Update 21:787–808. https://doi.org/10.1093/humupd/dmv036

Shea LD, Woodruff TK, Shikanov A (2014) Bioengineering the ovarian follicle microenvironment. Annu Rev Biomed Eng 16:29–52. https://doi.org/10.1146/annurev-bioeng-071813-105131

Shelling AN (2010) Premature ovarian failure. Reproduction 140:633–641. https://doi.org/10.1530/REP-09-0567

Skory RM, Xu Y, Shea LD, Woodruff TK (2015) Engineering the ovarian cycle using in vitro follicle culture. Hum Reprod 30:1386–1395. https://doi.org/10.1093/humrep/dev052

Torrealday S, Kodaman P, Pal L (2017) Premature ovarian insufficiency—an update on recent advances in understanding and management. F1000Res 6. https://doi.org/10.12688/f1000research.11948.1

Xiao S, Zhang J, Romero MM, Smith KN, Shea LD, Woodruff TK (2015) In vitro follicle growth supports human oocyte meiotic maturation. Sci Rep 5:17323. https://doi.org/10.1038/srep17323

12 Decellularization Methods of Uterus in Tissue Engineering

Hannes Campo, Sara López-Martínez, and Irene Cervelló

Abstract

A new field of investigation which aims to design tissues and organs similar to their native origin has been developed recently, named as regenerative medicine (tissue engineering and bio-engineering). Uterus is the main organ for regeneration and contributes in the fertility. At an ultimate level, the uterus plays a role in embryo implantation, sperm migration and fetal nutrition. Uterine congenital anomalies, attained uterine lesions and immune system disorders may affect such uterine functions preventing successful pregnancy. Due to following reasons, it is essential to consider regenerative medicine as a new approach for the treatment of uterine dysfunctions to overcome the failures that cannot be treated with clinical medication.

Keywords

Decellularization · Uterus · Tissue engineering · Infertility · Reproduction

H. Campo · S. López-Martínez · I. Cervelló (✉)
Fundación Instituto Valenciano de Infertilidad (FIVI), Instituto Universitario IVI/INCLIVA, Valencia, Spain
e-mail: Irene.Cervello@ivirma.com

12.1 History

The uterus is the largest organ of the female reproductive system, consisting of the fundus (superior to the fallopian tubes), corpus and cervix. This muscular, inverted pear-shaped organ evacuates the unfertilized egg during menstruation or nourishes the growing embryo until delivery. There is a large variety of uterus forms in mammals; the human uterus is fused in one organ (simplex), while other forms include duplex (rabbit, mouse, rat) and bicornuate (dog and pig) anatomy. The uterine wall is comprised of three histological layers, namely the outer perimetrium, middle myometrium and inner endometrium. The thickest layer is the myometrium, having three layers of smooth muscle cells. These cells are circularly or spirally organized allowing for elongation and distention of the uterus during the pregnancy. The most luminal and complex layer is the endometrium and is responsible for embryo implantation and development. This mucosal layer is highly responsive to the female sex hormones and is composed mainly by epithelial (luminal and glandular) and stromal compartments (Simón et al. 2009; Speroff and Fritz 2005).

Important advances have been demonstrated in different organs like liver (Baptista et al. 2011), kidney (Arenas-Herrera et al. 2013), pancreas (Berney and Berishvili 2015) and heart (Kim et al. 2016; Sánchez et al. 2015).

A. Kajbafzadeh (ed.), *Decellularization Methods of Tissue and Whole Organ in Tissue Engineering*, Advances in Experimental Medicine and Biology 1345,
https://doi.org/10.1007/978-3-030-82735-9_12

Concerning the uterus, it is classified as a non-vital organ such as the spleen, appendix, testicles, ovaries and eyes. But, transplantation involving these non-vital organs has been previously performed in humans to improve the quality of life (Kisu et al. 2013). Consequently, the uterus is not outside of the (r)evolution of traditional medicine showing huge progresses in the field of regenerative medicine investigating the improvement of quality of life ,as we are going to explain along this chapter.

12.2 Materials and Methods

Decellularized tissues have been proposed as the ideal biological scaffold for the development and homeostasis of cells, tissues and organs (Badylak 2007; Crapo et al. 2011). In order to obtain these acellular three-dimensional scaffolds, various decellularization (DC) techniques and protocols have been developed. It is necessary to adapt these to the specific organ/tissue characteristics and final objective (in vitro models, regeneration of tissues, whole organs replacements). It has been demonstrated that one DC technique does not always translate well between different organs or even the same organs of different species. Hence, optimization and comparison of protocols are necessary (Hellström et al. 2014; Santoso et al. 2014).

Tissues can be decellularized using (a combination of) the three main types of DC methods: These are physical, enzymatic and chemical methods, and extensive reviews have been published describing their respective advantages and disadvantages (Badylak et al. 2011). Physical decellularization techniques such as freeze–thawing and high hydrostatic pressure (HHP) are usually ineffective on its own. For this, they are commonly combined with the other two methods and can be applied via two ways, namely immersion and perfusion; a summary of relevant publications is given in Table 12.1. Furthermore, the methods used to validate the decellularization of uterine tissues are given in Fig. 12.1.

The most common approach, which is mainly used for small fragments and thin tissues, is to immerse and agitate them in the decellularization reagents. This has been proven effective for virtually all tissues, including all uterine histological layers. Young et al. were among the first to use these methodologies to decellularize fragments of the human and rat myometrium (Young and Goloman 2013). Strips (2 × 2 x 10 mm) of the pregnant human myometrium were obtained after Cesarean delivery, while rectangular sects. (1.5 × 2.0 cm) were obtained from timed pregnant rats (days 20–21). Cells were destroyed by agitating the fragments in 70% ethanol at room temperature for 24 h, followed by a 1-h wash in distilled water. They were also the only protocol using trypsin (1X in 0.25% EDTA, 3 or 24 h) in order to remove cellular debris. Complete decellularization was only demonstrated using a Masson's trichrome stain.

The first study comparing various DC techniques in the whole uterus was performed by Santoso et al. Rectangular cuts (15 mm × 65 mm) of the complete uterine wall (including peri-, myo- and endometrium) were subjected to various DC protocols, using the ionic detergent sodium dodecyl sulfate (SDS), the non-ionic detergent Triton X-100 and HHP (Santoso et al. 2014). A combination of concentrations and incubation times in solely SDS or Triton X-100 was investigated, and high hydrostatic pressure (980 MPa) was applied at different temperatures each being subjected to a different pressure gradient. Afterward, all tissues were washed extensively for one week at 4 °C in a DNase I solution containing antibiotics. After histological analysis (hematoxylin and eosin (HE), Masson's trichrome (MT) and Verhoeff's Van Gieson (VVG)), the best decellularization methods were selected for further analysis. These were 1% SDS for 1 h and HHP at 30 °C, and all Triton X-100 conditions were excluded, seeing they were unable to fully decellularize the tissues. Both protocols were able to destroy the native DNA. In contrast to HHP, a negative effect of SDS on the collagen fibers was observed but without significantly affecting ultrastructure. Furthermore, the HPP-treated samples showed increased mechanical strength. However, in a further article from the same group, the SDS-based protocol

Table 12.1 Summary of all relevant publications in the de- and recellularization of the uterus

Decellularization				In vitro tissue-engineered construct			In vivo experiments		Results (in vitro and/or in vivo)	References
Tissue	*Species*	*Method*	*Chemicals*	*Cells used*	*Time*	*Culture conditions*	*Implanted species*	*Time*		
Myometrial segments	Rat and human	Immersion	24 h ethanol (70%, RT), 1 h H_2O, 3 or 24 h Trypsin (0.25% in 1X EDTA)	Human and rat myocytes (4–5 × 10^6 cells/mL)	34 days	Non-static culture in DMEM with 10% FBS, Penicillin 10,000 U/mL, Streptomycin 10,000 mg/mL	/	/	Allo-neo myometrium: poor surface over growth, no bundle formation. Xeno-neo-myometrium: good cellularity, cell viability and showed coordinated contractions	Young and Goloman (2013)
Full thickness uterine segments (15 mm x 5 mm)	Rat	Immersion	1 h SDS (0.1% in PBS), 1 h SDS (1% in PBS), 2 h SDS (1% in PBS), 24 h Triton X- 100 (1% in PBS), 24 h Triton X-100 (3% in PBS), 48 h Triton X- 100 (3% in PBS). HHP: 10 min at 980 MPa (10 °C or 30 °C). All were washed for 1 week at 4 °C in 0.9% NaCl, 0.05 M $MgCl_2 \cdot 6H_2O$, 0.2 mg/ml DNase I and 1% P/S	/	/	/	Sprague Dawley rats	1 month	HPP and SDS result in successful DC with the former having a less detrimental effect on collagen fibers. Demonstration of a functional uteri after partial reconstruction in Sprague Dawley rats	Santoso et al. (2014)
Whole uterus	Rat	Perfusion	24 h SDS (0.01%, 4 °C), 24 h SDS (0.1%, 4 °C), 24 h SDS (1%, RT), 15 min H_2Od, 30 min Triton X-100 (1%), washed extensively with sterile PBS	Rat neonatal, adult uterine and mesenchymal stem cells (7,9 × 10^7)	3 days	Dynamic culture using smooth muscle cell basal medium 2 with 0.5 ng/mL EGF, 5% FBS, 2 ng/mL BFGF and 5 mg/mL insulin	Rat	28 or 90 days	SDS-based protocol maintained 3D architecture of the rat uterus. Tissue-engineered constructs were created in vitro which yielded structurally and functionally competent uterine tissues in vivo	Miyazaki and Maruyama (2014)

(continued)

Table 12.1 (continued)

Decellularization				In vitro tissue-engineered construct			In vivo experiments		Results (in vitro and/or in vivo)	References
Tissue	*Species*	*Method*	*Chemicals*	*Cells used*	*Time*	*Culture conditions*	*Implanted species*	*Time*		
Whole uterus	Rat	Perfusion	Protocol 1 (P1, 5 cycles): 4 h DMSO (4% in PBS + A), 4 h Triton X-100 (1% in PBS + A), 30 min PBS + A (+ stored O/N). Protocol 2 (P2): identical to P1 but dilutions made in dH_2O + sodium azide (0.05%; H_2Od + A), freeze–thaw cycle between cycles 2 and 3. Protocol 3 (P3, 5 cycles): 6 h SDC (2% in H_2Od + A), 2 h H_2Od + A, stored O/N in H_2Od + A at RT	/	/	/	/	/	Biochemical and mechanical comparison of three different DC protocols. P1 yielded a compact, weaker scaffold with P3 resulting in a porous scaffold resembling the native uterus	Hellström et al. (2014)
Whole uterus	Rat	Perfusion	P1, P2 and P3 as described above	Rat uterine primary cells and rat GFP-MSCs	3 days	Standard culture conditions + fed daily with EM medium (DMEM + 10 mM HEPES + 1X Anti-Anti + 10% fetal calf serum)	Rat	6 to 8–9 weeks	Superior repopulation of injected cells in P1 and P2 scaffolds, no placentation over any of the grafts but gave and gave proper support during pregnancy	Hellström et al. (2016)
Full thickness uterine segments	Mice	Immersion	1 h SDS (1% in PBS) + washed for 1 week at 4 °C in 0.9% NaCl, 0.05 M $MgCl_2 \cdot 6H_2O$, 0.2 mg/ml DNase I and 1% P/S	/	/	/	Mouse	Up to 28 days	Uterine epithelium rapidly initiates regeneration over transplanted DUM, independently of female sex steroids. An important role for STAT3 is demonstrated	Hiraoka et al. (2016)

(continued)

Table 12.1 (continued)

Decellularization				In vitro tissue-engineered construct			In vivo experiments		Results (in vitro and/or in vivo)	References
Tissue	*Species*	*Method*	*Chemicals*	*Cells used*	*Time*	*Culture conditions*	*Implanted species*	*Time*		
Endometrial segments	Human	Immersion	48 h 0.25% Triton X-100 + 0.25%SDC (37 °C), 4 days DMEM/F12 (4 °C), 24 h 100 µg/ml ribonuclease and 150 IU/ml DNase I (37 °C), 24 h DMEM/F12 (4 °C)	Primary culture human stromal and epithelial cells	8 weeks before integration in microfluidic culture (28 days)	Microfluidic co-culture (EVATAR) in standard growth medium (DMEM/F12, 10% FBS and 1% P/S)	/	/	28-day culture of endometrial construct connected for murine ovary human fallopian tube, ectocervix and liver culture resulted in tissue specific cells with active proliferation	Xiao et al. (2017)
Endometrial segments	Human	Immersion	48 h 0.25% Triton X-100 + 0.25%SDC (37 °C), 4 days DMEM/F12 (4 °C), 24 h 100 µg/ml ribonuclease and 150 IU/ml DNase I (37 °C), 24 h DMEM/F12 (4 °C)	Primary culture human stromal and epithelial cells	28 days	Static culture in DMEM/F12, 10% FBS and 1% P/S with stepwise hormone protocol	/	/	Endometrial three-dimensional co-culture favored stromal cells, producing decidualization markers during the 28-day hormone treatment	Olalekan et al. (2017)
Whole uterus	Pig	Perfusion	2 cycles: 1 h PBS, 18 h SDS (0,1%), 30 min H_2Od, 30 min Triton X-100 (1%), 5 h PBS	Human side population endometrial stem cell lines (ICE 6&7)	12 days	Static culture in DMEM, 10% FBS, 2 mM L-glutamine, 1/1000 Streptomycin/Penicillin/amphotericin	/	/	Mild SDS and Triton X-100-based perfusion protocol yields vascularized scaffold, endometrial disks are compatible with endometrial stem cells, preserving cell function and phenotype	Campo et al. (2017)
Fragmented uterine horn	Bovine	Immersion	2–4 weeks SDS (0,1%, 4 °C)	Human bone marrow-derived mesenchymal stem cells	28 days	static culture in 1 × low glucose DMEM + 10% FBS, HEPES buffer, l-glutamine, 10 units antibiotic and antimyotic	/	/	Large surface tissue papers were created that were structurally stable with adhesion, proliferation and infiltration of MSCs shown	Jakus et al. (2017)

Abbreviations *BFGF* basic fibroblast growth factor; *EGF* epidermal growth factor; *SDS* sodium dodecyl sulfate; *O/N* overnight; *RT* room temperature; *HPP*: high hydrostatic pressure; *DUM* decellularized uterine matrix; *P/S* penicillin and streptomycin; *FBS* fetal bovine serum; *DMSO* dimethyl sulfoxide; H_2Od distilled water

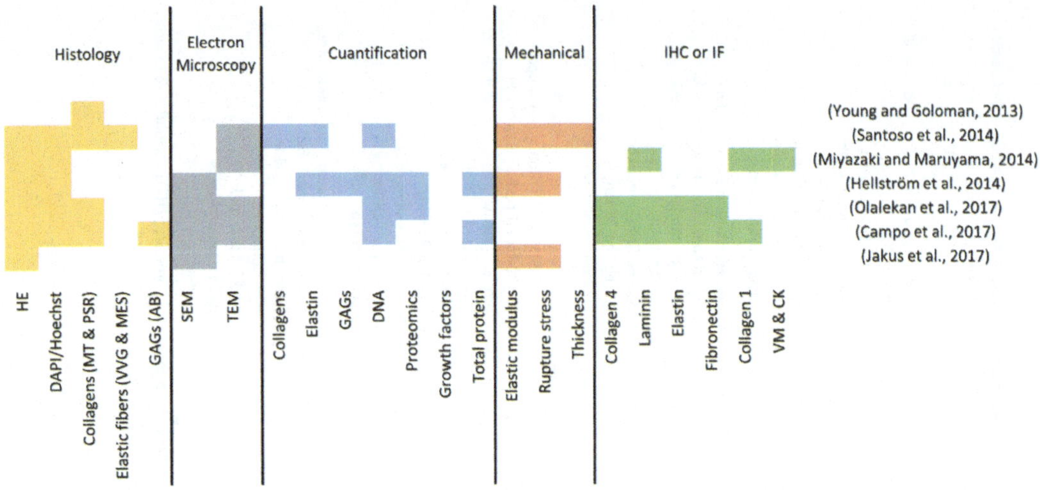

Fig. 12.1 Different methods used to validate the decellularization of the uterus and uterine tissues. Abbreviations: HE: hematoxylin and eosin; DAPI: 4′, 6-diamidino-2-phenylindole; MT: Masson's trichrome; VVG: Verhoeff's Van Gieson; AB: Alcian blue; SEM: scanning electron microscopy; TEM: transmission electron microscopy; GAGs: glycosaminoglycans; PSR: picrosirius red; MES: Miller's elastic stain; IHC: immunohistochemistry; IF: immunofluorescence; VM: vimentin; CK: cytokeratin

was used, suggesting that the convenience using the more commonly used SDS outweighs the possible benefits of HPP (Hiraoka et al. 2016).

Solely, endometrial tissue was also decellularized; Olalekan et al. developed a DC protocol to be used in 3D models of the endometrium (Olalekan et al. 2017; Xiao et al. 2017). Here, 0.5 mm thin sections were agitated in 0.25% of Triton X-100 and 0.25% sodium deoxycholate (SDC) at 37 °C for 48 h. After washing for 72 h with Dubelcco's modified Eagle medium (DMEM) at 4 °C, nuclear material was removed using 100 μg/ml ribonuclease and 150 IU/ml DNase I. Histology (HE and MT) showed an acellular scaffold, which was further corroborated by DNA quantification. Electron microscopy showed a fibrous collagen-rich scaffold, while a precise proteomic analysis showed a scaffold rich in proteins related to cell adhesion, cell matrix and cytoskeletal proteins.

Even though immersion and agitation are convenient methods of DC, samples are limited by the penetration of the DC agents in the tissue. This means that the possible samples are restricted by a maximum thickness. For this, perfusion-based DC techniques provide a solution for thick samples. Here, the vasculature of whole organs is used to deliver DC reagents, remove the cellular material and provide cells and nutrients for recellularization (RC) purposes (Ott et al. 2008). The first fully vascularized three-dimensional uterine scaffold in rats was published by Miyazaki et al.; via the aorta, rising concentrations of SDS were perfused at 50 ml/h through the organ to remove cellular material (Miyazaki and Maruyama 2014). Washing steps to remove SDS included the use of distilled water and Triton X-100 (1%). Lastly, the organs were extensively washed in PBS and stored at 4 °C for up to a week in PBS with antibiotics and antimycotics. Decellularization was demonstrated by HE, nuclear staining and immunofluorescence for α-smooth muscle actin, vimentin, cytokeratin, laminin and collagen I. Electron microscopy showed an intact ultrastructure and vasculature. In the same year, an extensive comparative study of three whole rat uterus decellularization protocols was published by Hellström et al. Perfusion was set at 8 ml/min (Hellström et al. 2014). The first DC protocol (P1) used five cycles consisting of 4-h perfusion with 4% DMSO, 4 h with 1% Triton X-100 (both diluted in PBS+0.05% sodium azide and PBS +A) and 30 min with PBS+A. Between the cycles, the organ was left unperfused at RT in PBS+A. The second protocol (P2) was very similar, the dilutions were made in distilled water, and a freeze–thaw procedure (−80 °C) was introduced between cycles 2 and 3. Lastly, P3 (protocol 3) used five cycles of 6-h perfusion with 2% SDC (in dH_2O+A) and 2-h perfusion with dH_2O+A. A quantitative analysis of the ECM and its biochemical composition following vascular decellularization was done. Differences were observed in various levels. For example, using unbuffered perfusion medium (P2) removed significantly more DNA than the other protocols. P3 on the other hand preserved more of the important structural ECM proteins such as elastin and provided the most mechanically strong scaffold.

Lastly, in order to establish a DC protocol for large reproductive organs, our laboratory created a robust protocol for the pig uterus, lasting only 49 h in total (Campo et al. 2017). Furthermore, the effect of a freeze–thaw cycle before decellularization was assessed. To procure the large vascularized scaffold, two identical perfusion cycles at a physiological rate of 15 ml/min were employed: After an initial flush of PBS for one hour, 0.1% SDS was perfused for 18 h, followed by wash steps for 30 min with distilled water, half an hour of 1% Triton X-100 and 5 h of PBS. Decellularization was demonstrated by histology (HE, DAPI, MT) and quantification of DNA and proteins. The collagen-rich scaffold was further analyzed with MT and Alcian blue staining and immunofluorescence of collagens I and IV, elastin, laminin and fibronectin. The ultrastructure showed a fibrous stroma with intact collagen fibers and an intact endometrial surface and vascular structures; the latter was further corroborated by a vascular corrosion cast, where capillary structures were present in all DC conditions. This protocol was slightly adapted in a further study in a different animal model, namely the rabbit model. Here, non-synchronous (NS, with a non-proliferating endometrium) and synchronous rabbit uteri (S, 72 h after ovarian stimulation and proliferating endometrium) were perfused at a flow rate of 8 ml/min (Campo et al. 2019). The same SDS and Triton X-100-based cycles were used, and finally, a perfusion with 2 μg/mL DNase 1 solution was done to improve the removal of residual DNA. After testing DC efficiency, the acellular endometrium was separated, via microdissection, lyophilized and milled. Tissue- and proliferation-specific NS and S hydrogels were made from these powders, yielding interesting biomaterials for possibly treating endometrial pathologies that can also serve as a future platform for three-dimensional cell culture (Campo et al. 2019).

These publications show that it is possible to decellularize uterine tissues, where different protocols successfully remove nearly all cellular material in tissues and whole organs while retaining the hierarchical structures of the complex extracellular matrix.

12.3 Cell Seeding

Ideally, these acellular three-dimensional bioscaffolds will be recellularized with several types of cells in order to create tissue-engineered constructs that are re-implantable in future recipients (Peloso et al. 2015) or to mimic organ-like microenvironment. To recellularize these extracellular matrix (ECM) scaffolds, several techniques can be used; they can be simply covered with cells and allowed to adhere, proliferate and penetrate, they can be introduced directly into the scaffold by injection, or it could be delivered via perfusion (only for whole organs). For reproductive tissue engineering (REPROTEN) purposes (Amorim 2017), only the covering and injection with cells techniques have been published.

The original goal of tissue engineering as described by Langer and Vacanti was to "restore, maintain or improve function of tissues or whole organs" (Langer and Vacanti 1993). However, it can also play a transformative role for in vitro applications such as drug testing, disease modeling and precision medicine (Wobma and Vunjak-Novakovic 2016). In the pursuit of restoring uterine defects before pregnancy, Young et al. cultured human and rat myocytes in vitro on (autologous and non-autologous) myometrial decellularized scaffolds. Cells were allowed to adhere first by covering scaffold "hammocks" with media containing 4–5 × 10^6 cells/mL and kept still for one hour, followed by culture in agitation. This way, allo-neo-myometrium (same species) and xeno-neo-myometrium (human and rat components) were made and were able to remodel over artificially made defects. The rat decellularized myometrium was the preferred environment for both cell types, showing improved adhesion and penetration of even human cells. This xeno-neo-myometrium formed typical myometrial-bundles structures that showed coordinated contractions. In an effort to closely mimic the in vivo conditions of reproductive tissues, a complex microfluidic systems (EVATAR) integrating female reproductive tract tissues (murine ovary, human fallopian tube, endometrium, ectocervix) and liver (Xiao et al. 2017) was created. In this system, endometrial constructs were made by covering the scaffold with approximately one million expanded endometrial cells in 50 μl of culture media. In the EVATAR setup, it was influenced by the female sex steroids produced by the ovarian module, and the endometrial cells expressed typical specific markers such as vimentin, cytokeratin, Ki67, estrogen receptors and progesterone receptors. In a different publication, this construct was subjected to a stepwise hormone protocol (Olalekan et al. 2017). Despite using endometrial tissue from different developmental stages, it was stated that more investigation was needed to assess the possible effects of the innate differences of the developing endometrium. Furthermore, the culture conditions clearly favored stromal cells. Despite these constraints, the construct produced and expressed typical decidualization markers such as prolactin and insulin-like growth factor-binding protein 1 (IGFBP-1).

To create a true three-dimensional environment, organoid-like structures were also created using human endometrial stem cell lines (Cervelló et al. 2011) and acellular endometrial disks, obtained from whole decellularized pig uteri (Campo et al. 2017). Disks measuring 100 microns thick and 5 mm wide were covered with a physiological ratio of stromal, and epithelial side population (SP) stem cells from human endometrial origin (4/5 ICE6 and 1/5 ICE7, 0.5 million in total) were cultured in hypoxic conditions for up to 12 days. After 4 days, these covered disks began to roll up, forming an organoid-like structure, containing cells that are closely interacting with the ECM and expressing typical endometrial markers (cytokeratin and vimentin).

To create ECM "inks," lyophilized powder of decellularized uterus (65 vol%) was mixed with PLGA polymer (35 vol%), and a unique surface topology was observed. These hybrid scaffolds ("tissue papers") were stable, and adhesion, proliferation and infiltration of mesenchymal stem cells (MSCs) were demonstrated (Jakus et al. 2017).

The injection of cells into the bioscaffolds is usually done to create a transplantable tissue-engineered construct. In a recent publication, Hellström et al. removed patches (20 × 5 mm) from the whole decellularized organ, 7.3 × 10^6 cells (0.66% isolated rat endometrial cells and 99.33% GFP-MSCs) per patch were injected and were incubated statically for 3 days. A typical gene-expression profile for GFP-MSCs in vitro before transplantation was observed (Hellström et al. 2016). Miyazaki et al. used a similar approach but with some key differences. Here, a mixture of neonatal uterine cells, adult uterine cells and rat mesenchymal stem cells was injected in the whole decellularized uterus matrix (DUM). Dynamic culture conditions were established via perfusion of oxygenated culture medium via the aortic tubing for 3 days. Recellularization was demonstrated by HE staining and immunofluorescent staining of vimentin, cytokeratin, α-smooth muscle actin and CD31 (Miyazaki and Maruyama 2014).

While much progress has been made in the recellularization of small sections mainly for in vitro models and also for pilot transplantation experiments, the recellularization of whole organs represents the next frontier/main hurdle in tissue engineering. Such as with other organ systems, there is still much investigation to be done to increase the RC efficiency and vascularization to one day regenerate whole transplantable organs (Pellegata et al. 2018).

12.4 Clinical Applications

Many uterine pathologies exist that result in reproductive dysfunction, while most of these can be treated adequately, some are still poorly treatable. One main example is congenital or acquired absolute uterine factor infertility (AUFI), affecting 1 in every 500 women (Milliez 2009). Congenital AUFI includes pathologies such as Mayer–Rokitansky–Kuster–Hauser (MRKH) syndrome, uterine hypoplasia and uterine malformation. MRKH syndrome or Müllerian agenesis is characterized by the absence of the uterus and upper two-third of the vagina (Bombard and Mousa 2014). Acquired AUFI on the other hand is the result of hysterectomy due to malignant uterine tumor, benign diseases (e.g., leiomyoma and adenomyosis), postpartum hemorrhage and loss of fertility due to intrauterine adhesions (Asherman's syndrome).

The only strategy available until recently was gestational surrogacy, but is far from ideal because of economic, legal and ethical reasons and is not practiced in many countries (Shenfield et al. 2005). The only real cure to restore fertility can be found in allogeneic uterus transplantation, and a proof of concept for this was published in 2015 by the Brännström group, with the first successful live birth after uterus transplantation (Brännström et al. 2015). The long-term effect of transplantations, tissue rejection and immunosuppressant-related side effects remains an issue (Sayegh and Carpenter 2004). Here, reproductive tissue engineering could play a transformative role by creating transplantable whole organs, effectively circumventing these issues.

Miyazaki and Maruyama were the first to test the in vivo viability of these biomaterials. Here, the DUM (without recellularization) was used to assess the potential of the scaffold to regenerate whole uterine wall injuries in rat models (Miyazaki and Maruyama 2014). Compared to the "excision" group, where the damages were left unrepaired, a well recellularized and thicker graft site was observed after 28 and 90 days. Furthermore, the regenerated horns supported the pregnancy, but there was no mention of embryo implantation or placentation on the graft sites. Interesting results were also shown concerning the epithelial regeneration, where after 90 days an organized epithelium with glands was observed. Even more, an intact epithelial layer was found after 28 days, suggesting the improved capacity of the endometrium to re-epithelialize, much like what happens after menstrual shedding. After shedding, torn surfaces with gland openings without epithelial layer are left, and during this phase, the re-epithelialization occurs (Maybin and Critchley 2015). Hiraoka et al. investigated this further, demonstrating that re-epithelialization takes place within a week and showing the important role of STAT3 in this regeneration process. Furthermore, using their DUM developed in previous investigation, they demonstrated fetal development over patched areas (Hiraoka et al. 2016). Published in the same year, Hellström et al. recellularized in vitro DUMs originating from three different protocols with primary uterine cells and green fluorescent protein-labeled bone marrow-derived mesenchymal stem cells (Hellström et al. 2016). Here, successful implantation at the transplanted DUM construct site was also demonstrated.

While these studies demonstrate that the expertise in this field is advancing rapidly, much investigation is still needed, from the tissue engineering techniques such as culturing uterine cells and creating novel 3D scaffolds to the chirurgical know-how to transplant these engineered patches and whole organs.

In conclusion, decellularized uterine constructs provide a promising avenue in creating recellularized and transplantable constructs. These investigations add to the regenerative approaches using other biomaterial such as collagen scaffolds loaded with growth factors (Li et al. 2011) or cells (Ding et al. 2014; Song et al. 2015), collagen hydrogels (Xu et al. 2017), myofibroblast-rich tissue (Campbell et al. 2008) or cell sheet engineering (Takagi et al. 2014).

12.5 Limitations

Although the decellularization methodologies of the uterus and uterine tissues are advanced and show promise, nowadays the focus is on improving the main limiting factor in generating transplantable tissue-engineered uterine constructs, namely the recellularization efficiency. Despite many efforts to recreate the promising results in other organ systems (Ott et al. 2008) and the many encouraging publications in both small and large animal models (Campo et al. 2017; Hellström et al. 2016; Hiraoka et al. 2016; Miyazaki and Maruyama 2014), this has not been achieved yet for the uterus.

Lastly, there is one other limitation that needs to be taken into consideration, which are the morphological differences of the uterus in mammals. Where different studies can use xenogeneic donors such as the pig this is not the case here, only the human uterus is large enough and of the simplex form. In other words, though DC protocols are established for mouse, rat, pig and rabbit uteri, it is paramount that the DC of the human uterus will be developed in order to one day regenerate and transplant the entire organ to humans.

12.6 Conclusion

This chapter describes the fundamental concepts of tissue and whole-organ engineering related to the uterus, including methods for decellularization, characterization of the extracellular matrix as a scaffold, types of cell seeding, techniques for the recellularization process and clinical applications.

Nowadays, the decellularization techniques and protocols can provide an acellular, natural and three-dimensional biologic scaffold material that can be used with or without several cell populations to create re-implantable tissue-engineered constructs or to mimic organ-like microenvironment. However, there are still some cons to consider like the standardization of scaffold production, the different materials or hydrogels, how to assess scaffold function and the potential beneficial impact in the damaged organ or tissue.

Preliminary studies with animal models have provided encouraging results as a proof of concept in the reproductive field (as described along this chapter). But, some aspects of these new technologies should also be clarified to facilitate the translation of tissue engineering in the laboratory to the clinic. Further investigation is required to establish the contribution of these scaffolds to organ function, and it should be determined which diseases are more likely to be successfully treated with success.

References

Amorim CA (2017) Special issue devoted to a new field of regenerative medicine: reproductive tissue engineering. Ann Biomed Eng 45:1589–1591. https://doi.org/10.1007/s10439-017-1862-0

Arenas-Herrera JE, Ko IK, Atala A, Yoo JJ (2013) Decellularization for whole organ bioengineering. Biomed Mater Bristol Engl 8:014106. https://doi.org/10.1088/1748-6041/8/1/014106

Badylak SF (2007) The extracellular matrix as a biologic scaffold material. Biomaterials 28:3587–3593. https://doi.org/10.1016/j.biomaterials.2007.04.043

Badylak SF, Taylor D, Uygun K (2011) Whole-organ tissue engineering: decellularization and recellularization of three-dimensional matrix scaffolds. Annu Rev Biomed Eng 13:27–53. https://doi.org/10.1146/annurev-bioeng-071910-124743

Baptista PM, Siddiqui MM, Lozier G, Rodriguez SR, Atala A, Soker S (2011) The use of whole organ decellularization for the generation of a vascularized liver organoid. Hepatol Baltim Md 53:604–617. https://doi.org/10.1002/hep.24067

Berney T, Berishvili E (2015) Toward clinical application of the bioartificial pancreas. Transplantation 99:2241–2242. https://doi.org/10.1097/TP.0000000000000865

Bombard DS, Mousa SA (2014) Mayer–Rokitansky–Kuster–Hauser syndrome: complications, diagnosis and possible treatment options: a review. Gynecol Endocrinol 30:618–623

Brännström M, Johannesson L, Bokström H, Kvarnström N, Mölne J, Dahm-Kähler P, Enskog A, Milenkovic M, Ekberg J, Diaz-Garcia C (2015) Livebirth after uterus transplantation. The Lancet 385:607–616

Campbell GR, Turnbull G, Xiang L, Haines M, Armstrong S, Rolfe BE, Campbell JH (2008) The peritoneal cavity as a bioreactor for tissue engineering visceral organs: bladder, uterus and vas deferens. J Tissue Eng Regen Med 2:50–60

Campo H, Baptista PM, López-Pérez N, Faus A, Cervelló I, Simón C (2017) De- and recellularization of the pig uterus: a bioengineering pilot study. Biol Reprod 96:34–45. https://doi.org/10.1095/biolreprod.116.143396

Campo H, García-Domínguez X, López-Martínez S, Faus A, Vicente Antón JS, Marco-Jiménez F, Cervelló I (2019) Tissue-specific decellularized endometrial substratum mimicking different physiological conditions influences in vitro embryo development in a rabbit model. Acta Biomater. https://doi.org/10.1016/j.actbio.2019.03.004

Cervelló I, Mas A, Gil-Sanchis C, Peris L, Faus A, Saunders PTK, Critchley HOD, Simón C (2011) Reconstruction of endometrium from human endometrial side population cell lines. PLoS One 6. https://doi.org/10.1371/journal.pone.0021221

Crapo PM, Gilbert TW, Badylak SF (2011) An overview of tissue and whole organ decellularization processes. Biomaterials 32:3233–3243. https://doi.org/10.1016/j.biomaterials.2011.01.057

Ding L, Li X, Sun H, Su J, Lin N, Péault B, Song T, Yang J, Dai J, Hu Y (2014) Transplantation of bone marrow mesenchymal stem cells on collagen scaffolds for the functional regeneration of injured rat uterus. Biomaterials 35:4888–4900. https://doi.org/10.1016/j.biomaterials.2014.02.046

Hellström M, El-Akouri RR, Sihlbom C, Olsson BM, Lengqvist J, Bäckdahl H, Johansson BR, Olausson M, Sumitran-Holgersson S, Brännström M (2014) Towards the development of a bioengineered uterus: comparison of different protocols for rat uterus decellularization. Acta Biomater 10:5034–5042. https://doi.org/10.1016/j.actbio.2014.08.018

Hellström M, Moreno-Moya JM, Bandstein S, Bom E, Akouri RR, Miyazaki K, Maruyama T, Brännström M (2016) Bioengineered uterine tissue supports pregnancy in a rat model. Fertil Steril 106:487-496.e1. https://doi.org/10.1016/j.fertnstert.2016.03.048

Hiraoka T, Hirota Y, Saito-Fujita T, Matsuo M, Egashira M, Matsumoto L, Haraguchi H, Dey SK, Furukawa KS, Fujii T, Osuga Y (2016) STAT3 accelerates uterine epithelial regeneration in a mouse model of decellularized uterine matrix transplantation. JCI Insight 1. https://doi.org/10.1172/jci.insight.87591

Jakus AE, Laronda MM, Rashedi AS, Robinson CM, Lee C, Jordan SW, Orwig KE, Woodruff TK, Shah RN (2017) "Tissue Papers" from organ-specific decellularized extracellular matrices. Adv Funct Mater 27. https://doi.org/10.1002/adfm.201700992

Kim JJ, Hou L, Huang NF (2016) Vascularization of three-dimensional engineered tissues for regenerative medicine applications. Acta Biomater 41:17–26. https://doi.org/10.1016/j.actbio.2016.06.001

Kisu I, Banno K, Mihara M, Suganuma N, Aoki D (2013) Current status of uterus transplantation in primates and issues for clinical application. Fertil Steril 100:280–294. https://doi.org/10.1016/j.fertnstert.2013.03.004

Langer R, Vacanti JP (1993) Tissue engineering. Science 260:920–926

Li X, Sun H, Lin N, Hou X, Wang J, Zhou B, Xu P, Xiao Z, Chen B, Dai J, Hu Y (2011) Regeneration of uterine horns in rats by collagen scaffolds loaded with collagen-binding human basic fibroblast growth factor. Biomaterials 32:8172–8181. https://doi.org/10.1016/j.biomaterials.2011.07.050

Maybin JA, Critchley HO (2015) Menstrual physiology: implications for endometrial pathology and beyond. Hum Reprod Update 21:748–761

Milliez J (2009) Uterine transplantation: FIGO committee for the ethical aspects of human reproduction and women's health. Int J Gynecol Obstet 106:270–270

Miyazaki K, Maruyama T (2014) Partial regeneration and reconstruction of the rat uterus through recellularization of a decellularized uterine matrix. Biomaterials 35:8791–8800. https://doi.org/10.1016/j.biomaterials.2014.06.052

Olalekan SA, Burdette JE, Getsios S, Woodruff TK, Kim JJ (2017) Development of a novel human recellularized endometrium that responds to a 28-day hormone treatment. Biol Reprod 96:971–981. https://doi.org/10.1093/biolre/iox039

Ott HC, Matthiesen TS, Goh S-K, Black LD, Kren SM, Netoff TI, Taylor DA (2008) Perfusion-decellularized matrix: using nature's platform to engineer a bioartificial heart. Nat Med 14:213–221. https://doi.org/10.1038/nm1684

Pellegata AF, Tedeschi AM, De Coppi P (2018) Whole organ tissue vascularization: engineering the tree to develop the fruits. Front Bioeng Biotechnol 6. https://doi.org/10.3389/fbioe.2018.00056

Peloso A, Dhal A, Zambon JP, Li P, Orlando G, Atala A, Soker S (2015) Current achievements and future perspectives in whole-organ bioengineering. Stem Cell Res Ther 6.https://doi.org/10.1186/s13287-015-0089-y

Sánchez PL, Fernández-Santos ME, Costanza S, Climent AM, Moscoso I, Gonzalez-Nicolas MA, Sanz-Ruiz R, Rodríguez H, Kren SM, Garrido G, Escalante JL, Bermejo J, Elizaga J, Menarguez J, Yotti R, Pérez del Villar C, Espinosa MA, Guillem MS, Willerson JT, Bernad A, Matesanz R, Taylor DA, Fernández-Avilés F (2015) Acellular human heart matrix: a critical step toward whole heart grafts. Biomaterials 61:279–289. https://doi.org/10.1016/j.biomaterials.2015.04.056

Santoso EG, Yoshida K, Hirota Y, Aizawa M, Yoshino O, Kishida A, Osuga Y, Saito S, Ushida T, Furukawa KS (2014) Application of detergents or high hydrostatic pressure as decellularization processes in uterine tissues and their subsequent effects on in vivo uterine regeneration in murine models. PLOS One 9, e103201. https://doi.org/10.1371/journal.pone.0103201

Sayegh MH, Carpenter CB (2004) Transplantation 50 years later—progress, challenges, and promises

Simón C, Horcajadas J, Pellicer A, García-Velasco J (2009) El Endometrio Humano. B. Aires Argent. Editor. Panam

Song T, Zhao X, Sun H, Li X, Lin N, Ding L, Dai J, Hu Y (2015) Regeneration of uterine horns in rats using collagen scaffolds loaded with human embryonic stem cell-derived endometrium-like cells. Tissue Eng Part A 21:353–361. https://doi.org/10.1089/ten.TEA.2014.0052

Speroff L, Fritz MA (2005) Clinical gynecologic endocrinology and infertility. Lippincott Williams & Wilkins

Takagi S, Shimizu T, Kuramoto G, Ishitani K, Matsui H, Yamato M, Okano T (2014) Reconstruction of functional endometrium-like tissue in vitro and in vivo using cell sheet engineering. Biochem Biophys Res Commun 446:335–340

Wobma H, Vunjak-Novakovic G (2016) Tissue engineering and regenerative medicine 2015: a year in review. Tissue Eng Part B Rev 22:101–113. https://doi.org/10.1089/ten.TEB.2015.0535

Xiao S, Coppeta JR, Rogers HB, Isenberg BC, Zhu J, Olalekan SA, McKinnon KE, Dokic D, Rashedi AS, Haisenleder DJ, Malpani SS, Arnold-Murray CA, Chen K, Jiang M, Bai L, Nguyen CT, Zhang J, Laronda MM, Hope TJ, Maniar KP, Pavone ME, Avram MJ, Sefton EC, Getsios S, Burdette JE, Kim JJ, Borenstein JT, Woodruff TK (2017) A microfluidic culture model of the human reproductive tract and 28-day menstrual cycle. Nat Commun 8:14584. https://doi.org/10.1038/ncomms14584

Xu L, Ding L, Wang L, Cao Y, Zhu H, Lu J, Li X, Song T, Hu Y, Dai J (2017) Umbilical cord-derived mesenchymal stem cells on scaffolds facilitate collagen degradation via upregulation of MMP-9 in rat uterine scars. Stem Cell Res Ther 8:84

Young RC, Goloman G (2013) Allo- and Xeno-reassembly of human and rat myometrium from cells and Scaffolds. Tissue Eng Part A 19:2112–2119. https://doi.org/10.1089/ten.tea.2012.0549

13 Decellularization Methods of Vagina and Cervix in Tissue Engineering

Sara López-Martínez, Hannes Campo, and Irene Cervelló

Abstract

The vagina is a fibromuscular elastic tubular tract that connects the cervix with the outer genitals and has an important function discharging uterine secretions, sexual intercourse and acts as the passage for the full-term fetus. Currently, a new field of investigation which aims to design tissues and organs similar to their native origin has been developed recently and was named regenerative medicine (tissue engineering and bioengineering). Malformations in cervix tissue represent a hard challenge for medicine. Experts in bioengineering have tried to reconstruct vaginas or cervix with the aim to achieve cervicovaginal disorders, most of them with congenital cause. However, only few research groups have launched themselves upon the decellularization. The aim of this chapter is investigating the decellularization methods for cervix and vaginal tissues.

Keywords

Vagina · Cervix · Tissue engineering · Decellularization

S. López-Martínez · H. Campo · I. Cervelló (✉)
Fundación Instituto Valenciano de Infertilidad (FIVI), Instituto de Investigación Sanitaria La Fe, Valencia, Spain
e-mail: Irene.Cervello@ivirma.com

13.1 History

The vaginal wall is composed of three layers: a stratified squamous non-keratinized epithelium with an underlying lamina propria of connective tissue, a muscular layer of smooth muscle fibers and a dense connective tissue that blends with the surrounding fascia (Mulhall et al. 2011). A key feature of the vagina is its elasticity, permitting elongation during sexual intercourse and the passing of the baby during birth. These characteristics depend greatly on the vaginal extracellular matrix (ECM), which composition contains various structural proteins such as collagen, elastin and microfibrils (Gartner and Hiatt 2006). Likewise, the cervix is a cylindrical anatomical structure with a central canal, the endocervical canal, through which spermatozoids enter the uterine cavity (Jordan et al. 2006). The wall of the cervix is thick and is composed of a dense fibroelastic connective structure housing smooth muscle and cervical glands (Gartner and Hiatt 2006). The cervix plays an important role in fertility and during pregnancy supporting the fetus (Kuo et al. 2017). Important advances have been demonstrated in different organs (Baptista et al. 2011; Arenas-Herrera et al. 2013; Berney and Berishvili 2015; Sánchez et al. 2015; Kim et al. 2016), and the vagina and the cervix are also participating in this new challenge as we detailed below. The first article in which vagina is decellularized to obtain an acellular vaginal

A. Kajbafzadeh (ed.), *Decellularization Methods of Tissue and Whole Organ in Tissue Engineering*, Advances in Experimental Medicine and Biology 1345,
https://doi.org/10.1007/978-3-030-82735-9_13

matrix (AVM) was published in 2017 by Zhang et al. (2017). In the same way, they followed in other organs the purpose of generating decellularized tissues from cervix and vagina is to obtain cell-free tissue-derived scaffolds with a tissue-specific cocktail of bioactive growth factors and structural proteins. These scaffolds must have a low antigenicity but at the same time must be able to induce chemotaxis and proliferation over endogenous or exogenous cells. Finally, the biocompatibility of these decellularized tissues among mammalian species is essential, allowing xenogeneic donors and consequently a quick and big source of material that is necessary for this kind of interventions.

13.2 Materials and Methods

Up to date, the successful decellularization of the vagina has been reported by two studies, in both cases using the pig as tissue source (Zhang et al. 2017; Greco et al. 2018). A summary of principal aspects of these studies can be found in Table 13.1.

In the first of the reported protocols, the fully decellularization of porcine vagina was achieved in 10 days (Zhang et al. 2017). The selected DC method consisted in the immersion of the tissue is subjected to a broad range of chemical and enzymatic treatments. First, tunica serosa and tunica muscularis were manually removed, and protease activity was inhibited with phenylmethylsulfonyl fluoride (PMSF) before starting decellularization. The main detergents used were 0.1% sodium dodecyl sulfate (SDS) and 1% Triton X-100, both commonly used decellularizing solutions. Then, the tissue was exposed to an enzymatic treatment with deoxyribonuclease I and ribonuclease A to remove the remaining DNA/RNA. Disinfection was done using peroxyacetic acid (PAA) and ethanol. Finally, AVM was freeze-dried, sealed into packages and sterilized one final time using gamma irradiation. After the decellularization process, the vaginal mucosa matrix became white and translucent, as described for other organs. The protocol showed good decellularization, following the guidelines suggested by Crapo et al. with a final amount of remnant DNA <50 ng/mg ECM dry weight, no presence of bands in gel electrophoresis and no presence of nuclei and cells after histological analysis (Crapo et al. 2011). A vital feature of DC matrix compared with other biomaterials is the presence of natural growth factors. In this case, AVM showed the retention of two important growth factors: fibroblast growth factor (FGF), a very effective angiogenic cytokine and an important differentiation inducing factor, and platelet-derived growth factor BB (PDGF-BB), which plays a critical role in cell proliferation, angiogenesis and fibrosis (Zhang et al. 2017).

Another important aspect is the amount of tissue recovered from the DC process. The natural human vagina measures around 6–10 cm and up to 11–12 cm long during sexual arousal (Pendergrass et al. 2003; Barnhart et al. 2006). With their presented approach, the average size of dried AVM was 7.6 cm wide and 11.6 cm long (Zhang et al. 2017), allowing for the use of one unique piece to reconstruct the vagina.

Recently, a milder and shorter approach published by Greco et al. used 0.25% Triton X-100 and 0.25% sodium deoxycholate (Greco et al. 2018). Here, complete decellularization of the porcine vagina was achieved within 5 days, using a modified protocol previously used for tracheae (Lange et al. 2017). As the protocol described above, a combination of detergents, DNase and RNase were used, and sterilization was done using antibiotics and antimycotics. Cells were also ruptured by osmotic forces by using (hypotonic) deionized water. After decellularization of the entire vagina, the mucosal layer is stripped from the muscularis and adventitia layers, sterilized again with peracetic acid and stored. Analysis of protocol efficiency was done by histology (H&E) and DNA quantification (40% of DNA remained compared with fresh tissue), showing a good decellularization. The mild DC allowed for the preservation of basement membrane proteins and the conservation of collagen and elastin, and a conservation of 50% of GAGs was also measured (Greco et al. 2018).

Table 13.1 Summary of all relevant publications in the decellularization and recellularization of the vagina

Decellularization				In vitro tissue-engineered construct			In vivo experiments		Results	References
Tissue	*Species*	*Method*	*Chemicals*	*Cells used*	*Time*	*Culture conditions*	*Implanted species*	*Time*		
Vagina	Pig	Immersion	2 days in 0.1 mM/L PMSF in dH_2O at 4 °C, 12 h 0.01% SDS in hypotonic tris buffer pH8.0 with protease inhibitor at 4 °C. Seven days in Triton X-100 with protease inhibitor. 12 h in 50 U/mL deoxyribonuclease I and 1 U/mL ribonuclease A in 50 mm Tris–HCl, pH 7.5 at 37 °C. 2H in 0.1% PAA and 20% ethanol, a cycle of freeze-dried and sterilization with gamma irradiation (25KGY)	/	/	/	Rat	90 days	Vaginal reconstruction feasible. Faster epithelialization and growth of nerves fibers in AVM over SIS. Organized muscle bundles and ER distribution pattern closer to native vagina. Frequency-dependent contractions after electrical field stimulation similar to native vagina	Zhang et al. (2017)
Vagina	Pig	Immersion	24 h in 0,25% Triton X-100 in PBS and 0.25% sodium deoxycholate in PBS at RT, two washes of 20 min in dH_2O at 4 °C, 24 h enzymatic digestion in 2000 U DNase, 0,1 g RNase in 2.5 mM $MgCl_2$, 0.5 mM $CaCl_2$, 1% antibiotic and antimycotic solution in PBS at RT. About 48-72 h in dH_2O (up to six washes)	Porcine vaginal epithelial cells and porcine AD-MSCs	10 days	Standard culture conditions of 5% CO_2 and 37 °C in KGM-Gold (Lonza) medium	/	/	Biocompatibility of scaffold showed good results in AD-MSCs and vaginal epithelial cells. AD-MSCs achieved differentiation of adipose, chondrocytes and osteoblasts lineage. Vaginal epithelial cells developed toward parabasal cells and intermediate cells	Greco et al. (2018)

AD-MSCs adipose-derived mesenchymal stem cells; *AVM* acellular vagina matrix; *dH_2O* distilled water; *ER* estrogen receptor; *PAA* peroxyacetic acid; *PBS* phosphate buffered saline; *PMFS* phenylmethylsulfonyl fluoride; *SDS* sodium dodecyl sulfate; *SIS* small intestinal submucosa

The difference in time and aggressiveness of both protocols showed here could be explained by an important point of difference of both studies, namely the size of the organs. Zhang et al. used pigs of approximately 100 kg, while Greco et al. used pig weighing around 35–55 kg. So, the time for achieving a successful DC is probably proportional to the size of the organ, pass from a time of 10 days for 100 kg to 5 days for 50 kg. So, the difference in the size of vaginas may be the clue in the time of choosing between a mild or more aggressive protocol.

13.3 Cell Seeding

The first question to answer after establishing successful DC is if the obtained vaginal scaffold is biocompatible and could improve the proliferation and behavior of natural vaginal and other cell components.

Greco et al. demonstrated the cytocompatibility and differentiation potential of porcine AVM using allogeneic porcine vaginal epithelial cells and adipose-derived mesenchymal stem cells (AD-MSCs), both from primary cultures (Greco et al. 2018). In this experiment, a three-dimensional (3D) in vitro platform was made by separately seeding both cellular types with a density of 250,000/cm^2 onto 1.2 cm^2 of AVM that was previously primed with culture medium. Sterile stainless-steel rings were used to make sure that the seeded cells remained on the matrix. Both setups were cultured at standard culture conditions of 5% CO_2 and 37 °C for up to 10 days. AVM allowed the proliferation of both types of cells. Cells grew on the scaffold, and after 3 days of culture, a well-established monolayer was formed and was maintained up to day 10. Furthermore, AVM achieved differentiation in both type of cells, showing markers of adipose, chondrocytes and osteoblasts lineages in AD-MSCs and the development of two different populations, parabasal cells and intermediate cells, in vaginal epithelial cells.

Zhang et al. on the other hand investigated the biocompatibility by implanting the acellular AVM subcutaneously for up to 56 days. By this "in situ" recellularization technique, they showed that porcine AVM induces a low immune reaction when implanted into rats and did not provoke a significant inflammatory response (Zhang et al. 2017). This data is similar to other DC tissues reported (Syazwani et al. 2014; Alberti and Xu 2016; Khorramirouz et al. 2018). When the AVM was used for reconstruction of a partial vaginectomy, a multilayered neovagina was observed, having organized muscle bundles after 90 days. This was not visible with the small intestinal submucosa (SIS) transplant and demonstrates the potential of DC vaginal tissues compared to non-tissue-specific solutions.

Despite these promising findings, the small amount of publications makes further in vitro studies of cell viability and behavior necessary in order to probe the regenerative potential of AVM. These studies should ideally be focused on probing the bioactivity of these porcine AVM in human cells from vaginal primary cultures or even with stem cells. Comparison between AVM and other non-specific ECM matrices could also be interesting to fully confirm and understand the potential of AVM.

13.4 Clinical Applications

The vagina and cervix can be affected by a variety of pathologies, which can both be congenital or acquired. Congenital disorders arise from abnormal development of the Mullerian ducts from where the entire female reproductive tract originates, including the cervix and vagina. The incidence of Mullerian anomalies in the general population is estimated at 2% (Pfeifer 2016). Within Mullerian anomalies, congenital agenesis, aplasia or dysgenesis of the uterine cervix or/and vagina in the presence of a functioning endometrium is a rare Müllerian duct anomaly (Fujimoto et al. 1997). Vagina aplasia, or the nonappearance of a vagina since the birth, is caused by disorders such as Mayer–Rokitansky–Kuster–Hauser syndrome (MRKHS), cloacal malformation and endocrine abnormalities (i.e., adrenal hyperplasia) (Kimberley et al. 2012). About 50% of female patients with vaginal agenesis also carry cervical agenesis (Mhaskar 2005). Moreover, a loss or damage of the cervicovaginal organs can also be produced by acquired disorders such as cancer or trauma (Raya-Rivera et al. 2014).

Cervicovaginal malformations are very difficult to treat and require surgical intervention, and extensive reconstruction is usually necessary to restore the normal anatomy of the internal genitalia. The solution for pediatric and teenage patients in the early days of cervicovaginal surgical interventions during the 1990s was a hysterectomy; a consequence of this was the loss of all reproductive chances for these patients. With the improvement of surgical techniques, these methods were gradually changed toward conservative approaches such as canalization methods (Fujimoto et al. 1997) and the use of autologous or allogeneic grafts for the creation of new tissue. Nowadays, a broad range of autologous grafts from non-vaginal sources has been tested for cervicovaginal reconstruction. One of the most popular approaches is the use of bowel which has the advantage of being a vascularized tissue (O'Connor et al. 2004; Hensle et al. 2006). However, this method produces hygienic issues and a higher incidence of neovaginal neoplasia (Raya-Rivera et al. 2014). Furthermore, the amount of tissue extracted for transplantation could be a limiting factor in this approach. Allogeneic tissues, such as amniotic membrane, have also been tested in patients. However, these solutions have the disadvantage of not being a sterile material and can possibly transmit diseases (Chakravarty et al. 2000; Alborzi et al. 2005; Mhaskar 2005).

In the last decades, the use of bioengineering has emerged as a compelling option for organ reconstruction and regeneration; this is also the case for cervicovaginal disorders. To date, there have been several research groups that explored different biomaterials and cell sources to create tissue-engineered constructs for cervicovaginal abnormalities. Examples are synthetic scaffolds as polyglycolic acid (de Filippo et al. 2003) or silk fibroin (Chang et al. 2017) in vagina. For cervical reconstruction, a well-researched option was done by the group of House et al. also with silk (House et al. 2010).

On the other hand, after realizing the pivotal role of the natural ECM in homeostasis, the decellularized matrix as scaffolds has become more important in the last years. In this respect, some commercial xenogeneic acellular matrices such as porcine acellular dermal matrix (ADM) or porcine SIS have been shown to be viable for cervicovaginal reconstructions, resulting in near normal sexual function (Zhu et al. 2013; Ding et al. 2014; Raya-Rivera et al. 2014). SIS grafts in particular have been widely used in vaginoplasty and have even showed to be resistant to infection (Ding et al. 2014). In 2014, four MRKHS patients underwent a vaginoplasty of SIS scaffolds recellularized with autologous vulvar cells. The follow-up after 8 years showed a vaginal structure with epithelium and muscle that was able to produce lubrication, orgasm and painless intercourse (Raya-Rivera et al. 2014).

Despite that, the capability of SIS grafts to improve on other techniques such as laparoscopic peritoneal vaginoplasty have been questioned (Ding et al. 2015). Furthermore, some experts think the use of a non-specific ECM for the reconstruction of cervix or vaginal scaffold could lead to a not ideal environment for tissue reconstruction in vivo and subsequently results to a limited functionality of the organ. Opening up a window to think about a transplantable tissue-specific engineered construct.

The theory of a tissue-specific cervicovaginal material could be more appropriate than the use of non-specific source, this was checked with the results obtained from the comparison of AVM with SIS in vaginal reconstruction in rats. The new AVM vagina got a faster and a more complete regeneration than SIS scaffold in epithelialization and growth of nerve fibers. Furthermore, AVM showed a multilayered architecture with organized muscle bundles and an estrogen receptor distribution pattern closer to a normal vagina. Notwithstanding, the most importantly achieved progress is the contractile function, acquired only in AVM grafts, which one was similar to native vagina (Zhang et al. 2017).

All in all, taking into account currently available information, it becomes apparent that tissue-specific biomaterials from natural sources represent a vital and promising avenue to achieve progress in cervicovaginal pathologies. Despite this, bioengineering approaches to obtain a functional bioengineered vagina are in its infancy with only a few reported studies. Even so, the use of decellularized vaginal tissue opens up incredible opportunities in the repair of vaginal defects and can be considered as the next logical step in clinical cervicovaginal reconstruction.

13.5 Limitations

The long-term goal of the decellularized cervix or vagina is to create a transplantable tissue-engineered scaffold that can be used for women with cervicovaginal pathologies. Preferably these engineered organs must be implanted without any immunological adverse reaction, even support vaginal bacterial flora development, and must presented the typical histological layers with complete normal function in terms of lubrication, orgasm and painless intercourse among others.

Among the possible tissue sources, it is logical to consider the use of human vagina as the ideal solution. However, human organs could be a difficult option due to its limited availability. Instead, porcine vagina might be considered as an appropriate source for vaginal reconstruction in humans due to their similar shape, demonstrated interspecies biocompatibility and high availability. In fact, nowadays acellular porcine products are widely used for clinical applications (Crapo et al. 2011).

However, further studies investigating the effect of xenogeneic AVM with human cells are necessary. Only one in vitro cytocompatibility study has been performed with allogeneic (porcine) cells (Greco et al. 2018).

Another limitation that this approach shares with other organs reconstruction techniques is the difficulty of completely recellularize the scaffold ex vivo. Firstly, it is difficult to recover enough tissue-specific cells in patients suffering from congenital disorders. While it could be possible to augment the cells obtained via primary culture of biopsies, it is likely that this will not be sufficient for the recellularization of entire vaginal bioscaffolds. Secondly, current cell seeding and conditioning techniques of non-cannulated DC tissues still need to be optimized to create a transplantable construct in a bioreactor (VeDepo et al. 2017). This lack of full recellularization of AVM is easily seen, for example, in the inability of producing tonic contractile responses in the same way as the native vagina (Zhang et al. 2017). In these cases, differentiation from autologous stem cells to recellularize scaffold may be a possibility.

Finally, the approaches mentioned above are developed focusing on the restoration of vaginal function during sexual intercourse, improving the quality of life of the patients, but they remain infertile. The functionality of these new organs during a natural childbirth is not investigated. Consequently, infertility in these patients, or at least the possibility of a natural delivery, continues to be an unsolvable problem.

13.6 Conclusions

When taking all information into account, it becomes apparent that tissue-specific biomaterials represent a vital and promising avenue to achieve progress. The use of a non-specific ECM as cervix or vaginal scaffold could lead to a not ideal environment for tissue reconstruction and subsequently results to a limited functionality of the organ (Jaureguito and Paulos 1996). A good way for understanding this is the "plastic principle" which stipulates that when a part of the body is lost, it should be replaced with a similar material (Zhang et al. 2017). This concept is clearly proven when AVM is compared with SIS in a vaginal reconstruction. The superiority of AVM over SIS scaffold is expressed in a faster and a more complete regeneration. The new AVM vagina shows organized muscle bundles, faster epithelialization and growth of nerve fibers and a multilayered architecture and estrogen receptor distribution pattern closer to a normal vagina. However, actually, the most important progress is the physiological contractile function acquired, only present in AVM grafts, which ones were similar to native vagina (Zhang et al. 2017).

References

Alberti KA, Xu Q (2016) Biocompatibility and degradation of tendon-derived scaffolds. Regen Biomater [Internet] 3:1–11

Alborzi S, Momtahan M, Parsanezhad ME, Yazdani M (2005) Successful treatment of cervical aplasia using a peritoneal graft. Int J Gynecol Obstet [internet] 88:299–302

Arenas-Herrera JE, Ko IK, Atala A, Yoo JJ (2013) Decellularization for whole organ bioengineering. Biomed Mater [Internet] 8:014106

Baptista PM, Siddiqui MM, Lozier G, Rodriguez SR, Atala A, Soker S (2011) The use of whole organ decellularization for the generation of a vascularized liver organoid. Hepatology [Internet] 53:604–617

Barnhart KT, Izquierdo A, Pretorius ES, Shera DM, Shabbout M, Shaunik A (2006) Baseline dimensions of the human vagina. Hum Reprod [Internet] 21:1618–1622 (Oxford University Press)

Berney T, Berishvili E (2015) Toward clinical application of the bioartificial pancreas. Transplantation [internet] 99:2241–2242

Chakravarty B, Konar H, Chowdhury NNR (2000) Pregnancies after reconstructive surgery for congenital cervicovaginal atresia. Am J Obstet Gynecol [internet] 183:421–423

Chang Y, Sun X, Li Q, Ding X, Liu H, Wang J (2017) Silk fibroin scaffold as a potential choice for female pelvic reconstruction: a study on the biocompatibility in abdominal wall, pelvic, and vagina. Microsc Res Tech [Internet] 80:291–297

Crapo PM, Gilbert TW, Badylak SF (2011) An overview of tissue and whole organ decellularization processes. Biomaterials [internet] 32:3233–3243

Ding J-X, Chen X-J, Zhang X-Y, Zhang Y, Hua K-Q (2014) Acellular porcine small intestinal submucosa graft for cervicovaginal reconstruction in eight patients with malformation of the uterine cervix. Hum Reprod [Internet] 29:677–682. https://doi.org/10.1093/humrep/det470

Ding J-X, Chen L-M, Zhang X-Y, Zhang Y, Hua K-Q (2015) Sexual and functional outcomes of vaginoplasty using acellular porcine small intestinal submucosa graft or laparoscopic peritoneal vaginoplasty: a comparative study. Hum Reprod [Internet] 30:581–589. https://doi.org/10.1093/humrep/deu341

cde Filippo RE, Yoo JJ, Atala A (2003) Engineering of vaginal tissue in vivo. Tissue Eng [internet] 9:301–306

Fujimoto VY, Miller JH, Klein NA, Soules MR (1997) Congenital cervical atresia: report of seven cases and review of the literature. Am J Obstet Gynecol [internet] 177:1419–1425

Gartner LP, Hiatt JL (2006) Color atlas of histology [Internet]. Lippincott Williams & Wilkins. Available from: https://books.google.es/books?id=KQ__HPy0XgoC&dq=Color+Atlas+of+Histology.+Baltimore,&hl=es&sa=X&ved=0ahUKEwjgkqWZ8N3fAhUS5uAKHXQdDlwQ6AEIUzAF

Greco KV, Jones LG, Obiri-Yeboa I, Ansari T (2018) Creation of an acellular vaginal matrix for potential vaginal augmentation and cloacal repair. J Pediatr Adolesc Gynecol [internet] 31:473–479

Hensle TW, Shabsigh A, Shabsigh R, Reiley EA, Meyer-Bahlburg HFL (2006) Sexual function following bowel vaginoplasty. J Urol [internet] 175:2283–2286

House M, Sanchez CC, Rice WL, Socrate S, Kaplan DL (2010) Cervical tissue engineering using silk scaffolds and human cervical cells. Tissue Eng Part A [internet] 16:2101–2112

Jaureguito JW, Paulos LE (1996) Why grafts fail. Clin Orthop Relat Res [Internet], pp 25–41. Available from: http://www.ncbi.nlm.nih.gov/pubmed/8998884

Jordan JA, Singer A, Jones HW, Shafi MI (2006) Wiley InterScience (Online service). The cervix [Internet]. Blackwell. Available from: https://www.wiley.com/en-us/The+Cervix%2C+2nd+Edition-p-9781405131377

Khorramirouz R, Go JL, Noble C, Jana S, Maxson E, Lerman A, Young MD (2018) A novel surgical technique for a rat subcutaneous implantation of a tissue engineered scaffold. Acta Histochem [internet] 120:282–291

Kim JJ, Hou L, Huang NF (2016) Vascularization of three-dimensional engineered tissues for regenerative medicine applications. Acta Biomater [internet] 41:17–26

Kimberley N, Hutson JM, Southwell BR, Grover SR (2012) Vaginal agenesis, the hymen, and associated anomalies. J Pediatr Adolesc Gynecol [internet] 25:54–58

Kuo C-Y, Baker H, Fries MH, Yoo JJ, Kim PCW, Fisher JP (2017) Bioengineering strategies to treat female infertility. Tissue Eng Part B Rev [internet] 23:294–306

Lange P, Greco K, Partington L, Carvalho C, Oliani S, Birchall MA, Sibbons PD, Lowdell MW, Ansari T (2017) Pilot study of a novel vacuum-assisted method for decellularization of tracheae for clinical tissue engineering applications. J Tissue Eng Regen Med [internet] 11:800–811

Mhaskar R (2005) Amniotic membrane for cervical reconstruction. Int J Gynaecol Obstet [internet] 90:123–127

Mulhall JP, Incrocci L, Goldstein I, Rosen R (2011) Cancer and sexual health [Internet]. Humana Press. Available from: https://books.google.es/books?id=GpIadil3YsQC&pg=PA13&redir_esc=y&hl=es#v=onepage&q&f=false

O'Connor JL, DeMarco RT, Pope JC, Adams MC, Brock JW (2004) Bowel vaginoplasty in children: a retrospective review. J Pediatr Surg [internet] 39:1205–1208

Pendergrass PB, Belovicz MW, Reeves CA (2003) Surface area of the human vagina as measured from vinyl polysiloxane casts. Gynecol Obstet Invest [Internet] 55:110–113 (Karger Publishers)

Pfeifer SM (2016) Congenital Müllerian Anomalies: Diagnosis and Management. In: Pfeifer SM (ed). Springer International Publishing

Raya-Rivera AM, Esquiliano D, Fierro-Pastrana R, López-Bayghen E, Valencia P, Ordorica-Flores R, Soker S, Yoo JJ, Atala A (2014) Tissue-engineered autologous vaginal organs in patients: a pilot cohort study. Lancet [Internet] 384:329–336 (Elsevier)

Sánchez PL, Fernández-Santos ME, Costanza S, Climent AM, Moscoso I, Gonzalez-Nicolas MA, Sanz-Ruiz R, Rodríguez H, Kren SM, Garrido G et al (2015) Acellular human heart matrix: a critical step toward whole heart grafts. Biomaterials [internet] 61:279–289

Syazwani N, Azhim A, Morimoto Y, Furukawa K, Ushida T (2014) Immune response of implanted aortic scaffolds decellularized by Sonication Treatment. Springer, Cham, pp 275–278. Available from: http://link.springer.com/10.1007/978-3-319-02913-9_70.

VeDepo MC, Detamore MS, Hopkins RA, Converse GL (2017) Recellularization of decellularized heart valves: Progress toward the tissue-engineered heart valve. J Tissue Eng [Internet] 8, 2041731417726327. SAGE Publications.

Zhang J-K, Du R-X, Zhang L, Li Y-N, Zhang M-L, Zhao S, Huang X-H, Xu Y-F (2017) A new material for tissue engineered vagina reconstruction: acellular porcine vagina matrix. J Biomed Mater Res A [internet] 105:1949–1959

Zhu L, Zhou H, Sun Z, Lou W, Lang J (2013) Anatomic and sexual outcomes after vaginoplasty using tissue-engineered biomaterial graft in patients with Mayer-Rokitansky-Küster-Hauser syndrome: a new minimally invasive and effective surgery. J Sex Med [internet] 10:1652–1658

Decellularization of Male Reproductive Tissue

14

Joery De Kock and Yoni Baert

Abstract

Decellularized testicular matrix (DTM) enables researchers to focus on the specific composition of the testicular extracellular matrix (ECM) and elucidate its role in spermatogenesis. Furthermore, it provides the natural architectural arrangement that could guide the reorganization of dissociated testicular cells in vitro. This is a key consideration as the presence of an authentic nutritive and endocrine support has been proven to be essential for in vitro spermatogenesis, at least in the mouse (Oliver and Stukenborg in Andrology 8:825–834, 2020; Richer et al. in Andrology 12741, 2019). Hence, scaffolds of DTM could be harnessed for the development of a human in vitro spermatogenesis culture system, which is a missing link in male fertility preservation and could be a possible treatment for nonobstructive azoospermia (Gassei and Orwig in Steril 105:256–266, 2016).

J. De Kock
In Vitro Toxicology and Dermato-Cosmetology (IVTD), Vrije Universiteit Brussel (VUB), Laarbeeklaan 103, 1090 Brussels, Belgium

Y. Baert (✉)
Biology of the Testis (BITE) Research Group, Department of Reproduction, Genetics and Regenerative Medicine, Vrije Universiteit Brussel (VUB), Laarbeeklaan 103, 1090 Brussels, Belgium
e-mail: yoni.baert@vub.be

Keywords

Reproductive system · Scaffold · Tissue engineering · Testis

14.1 History

14.1.1 Decellularization of Testicular Tissue

We were the first to derive DTM and develop natural cytocompatible scaffolds from the human testis (Baert et al. 2015; Baert and Goossens 2018). Mechanically agitating testicular fragments for 24 h in 1% (w/v) sodium dodecyl sulfate (SDS) detergent was more effective in decellularizing and removing DNA from the tissue while maintaining important ECM proteins compared to treatment in 1% (v/v) Triton X-100. Analyzing these acellular fragments by LC–MS/MS gave us an exclusive look into the testicular ECM and revealed 102 proteins specifically assigned to the extracellular region (Baert et al. 2015). The mass spectrometry data were deposited to the ProteomeXchange with identifier PXD001524. Following recellularization of 90-μm-thick scaffolds with adult or pubertal human testicular cells at the air-media interphase, testicular somatic and germ cells attached to the scaffold and created 3D mini-testicular tissues referred to as testicular organoids (Baert et al.

A. Kajbafzadeh (ed.), *Decellularization Methods of Tissue and Whole Organ in Tissue Engineering*, Advances in Experimental Medicine and Biology 1345,
https://doi.org/10.1007/978-3-030-82735-9_14

2017a, b). The testicular organoids were composed of proliferative spermatogonia and functional niche cells secreting testosterone and inhibin B for at least 4 weeks. However, the testicular organoids did not show the critical testicular compartmentalization and spermatogenesis was not studied. Nevertheless, soon after these first promising results, more publications on DTM from other research groups followed. Vermeulen and colleagues compared several decellularization protocols for prepubertal porcine testicular fragments. Following recellularization with adult human Sertoli cells, 0.01% SDS + 1% Triton X-100 and 0.05% trypsin + 0.02% EDTA + 3% Triton X-100-derived scaffolds offered the best compromise in terms of DNA elimination and ECM preservation, while securing attachment, proliferation, and functionality of the Sertoli cells (Vermeulen et al. 2018). In a follow-up paper, hydrogels were derived from their DTM and used to generate porcine testicular organoids by encapsulating piglet testicular cell suspensions. The organoids were maintained for 45 days and consisted of tubule-like structures surrounded by interstitial cells. However, germ cell maturation was not achieved (Vermeulen et al. 2019). Another study fabricated testis-derived scaffolds from ram testicular tissue (Rezaei Topraggaleh et al. 2019). Testicular ECM was successfully extracted by treating testicular fragments with a serial combination of 1% Triton X-100 and 1% SDS for 48 h, digested using an acid-pepsin solution digestion to form a hydrogel and subsequently lyophilized to generate scaffolds. The resulting macroporous scaffolds were inoculated with neonatal mouse testicular cells and supported formation of organoids that lacked the testicular architecture but nevertheless produced hormones and formed post-meiotic cells (Rezaei Topraggaleh et al. 2019). Yang et al. cultured CD326-purified spermatogonial stem cells on DTM-coated plates. DTM hydrogels were prepared by stirring for 18 h in 1% SDS followed lyophilization and pepsin-digestion. It was found that the viability of the spermatogonial stem cells inoculated onto DTM hydrogel scaffolds was significantly higher than those inoculated on Matrigel or laminin. Also, it was observed that intracellular gene expression and DNA-imprinting patterns of these cells were similar to that of native spermatogonial stem cells. Importantly, DTM promoted spermatogonial stem cells to proliferate and differentiate into round spermatids without requiring additional supportive cells (Yang et al. 2020). Movassagh et al. cultured human stem cells on sheep DTM (Ashouri Movassagh et al. 2020). They observed that after the initiation of differentiation, pre-meiotic, meiotic, and post-meiotic related genes were significantly higher expressed in the cells cultured in DTM substrate compared to 2D conditions.

14.2 Decellularization of Whole Testis

Instead of decellularizing fragments, a number of studies looked into the decellularization of the whole organ by perfusion. Gharenaz et al. used a combination of SDS and Triton X-100 to remove up to 98% of the DNA from mouse testis while preserving glucosaminoglycans, collagens, fibronectin and laminins. It could also be shown that spermatogonial stem cells proliferated and differentiated in to spermatocytes after being injected in these murine decellularized testicular scaffolds (Majidi Gharenaz et al. 2020). Kargar-Abarghouei et al. perfused frozen-thawed rat testes with 1% Triton X-100 through the ductus deferens for 4 h followed by 1% SDS for 24 h. ECM components were well preserved and the cells were completely removed after decellularization. In-vitro studies excluded soluble toxicity and confirmed cell compatibility of their decellularized scaffolds. In addition, mesenchymal stem cell-laden and cell-free scaffolds were transplanted to different positions in acceptor rats. Although testicular-like cells were observed in scaffolds located in the mesentery and liver, they could not differentiate into post-meiotic cells (Kargar-Abarghouei et al. 2018). More recently, Akbarzadeh et al. reported a whole-organ decellularization protocol for a human-sized ovine testis using the testicular arteries as entry point. Organ perfusion with 1% SDS for 6–8 h demonstrated the most desirable outcomes regarding decellularization and ECM preservation. Extensive washing prevented residual toxicity as shown in their MTT analysis, and MRI after contrast solution perfusion showed a three-dimensional testicular bioscaffold resembling the properties of the native organ. Noteworthy, the cell attachment properties were not addressed in this study (Akbarzadeh et al. 2019). The group of Sabetkish obtained entire immature human testicles from patients with testicular feminization syndrome and perfused those with 1% Triton X-100 and 1% SDS, each for 3 h in 5 consecutive cycles. Next, they investigated the implantation of human DTM in C57BL/6 mice as a first stepping stone for further testis tissue engineering. Strikingly, they confirmed the presence of spermatogonial stem-like cells on these human decellularized implants that well differentiated subsequently over time (Sabetkish et al. 2020).

14.3 Conclusion

In the most recent years, more and more attention has been put on the opportunities of DTM, especially for in vitro differentiation studies. SDS and Triton X-100 alone, but also combinations with other enzymatic and non-enzymatic reagents resulted in biocompatible scaffolds. However, more research with a focus on optimizing the recellularization protocols is needed to re-establish the naïve cell–cell and cell–matrix attachments, which in turn will facilitate efficient spermatogenesis in DTM scaffolds.

Acknowledgements Y. Baert is a postdoctoral fellow of the Research Foundation—Flanders (FWO; 62930).

References

Akbarzadeh A, Kianmanesh M, Fendereski K, Ebadi M, Daryabari SS, Masoomi A, Ghazisaeedi F, Seyyed Hossein Beigi R, Sheikh R, Kajbafzadeh A-M (2019) Decellularised whole ovine testis as a potential bio-scaffold for tissue engineering. Reprod Fertil Dev 31:1665

Ashouri Movassagh S, Ashouri Movassagh S, Banitalebi Dehkordi M, Pourmand G, Gholami K, Talebi A, Esfandyari S, Jabari A, Samadian A, Abbasi M (2020) Isolation, identification and differentiation of human spermatogonial cells on three-dimensional decellularized sheep testis. Acta Histochem 122:151623

Baert Y, Stukenborg J-B, Landreh M, De Kock J, Jornvall H, Soder O, Goossens E (2015) Derivation and characterization of a cytocompatible scaffold from human testis. Hum Reprod 30:256–267

Baert Y, Goossens E (2018) Preparation of Scaffolds from decellularized testicular matrix. Humana Press, Totowa, NJ, pp 1–7

Baert Y, De Kock J, Alves-Lopes J P, Söder O, Stukenborg J-B, Goossens E (2017a) Primary human testicular cells self-organize into organoids with testicular properties. Stem Cell Reports 8:30–8

Baert Y, Rombaut C, Goossens E (2017b) Scaffold-based and Scaffold-free testicular organoids from primary human testicular cells. In: Methods in molecular biology (Clifton, N.J.) vol 1283, pp 161–9

Gassei K, Orwig KE (2016) Experimental methods to preserve male fertility and treat male factor infertility. Fertil Steril 105:256–266

Kargar-Abarghouei E, Vojdani Z, Hassanpour A, Alaee S, Talaei-Khozani T (2018) Characterization, recellularization, and transplantation of rat decellularized testis scaffold with bone marrow-derived mesenchymal stem cells. Stem Cell Res Ther 9:324

Majidi Gharenaz N, Movahedin M, Mazaheri Z (2020) Three-dimensional culture of mouse spermatogonial stem cells using a decellularised testicular scaffold. Cell J 21:410–418

Oliver E, Stukenborg JB (2020) Rebuilding the human testis in vitro. Andrology 8:825–834

Rezaei Topraggaleh T, Rezazadeh Valojerdi M, Montazeri L, Baharvand H (2019) A testis-derived macroporous 3D scaffold as a platform for the generation of mouse testicular organoids. Biomater Sci 7:1422–1436

Richer G, Baert Y and Goossens E (2019) In-vitro spermatogenesis through testis modelling: towards the generation of testicular organoids. Andrology 12741

Sabetkish S, Kajbafzadeh A-M, Sabetkish N (2020) Recellularization of testicular feminization testis in C57bl6 as a natural bioreactor for creation of cellularized seminiferous tubules: an experimental study. Cell Tissue Bank 6

Vermeulen M, Vento F Del, Michele F de, Poels J, Wyns C (2018) Development of a cytocompatible Scaffold from pig immature testicular tissue allowing human Sertoli cell attachment, proliferation and functionality. Int J Mol Sci 19(1):227

Vermeulen M, Del Vento F, Kanbar M, Pyr dit Ruys S, Vertommen D, Poels J and Wyns C (2019) Generation of organized porcine testicular organoids in solubilized hydrogels from decellularized extracellular matrix. Int J Mol Sci 20:5476

Yang Y, Lin Q, Zhou C, Li Q, Li Z, Cao Z, Liang J, Li H, Mei J, Zhang Q, Xiang Q, Xue W, Huang Y (2020) A testis-derived hydrogel as an efficient feeder-free culture platform to promote mouse spermatogonial stem cell proliferation and differentiation. Front Cell Dev Biol 8:1–14

Decellularization of Skin Tissue

15

Naveen Kumar, Vineet Kumar, Sanjay Purohit,
Anil Kumar Gangwar, Sameer Shrivastava,
Swapan Kumar Maiti, Sonal Saxena,
Dayamon David Mathews,
Pawan Diwan Singh Raghuvanshi,
Ajit Kumar Singh, and Karam Pal Singh

Abstract

Biomaterials science encompasses elements of medicine, biology, chemistry, materials, and tissue engineering. They are engineered to interact with biological systems to treat, augment, repair, or replace lost tissue function. The choice of biomaterial depends on the procedure being performed, the severity of the patient's condition, and the surgeon's preference. Prostheses made from natural-derived biomaterials are often derived from decellularized extracellular matrix (ECM) of animal (xenograft) or human (allograft) origin. Advantages of using ECM include their resemblance in morphology and three-dimensional structures with that of tissue to be replaced. Due to this, scientists all over are now focusing on naturally derived biomaterials which have been shown to possess several advantages compared to synthetic ones, owing to their biocompatibility, biodegradability, and remodeling properties. Advantages of a naturally derived biomaterial enhance their application for replacement or restoration of damaged organs/tissues. They adequately support cell adhesion, migration, proliferation, and differentiation. Naturally derived biomaterials can induce extracellular matrix formation and tissue repair when implanted into a defect by enhancing attachment and migration of cells from surrounding environment. In the current chapter, we will focus on the natural and synthetic dermal matrix development and all of the progress in this field.

Keywords

Skin · Decellularization · Tissue engineering · Dermal Matrix

15.1 History

Biomaterials are those materials that are used in medical devices or in contact with biological systems (Ratner 2006). A biocompatible or suitable biomaterial for one application may not be biocompatible in another (Ratner 2006). An ideal implant should effectively repair the defect without eliciting an adverse host response while maintaining mechanical, as well as, biological integrity for a desired time ranging from a few weeks to even several years' durations. Further, they must be easily manufactured not only on macroscopic but also at the cellular level. However, on the odd occasion, these materials can cause immunological reactions in the host (Ratner et al. 1996).

N. Kumar (✉) · V. Kumar · S. Purohit ·
A. K. Gangwar · S. Shrivastava · S. K. Maiti ·
S. Saxena · D. D. Mathews · P. D. S. Raghuvanshi ·
A. K. Singh · K. P. Singh
Division of Surgery, ICAR-Indian Veterinary Research Institute, Izatnagar 243122, India

A. Kajbafzadeh (ed.), *Decellularization Methods of Tissue and Whole Organ in Tissue Engineering*, Advances in Experimental Medicine and Biology 1345,
https://doi.org/10.1007/978-3-030-82735-9_15

The success of implants mainly depends upon their source and nature. Biomaterials are classified broadly into two types; synthetic and natural biomaterials. Metals, ceramics, non-biodegradable polymers are classified as synthetic materials. Metal hip, Dacron, plastic intraocular lens are examples of synthetic biomaterials that have been commercialized and applied in clinical treatment. However, synthetic biomaterials are having some disadvantages. Their structure and composition are not similar to the native organ/tissue. Moreover, the biocompatibility and ability to induce tissue remodeling are low for synthetic biomaterials. There is a need for development of biocompatible materials that encourage integration while minimizing adverse responses, like tissue reaction and adhesion formation. The design and development of optimal biological soft materials for tissue engineering and surgical repair to reinforce or replace soft tissue remain a challenge. Varieties of scaffold materials are available, each with different physical properties and associated with a specific and unique host response. Scaffold materials can be either synthetic or natural. The most commonly used naturally occurring scaffold material is the structural protein collagen. The collagen is a naturally occurring highly conserved macromolecule that is ubiquitous among mammalian species and accounts for approximately 30% of all body proteins. Natural collagenous materials are being investigated in tissue engineering because of inherent low antigenicity and their ability to integrate with the surrounding tissue. To repair and reconstruct lost or damaged tissue and organs it is desirable to design and fabricate three-dimensional scaffolds so that newly regenerated tissue is similar in anatomical structure and function.

15.2 Properties of an Ideal Biomaterial

An ideal biomaterial should elicit either nil or minimal immune response and should allow cellular infiltration. Simultaneously, it must maintain its 3D structure required for performing the intended function. Materially, it should also provide mechanical integrity for association with the function of the reconstructed tissue. Ultimately, it should degrade and promote regeneration of healthy tissue rather than fibrous scarring. The physiological compatibility of biomaterials presumably remains the most critical factor that governs its approval for clinical use. Native ECM induces a more predictable restorative effect than synthetic biomaterials (Badylak et al. 2002), by promoting cellular infiltration, proliferation, and differentiation into structures similar to those of intact host tissue. Many of the first ECM biomaterials were used as prosthetics to provide structural support and mechanical functionality (Angell and Angell 1976; Dewanjee 1985; Liotta et al. 1978). For generating biomaterial, precreation of original structure and strength while reducing immunogenicity is of paramount importance. Collagen is the most abundant protein in ECM, which is remarkably preserved across species (Rest and Garrone 1991). For this reason, ECM invokes one of the weakest immune responses as compared to other biomaterials. Thus, bovine collagen is still one of the most commonly used and abundantly available xenographic materials for clinical applications (Badylak 2002). However, in hypersensitive patients or cases with extenuating circumstances, even collagen can provoke an immune reaction (Klein et al. 2001; Zhang et al. 2005; Dzemeshkevich et al. 1994). Proper treatments with detergents, terminal sterilization, and gamma irradiation or ethylene oxide gas treatment is enough to reduce the immune response to a very minimal level, much lower than that of synthetic meshes (Allman et al. 1980; Patino et al. 2003; Konstantinovic et al. 2005; O'Neill and Booth 1984).

15.3 Strategies for Developing Acellular Dermal Matrix from Different Skin Tissues

15.3.1 Decellularization

Autologous grafts have been considered as a gold standard for implantation. In autologous grafting, quantity is the most restricting factor. The need of patients can be met by a number of autografts. Even homografts are unavailable in sufficient amounts to satisfy the needs of patients. Usage of xenograft is limited due to the issue of immune reaction. In order to overcome this, decellularization can be carried out for processing homografts and xenografts. Decellularized tissues and organs have been used successfully for a variety of tissue engineering and regenerative medicine applications. The method of decellularization varies widely depending upon the tissue or organ of interest. Efficiency of cell removal varies with the tissue origin and specific decellularization method used. Each of these physical, chemical, and enzymatic treatments affects the biochemical composition, tissue ultrastructure, and mechanical behavior of the remaining ECM scaffold, which in turn alters the response of a host to the material. The ultimate aim of a decellularization protocol is to efficiently remove all the cellular and nuclear components while minimizing adverse effects on the composition, biological activity, and mechanical integrity of the remaining ECM (Gilbert et al. 2006).

Complex three-dimensional structure and the mechanical properties of the native material need to be retained by the tissue after undergoing decellularization. Decellularization protocol typically starts with the lysis of cell membrane by physical treatment or ionic solution. This is followed by separation of cellular components from the ECM using enzymatic treatments, solubilization of cytoplasmic and nuclear elements with detergents, and finally removal of cellular debris from the tissue. Effectiveness of these steps can be increased by coupling with mechanical agitation (Gilbert et al. 2006). In order to solubilize cytoplasmic portions of the cell and to remove nucleic acids, alkaline and acid treatments are used in decellularization protocol.

Decellularization carried out by either chemical, enzymatic or mechanical method leaves a material composed of extracellular matrix components. The acellular tissue retains its natural mechanical properties and prosthesis remodeling is promoted by neovascularization and recellularization by the host (Schmidt and Baier 2000). The extracellular matrices contain cells that have class I and II histocompatibility antigens capable of evoking rejection reactions. These cells also have glycoprotein that can be recognized by immune system of the host resulting in rejection reaction. Therefore, by elimination of these substances from ECM, rejection response can be prevented. However, complete removal of all antigens is considerably difficult (Malone et al. 1984). In many cases, residual cellular components and lipids could be unremoved, and these promote undesirable effects like calcification and host immune response resulting in inflammatory reactions in the recipients. Even after the removal of cellular and nuclear debris from the biomaterial, the ECM of the acellular tissue itself might evoke some sort of immune responses (Coito and Kupiec-Weglinsky 1996). Decellularization is a multistep process intended to remove all cellular components from a tissue or an organ leaving behind intact ECM. Many researchers have tested different decellularizing agents such as physical, chemical, and enzymatic methods. Since every decellularizing agent has a specific effect on cells and ECM. So, a combination of these agents is used generally to create an effective decellularization protocol which removes all cell components and recovers maximum ECM. Effectiveness of decellularization depends on the type of tissue or organ. A decellularizing and good agent for one tissue may not be the same for another (Schechter 1975). Furthermore, this cell derived ECM can be used as a matrix for cell culture.

Prevention of rejection of these substances is achievable using two strategies. Primary strategy employs reduction of the host immune responses while the focuses on reduction in the antigenicity of allograft or xenograft by crosslinking (Weadock et al. 1984). Yannas (1996a) demonstrated that although the immunogenicity of the collagen is measurable but the actual significance of such immunogenicity is extremely low due to small species difference among various types of collagen. It has been demonstrated that acellular grafts are less immunogenic and are better tolerated by allogenic hosts and are equally effective as isograft too (Gulati and Cole 1994). Thus, decellularization protocols which can effectively remove cellular components and also maintain the integrity of extracellular matrix are desired. One of the main objectives in utilizing natural biodegradable material is to induce the host, to replace the implanted constructive with tissue (Yoganathan 1995). Naturally derived xenografts have been investigated for their apparent ability to elicit a regenerative response in the host tissue rather than a standard healing response (Badylak 2002; Kropp et al. 2004).

Physical method of decellularization includes snap freezing (Gulati 1988), mechanical agitation and sonication, applying direct pressure (Freytes et al. 2004). For chemical decellularization, acid, and alkaline agents, non-ionic detergents like triton X-100, ionic detergents like sodium deoxycholate, sodium dodecyl sulphate, triton X-200, zwitter ionic detergents like 3-[(3-cholamidopropyl) dimethylammonio]-1-propanesulfonate (CHAPS), sulfobetaine-10 and -16 (SB10, SB-16), Tri (n-butyl) phosphate hypotonic and hypertonic solution are used. Trypsin, endonuclease, and exonuclease are used for enzymatic decellularization (Gilbert et al. 2006). Both cytoplasmic and nuclear cellular membranes are effectively solubilized by ionic detergents however, they tend to denature protein by disruption of protein-protein interaction. Cellular remnants can be effectively removed by sodium deoxycholate, but it badly affects the native tissue architecture in comparison to sodium dodecylsulphate. Hypotonic solutions such as deionized water are used to lyse the cells within tissue and organs by osmotic shock (Woods and Gratzer 2005). Peracetic acid (PAA) at a concentration of approximately 0.10–0.15% (w/v) has been used to decellularize porcine small intestinal submucosa. It effectively removes cellular components from this ECM structure and disinfection of the ECM was also possible simultaneously (Hodde and Hiles 2002). Each and every decellularization agent has specific affections on the cell and ECM. So a combination of different agents is used to create an effective decellularization protocol which removes all cell components and preserves maximum ECM. Effectiveness of decellularization protocol depends on the type of tissue or organ. An excellent agent in decellularization of one tissue need be ungood for another tissue. Besides this, desired ECM can be used as a matrix for cell culture (Table 15.1).

15.3.2 Acellular Matrices

Acellular matrix should provide an ideal biological and physiological environment to ensure homologous distribution of cell and extracellular matrix. It should also provide the appropriate size and morphology of the neo-tissue required. Collagen is a major constituent of connective tissue and thus it is a potentially useful biomaterial among all other protein oriented biomaterials. This protein has been well characterized and is ubiquitous across both animal and plant kingdoms (Rest and Garrone 1991). Further, it is biocompatible, nontoxic having cellular mobility, and is porous in texture. The collagen from bovine tendons contains monomeric, dimeric, trimeric, and higher polymeric forms. Owing to its above mentioned properties, collagen is the most widely used matrix material for tissue engineering in the dermis (Hardin-Young and Parenteau 2002). It is the major constituent of native dermal ECM and is used in reconstructive surgery and tissue engineering. Burke et al. (1981) used bovine collagen matrix which provides mechanical strength, prevent infection, and loss of moisture. Collagen-based scaffolds are under extensive investigation and used as matrices for tissue engineering as being natural matrix materials they mimic natural ECM structure and composition; emulate native stimulating effects of ECM on cells and allow the incorporation of growth factors. In modern medicine, natural and synthetic biomaterials play an indispensable role in the treatment of diseases and the improvement of health care. For tissue engineering applications, biomaterials frequently serve as a scaffold for a specific cell type (Kim et al. 2005).

All residual chemicals must be removed after decellularization to avoid an adverse host tissue response to the chemical (Gilbert et al. 2006). Various chemical and enzymatic techniques are used to remove cellular components. The

Table 15.1 Decellularization methods, their mode of action, and effects on ECM (Gilbert et al. 2006)

Method	Mode of action	Effects on ECM
Physical Mechanical agitation	Can cause cell lysis, but more commonly used to facilitate chemical exposure and cellular material removal	Aggressive agitation or sonication can disrupt ECM as the cellular material is removed
Non-ionic detergent Triton X-100	Disrupts lipid-lipid and lipid-protein interactions, while leaving protein-protein interactions intact	Mixed results; efficiency dependent on tissue, removes GAGs
Ionic detergent Sodium dodecyl sulphate (SDS)	Solubilize cytoplasmic and nuclear cellular membranes; tend to denature proteins	Removes nuclear remnants and cytoplasmic proteins; tends to disrupt native tissue structure, remove GAGs, and damage collagen
Ionic detergent Sodium deoxycholate		More disruptive to tissue structure than SDS
Zwitterionic detergent Tri(n-butyl) phosphate	Organic solvent that disrupts protein-protein interactions	Variable cell removal; loss of collagen content, although effect on mechanical properties was minimal
Hypotonic and Hypertonic solutions	Cell lysis by osmotic shock	Efficient for cell lysis, but does not effectively remove the cellular remnants
EDTA	Chelating agents that bind divalent metallic ions, thereby disrupting cell adhesion to ECM	No isolated exposure, typically used with enzymatic methods (e.g., trypsin)
Enzymes Trypsin	Cleaves peptide bonds on the C-side of Arg and Lys	Prolonged exposure can disrupt ECM structure, removes laminin, fibronectin, elastin, and GAGs
Enzymes Endonucleases	Catalyze the hydrolysis of the interior bonds of ribonucleotide and deoxyribonucleotide chains	Difficult to remove from the tissue and could invoke an immune response

removal of cells from the tissue leaves the complex mixture of structural and functional proteins that represent the ECM. Various factors affecting host response to the ECM scaffold following implantation include: the tissue from which ECM is harvested, species of origin, decellularization protocol and the method of sterilization. The sodium dodecyl sulphate treatment presented chronic rejection of arterial allograft and led to proliferation of elastin rich intima. Triton X-100 leads to complete loss of GAG and decrease in the laminin and fibronectin content of the tissue. It is the most widely used non-ionic detergent for decellularization (Grauss et al. 2005). Sodium deoxycholate is also very effective for removing cellular constituents. Treatment with hypotonic solution followed by hypertonic solution results in lysis but the resultant cellular remnants are unremoved from the tissues (Dahl et al. 2003). The trypsin is the most common proteolytic enzyme in decellularization protocols. It reduces the laminin and fibronectin content of the ECM. Prolonged exposure to trypsin also reduces GAGs.

The remaining ECM still supports endothelial cell growth in vitro despite the removal of ECM components (Grauss et al. 2005). The residual chemicals (particularly SDS) must be flushed from ECM after decellularization. The decellularized tissue requires multiple (more than six) washings along with agitation to completely remove the detergents (Cebotari et al. 2010). It has been observed that lipase treatment of rat skin followed by Triton X-100 could completely remove the cellular components from the dermis (Takami et al. 1996) and subcutaneously implanted acellular dermal matrix (ADM) in rats evoked no immunological reaction even after 20 weeks of implantation. Yannas et al. (1996b) demonstrated the immunogenicity of the collagen was measurable, but the actual significance of such immunogenicity was exceptionally low due to small species differences among various types of collagen. Immunogenicity of acellular graft was less, and it has better tolerance by allogenic host and equal effectiveness as that of isograft (Gulati and Cole 1990). Interspecies, rather than intraspecies immunogenicity of arterial ECM leads to inflammation, biodegradation, dilation, and ultimately chronic rejection of the graft (Allaire et al. 1997). Cartmell and Dunn (2000) reported the efficacy of an acellular matrix in the treatment of abdominal wall defect which largely depended upon its low antigenicity, capacity for rapid vascularization, and stability as abdominal wall tissue.

15.4 Development of Acellular Dermal Matrices from Different Species of Animals

In our laboratory, the protocols have been developed for making acellular dermal matrices from native skin of different species of animals. For each native skin a specific protocol has been developed for making them acellular which reduces their immunogenicity by the use of biological detergents (ionic, non-ionic, and zwitterionic detergents), enzymes (trypsin, pepsin, dispase), hypotonic, and hypertonic solution, EDTA, endonucleases and exonucleases. The materials were cut into 2×2 cm^2 sizes and were continuously agitated in ionic and non-ionic biological detergents in different concentrations on orbital shaker. It was observed that continuous agitation of these soft tissue materials resulted in complete decellularization. The parameters used for evaluation included:

15.4.1 Histopathological Evaluation

The tissue samples collected at different time intervals during standardization of decellularization protocols were subjected to histological examination. The samples were fixed in 10% formal saline solution, dehydrated in ethanol, cleared in xylene, and embedded in paraffin to get 5 micron thin sections. The sections were stained with hematoxylin and eosin staining. Masson's trichrome staining was done for assessing collagen fiber arrangement.

15.4.2 DNA Quantification

The DNA contents analysis before and after completion of protocols was done as per the method described by Gilbert et al. (2009).

15.4.3 Sodium Dodecyl Sulphate Polyacrylamide Gel Electrophoresis (SDS-PAGE)

The expression of protein bands was determined by the sodium dodecyl sulphate polyacrylamide gel electrophoresis as per Laemmli (1970).

15.5 Protocols for Preparing Acellular Dermal Matrices

Fresh ventral abdominal skin of pig, goat, sheep, and buffalo were procured from the local abattoir. The skin of rabbit and rat was collected from the animal euthanized for some other reasons. The skin was preserved in cold physiological saline just after the collection and later on rinsed with normal saline to remove the adhered blood. The maximum time period between the retrieval and the initiation of protocols was less than 4 h.

The tissue samples were cut into 2×2 cm^2 size pieces. The epidermis of the different species was removed using trypsin and sodium chloride in different concentrations and time intervals and was subjected to different protocols for the preparation of acellular dermal matrix. This phase was divided into two parts. Firstly, the protocols for de-epithelization of the rabbit skin were optimized, and then protocols to prepare acellular dermal matrix were optimized. For de-epithelization, the skin of different species was subjected to 0.25, 0.5, 1, 2% trypsin (pH 6) and 1 M, 2 M NaCl (pH 6) treatment to separate the epidermis. The observations were recorded from 48 to 108 h of post treatment. The samples were continuously agitated on a shaker.

The optimization of de-epidermis protocol was done on the basis of gross and histopathological studies. The optimized protocol was used to separate the epidermis and the remaining cellular dermis matrix was further used for optimization of acellular protocols for each species. Four protocols for the acellularity of dermal graft were tested. Trypsin enzyme and biological detergents combination were used. Two anionic (sodium deoxycholate and sodium dodecyl sulphate) and two non-ionic biological detergents (tritron-×100 and tween-20) with trypsin were used for making acellular dermis matrix. The tissue matrix in each solution was continuously agitated at 180 rpm at 37 °C in orbital shaker to provide better contact of tissue with treatments solution in each protocol.

15.5.1 Protocol-I (Trypsin–Sodium Deoxycholate Combination)

In this protocol, the skin was treated with trypsin (HiMedia Laboratories Pvt. Ltd., India, RM 6216_0) solution containing sodium salt of EDTA (Qualigens Fine Chemicals, Bombay, India, No. 18454) and sodium azide (Sigma Aldrich Co., St Louis, MO, USA, S2002) for 24–48 h. This was followed by treatment with sodium deoxycholate at room temperature (37 °C) for 12–24 h. It was followed by trypsin for 12–24 h. Again the biomaterials were treated with sodium deoxycholate. Finally, the tissue was thoroughly rinsed twice with distilled water for 15–30 min each.

15.5.2 Protocol-II (Trypsin-Sodium Dodecyl Sulphate Combination)

In this protocol, the trypsin and sodium dodecyl sulphate combination was used.

15.5.3 Protocol-III (Trypsin-Tritron-× 100 Combination)

In this protocol, the trypsin and tritron-× 100 combination was used.

15.5.4 Protocol-IV (Trypsin-Tween-20 Combination)

In this protocol, trypsin and tween-20 combination was used.

Each step in different protocols was optimized for different species keeping in view the overall thickness of the skin. All the solutions were filter sterilized and each procedure was performed aseptically. The time intervals, concentration, and pH of the chemicals were modified to obtain the desired result. The tissue matrices were preserved for histopathological and SEM examination.

The acellular dermal grafts prepared by different protocols were evaluated on the bases of the histological scorecard which include three parameters with different grading:

Cellularity: +++ = 90–100%, ++ = 80–90%, + = 70–80%, − = <70%

Compactness of collagen fibers: +++ = compact, ++ = mildly loose, + = moderately loose, − = heavy loose;

Collagen fiber morphology: +++ = Normal to the skin dermis, ++ = mildly thick, + = thin, − = very thin.

15.5.5 Optimization of Protocols for Rabbit Skin

15.5.5.1 Protocols for De-Epithelization

The histological picture of native rabbit skin is presented in Fig. 15.1a. The observations were recorded at 6 h intervals up to 48 h. On the basis of the gross and microscopic observations 0.5% trypsin concentration and 24 h time interval was optimized and adapted for further application in the next phase (Fig. 15.1b).

15.5.5.2 Protocols for Preparation of Acellular Dermal Matrix

The de-epithelized skin of rabbits (Fig. 15.1c) was subjected to anionic and non-ionic biological detergents with trypsin to extract out the cellular contents from the dermis. The time of the reaction and concentration of these solutions were adjusted to obtain the optimized acellular dermis. Four protocols were used to obtain the acellular dermal matrix from de-epithelized skin. Each protocol was run in triplicate.

In protocol-I, the de-epithelized rabbit skin was subjected to 0.5% trypsin treatment for 24 h. and later on it was further treated with 0.5% and 1% sodium deoxycholate for 12 h. Then it was again subjected to 0.5% trypsin treatment for 12 h and later on with same biological detergent for 12 h. The samples were washed thrice at every step to rinse the residuals of the previous step. In protocol-II the de-epithelized skin after treatment with 0.5% trypsin concentration for 24 h treated with sodium dodecyl sulphate (0.5% and 1%) for 12 h. The 0.5% trypsin was again used for 12 h. Then the samples were subjected to same time intervals and concentration of biological detergent as described previously. In protocol-III, the de-epithelized skin after 0.5% trypsin treatment for 24 h were treated with tritron-×100 (0.5% and 1%) for 12 h. The 0.5% trypsin was again used for 12 h. Then the samples were subjected to same time interval and

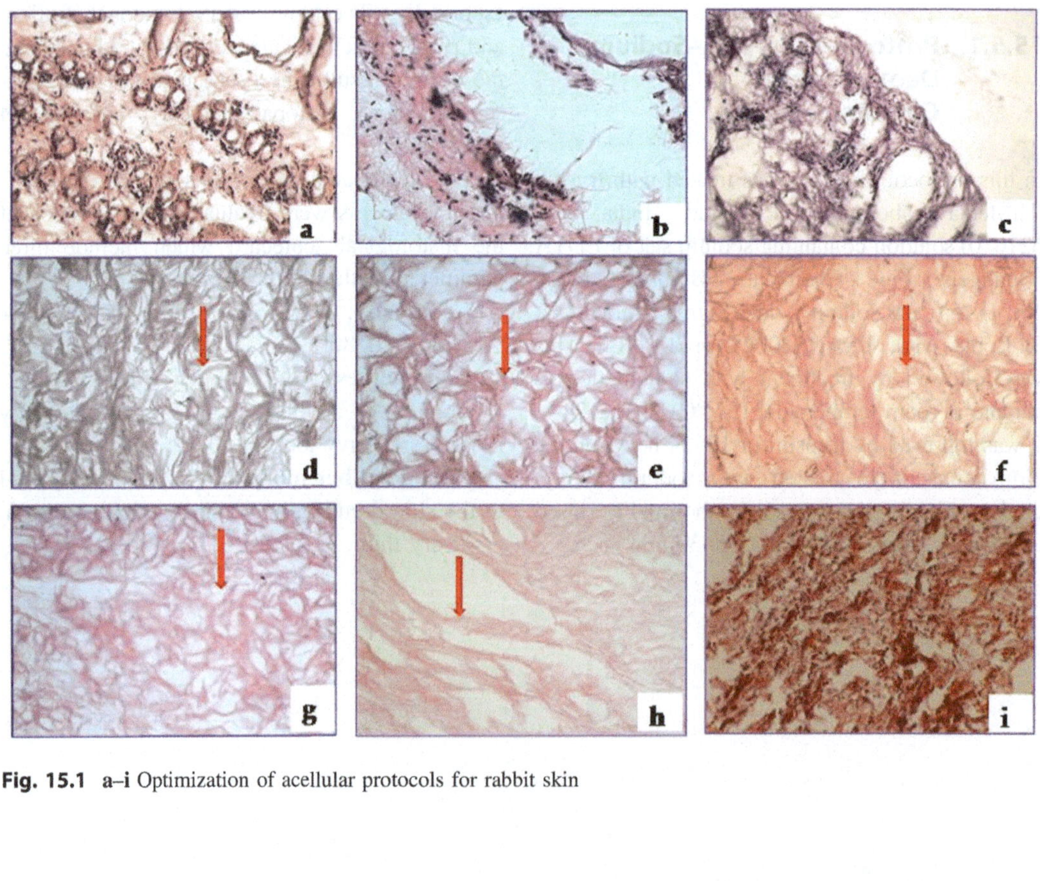

Fig. 15.1 **a–i** Optimization of acellular protocols for rabbit skin

concentration of the detergent treatment. In protocol-IV, the de-epithelized skin was subjected to 0.5% trypsin concentration for 24 h. Tween-20 (0.5% and 1%) was used for 12 h and 0.5% trypsin was again used for 12 h.

The protocols were further modified to obtain the desired results. The reaction time of 0.5% trypsin treatment after the biological detergents were increased to 24 h used instead of 12 h. The concentration of solutions and their reaction time intervals in the remaining steps were the same as previous protocols.

15.5.5.3 Microscopic Observations

In 0.5% sodium deoxycholate the scaffolds showed a 70–80% decrease in the cellular contents with mildly thin and moderately loose collagen fibers. However, in 1% concentration, the scaffolds were almost acellular with mildly thin and loose collagen fibers (Fig. 15.1d). In 0.5% and 1% sodium dodecyl sulphate the scaffolds were almost acellular with mildly thin and loose collagen fibers (Fig. 15.1e). However, moderate loosening of the collagen fibers were observed in 1% concentration of the detergent (Fig. 15.1f). In 0.5% and 1% tritron-×100 the scaffolds showed a 70–80% decrease in the cellular contents with mildly thin and moderately loose collagen fibers. However, the morphology of the collagen deteriorates in 1% concentration of the detergent. In 0.5% and 1% tween-20 the scaffold showed the acellularity below 70% with mild to moderate thin and moderately loose collagen fibers.

The protocols were further modified to obtain the desired results. The time interval of second trypsin treatment was increased from 12 to 24 h in these protocols. The 0.5% sodium deoxycholate showed a 80–90% decrease in the cellular contents with mildly thin and moderately loose collagen fibers. However, in 1% concentration, the scaffolds were almost 100% acellular with moderately thin and loose collagen fibers (Fig. 15.1g). In 0.5% and 1% sodium dodecyl sulphate the scaffolds were almost acellular with mildly thin and loose collagen fibers (Fig. 15.1 h). Moderate looseness of the collagen fibers were observed in 1% concentration of the detergent. The cellularity of the scaffolds decreased as compared to the unmodified protocols. In 0.5% and 1% tritron-×100 the scaffolds showed a 70–80% decreased in the cellular contents with mildly thin and moderately loose collagen fibers. However, the compactness and morphology of collagen scaffold deteriorate in the higher concentration of the detergents (Fig. 15.1i). In 0.5% and 1% tween-20 the scaffold showed 70–80% acellularity with mildly thin and heavily loose collagen fibers.

15.5.6 Optimization of Protocols for Pig Skin

15.5.6.1 Protocols for De-Epithelization

The pig skin was subjected to 2% trypsin (pH 6) treatment. Gross and microscopic observations revealed that the 2% trypsin at 48 h time interval epidermis was easily separated with dermis (Fig. 15.2a). The de-epithelized dermis (Fig. 15.2b) in histological section showed cellularity with retention of basement membrane (Fig. 15.2c) and mild looseness of the collagen fibers. This concentration and time interval of trypsin treatment were optimized and used in next phase.

15.5.6.2 Protocols for Preparation of Acellular Dermal Matrix

In protocol-I, the de-epithelized skin was subjected to 2% trypsin concentration for 24 h followed by 2% and 3% concentrations of sodium deoxycholate for 12 h. The 2% trypsin was again used for 12 h. Then the samples were subjected to same concentration of detergents for the same time intervals. In protocol-II sodium dodecyl sulphate was used as biological detergent. In protocol- III tritron-×100 and in protocol-IV tween-20 was used as biological detergent.

15.5.6.3 Microscopic Observations

Treatment with sodium deoxycholate in 2% concentration showed about 70% decrease in the cellular contents with mildly thick and loose of the collagen fibers (Fig. 15.2d). In 3% concentration, the scaffolds showed almost similar

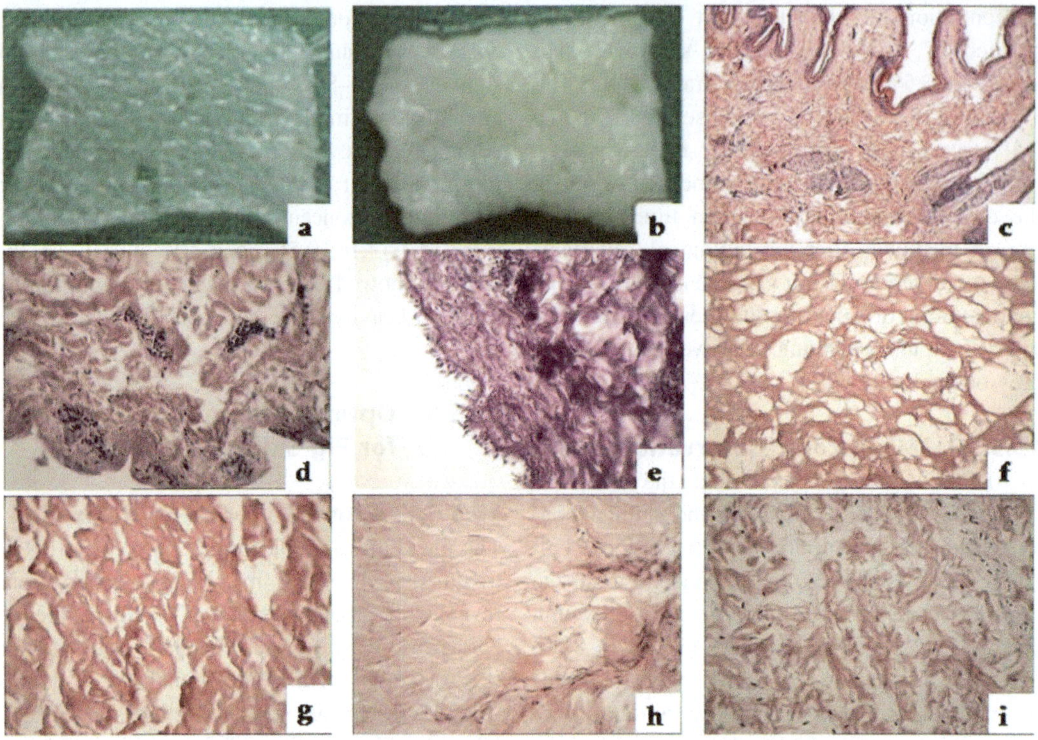

Fig. 15.2 **a–i** Optimization of acellular protocols for pig skin

cellular contents as 2% but, the collagen fibers became thin and moderately loose. Treatment with 2% sodium dodecyl sulphate revealed more than 70% decrease in the cellular contents with moderately loose and mildly thick collagen fibers (Fig. 15.2e). In 3% concentration the scaffolds became 100% acellular with mildly thick and loose collagen fibers (Fig. 15.2f). Treatment with 2% and 3% tritron-×100 showed a 90–100% decrease in the cellular contents with mildly thin and moderately loose collagen fibers (Fig. 15.2g) however, the morphology of collagen deteriorates in 3% concentration. Treatment with 2% and 3% tween-20 revealed reduction in cellularity below 70% (Fig. 15.2h), but the morphology and compactness of the collagen fibers were deteriorated in 3% concentration of detergents (Fig. 15.2i).

The concentration of the biological detergent was increased from 2 and 3% to 4 and 5% in the modified protocols. The results of the modified protocols showed more or less similar cellular contents as observed in previous protocols. However, the compactness and morphological changes revealed deterioration of the graft quality in their respective detergent groups.

15.5.7 Optimization of Protocols for Goat Skin

15.5.7.1 Protocols for De-Epithelization

The results of the gross and microscopic observations revealed that the treatment with 1% trypsin at 36 h time interval epidermis was easily separated with dermis (Fig. 15.3a). The de-epithelized dermis (Fig. 15.3b) in histological section showed cellularity with retained basement membrane (Fig. 15.3c) and mild looseness of the collagen fibers. This concentration and time interval of trypsin treatment were optimized and used in next phase.

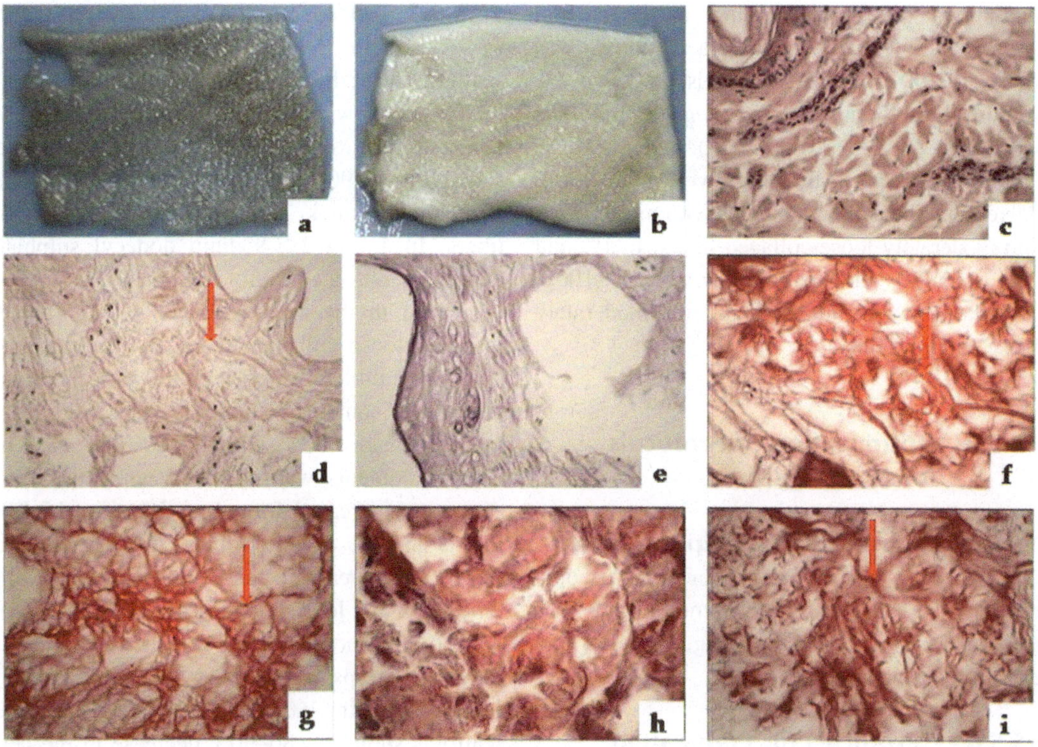

Fig. 15.3 **a–i** Optimization of acellular protocols for goat skin

15.5.7.2 Protocols for Preparation of Acellular Dermal Matrix

In protocol-I the de-epithelized skin (Fig. 15.3b) of goat was subjected to 1% trypsin concentration for 24 h. The 1% and 2% concentration of sodium deoxycholate was used for 12 h. The 1% trypsin was again used for 12 h. Then the samples were subjected to same time intervals and concentration of detergents. The samples were washed thrice at every step to rinse the residuals of previous steps. In protocol-II sodium dodecyl sulphate was used as biological detergent. In protocol-III tritron-×100 and in protocol-IV tween-20 was used as biological detergent.

The protocols were further modified to obtain the desired results. Time interval with 1% trypsin treatment was increased from 12 to 24 h. The concentration and time intervals in the remaining steps remained the same as described previously.

15.5.7.3 Microscopic Observations

Treatment with 1 and 2% sodium deoxycholate, the scaffolds showed a 90–100% decrease in the cellular contents with mildly thick and loose collagen fibers (Fig. 15.3d). However, moderate loosening of the collagen fibers was observed in 2% concentration. Treatment with 1% and 2% sodium dodecyl sulphate revealed that the scaffolds were almost acellular with mildly thin and loose collagen fibers (Fig. 15.3e). However, in 2% of the detergent treatment, thin collagen fibers were observed. In 1% and 2% tritron-×100 treatment the scaffolds showed a 80 to 90% decreased in the cellular contents with mildly thick and loose collagen fibers (Fig. 15.3f, g). Treatment with 1% and 2% of tween-20, the scaffold showed a 90–100% decrease in the cellular contents with mild to moderate thin and moderately loose collagen fibers (Fig. 15.3h).

In further modified protocols the 1 and 2% sodium deoxycholate treatment, showed more or less similar histological findings as observed in previous protocols, but the acellularity was nearer to 100% in these scaffolds. In 1 and 2% sodium dodecyl sulphate treatment the scaffolds were almost acellular (Fig. 15.3i). In 1% and 2% tritron-×100 the scaffolds showed a 90–100% decrease in the cellular contents with mildly thick and moderately loose collagen fibers. In 1% and 2% tween-20, the scaffold showed 90–100% acellularity with mildly thick and moderately loose collagen fibers.

15.5.8 Optimization of Protocols for Sheep Skin

15.5.8.1 Protocols for De-Epithelization

On the basis of the gross and microscopic observation 1% trypsin concentration and 36 h time interval was optimized (Fig. 15.4b, c) and used in next phase.

15.5.8.2 Protocols for Preparation of Acellular Dermal Matrix

The de-epithelialized sheep skin was subjected to protocols similar to that of goat.

15.5.8.3 Microscopic Observations

In 1% sodium deoxycholate concentration the scaffold showed a 70–80% decrease in the cellular contents with mildly thick and mildly loose collagen fibers. In 2% concentration, the scaffold showed a 80–90% decrease in cellular contents with mildly thick and mildly loose collagen fibers. In 1 and 2% sodium dodecyl sulphate concentration, the scaffold showed a 80–90% decreased in the cellular contents with mildly thin and loose collagen fibers. In 1% tritron-× 100 the scaffold showed a 70–80% decrease in the cellular contents with mildly thick and mildly loose collagen fibers. The 90–100% decrease in cellular contents was observed when the concentration increased from 1 to 2% (Fig. 15.4d). In 1% and 2% tween-20, the scaffold showed a 90–100% decrease in the cellular contents with mild thick and loose collagen fibers (Fig. 15.4e).

In further modified protocols the time interval for second trypsin treatment was increased from 12 to 24 h. In 1% sodium deoxycholate the scaffolds showed a 80–90% decrease in the cellular contents with mildly thick and mildly loose collagen fibers. In 2% concentration, the scaffold showed a 90–100% decrease in cellular contents with mildly thick and mildly loose collagen

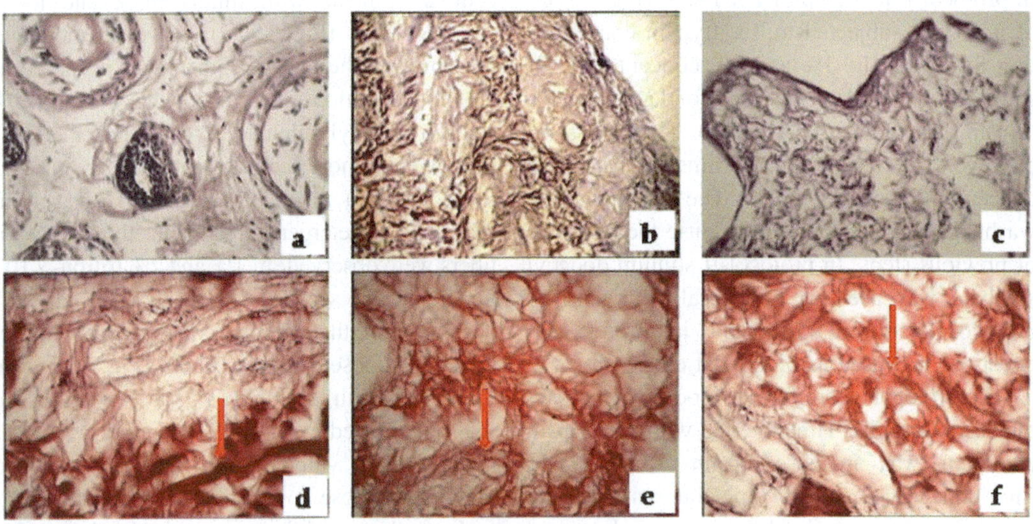

Fig. 15.4 a–i Optimization of acellular protocols for sheep skin

fibers (Fig. 15.4f). The 1 and 2% sodium deoxycholate and sodium dodecyl sulphate showed more or less similar histological findings as observed in previous protocols.

15.5.9 Optimization of Protocols for Buffalo Skin

15.5.9.1 Protocols for De-Epithelization

On the basis of the gross observations, the desirable removal of epidermis in a single layer was not possible with trypsin and NaCl solutions. The separation of epidermis was done with the help of dermatome (Fig. 15.5a) and the dermal tissue samples were adopted for further research in the next phase (Fig. 15.5b). Histological observations also showed that removal of epidermis was not possible with trypsin and NaCl solutions (Fig. 15.5c). Figure 15.5d shows removal of epidermis with dermatome.

15.5.9.2 Protocols for Preparation of Acellular Dermal Matrix

In protocol-I, the de-epithelized skin (Fig. 15.4) of bovine was subjected to 2% trypsin concentration for 24 h. The 2, 3, 4, and 5% concentration of sodium deoxycholate was used for 12 h. Then the samples were again treated with 2% trypsin for 12 h. In the next step, the samples were subjected to the same time and concentration of detergents treatment. The samples were washed thrice at every step to rinse the residuals of previous steps. In protocol-II sodium dodecyl sulphate was used as biological detergent. In protocol-III tritron-×100 and in protocol-IV tween-20 was used as biological detergent.

15.5.9.3 Microscopic Observations

Treatment with sodium deoxycholate and sodium dodecyl sulphate in 2 and 3% showed a 80–90% decrease in the cellular contents. The collagen fibers were mildly thin in 2%, became mildly thick, and loose in 3% solution. In 2% tritron-× 100 detergent the scaffolds showed a 80–90% decrease in the cellular contents (Fig. 15.5e) and in 3% solution scaffolds showed 90–100% decrease in the cellular contents (Fig. 15.5f). The collagen fibers were mildly thick and mildly loose in architecture. In 2% tween-20 biological detergent the scaffold showed a 80–90% decrease in the cellular contents (Fig. 15.5g) and 90–100% decrease in 3% solution (Fig. 15.5h). Mildly thick and moderately loose collagen fibers were seen in both the concentration.

In 4% and 5% concentration, of sodium deoxycholate the scaffolds showed a 90–100% decrease in the cellular contents. The collagen fibers were mildly thick and mildly loose in 4% concentration and became moderately loose in 5% concentration. In 4% and 5% concentrations of sodium deoxycholate, the scaffolds also showed a 90–100% decrease in the cellular contents. Collagen fibers were mildly thin and loose in 4% and mildly thick (Fig. 15.5i) and moderately loose in 5% concentration. In both the concentration tritron-×100 and tween-20 scaffolds showed a 90–100% decrease in the cellular contents with mildly thick collagen fibers. The collagen fibers were mildly loose in 4% and moderately loose in 5% concentration solution (Figs. 15.6 and 15.7).

15.5.10 Optimization of Protocols for Rat Skin

15.5.10.1 Protocols for De-Epithelization

Fresh skins were obtained from Sprague–Dawley rats of approximately 300–500 g immediately following sacrifice for other research work, and then stored at 4 °C in PBS (pH 7.4) containing 0.1% amikacin and 0.2025% ethylene diamine tetra acetic acid (EDTA). The skin was washed with PBS before being placed in hypertonic solution (605 mg tris base, 4 g sodium chloride, 202.5 mg EDTA in 100 ml PBS) at 37 °C for 8 h on a shaker. Following soaking in hypertonic solution, epidermal layer of skin was removed and then washed in PBS on a shaker for 1 h.

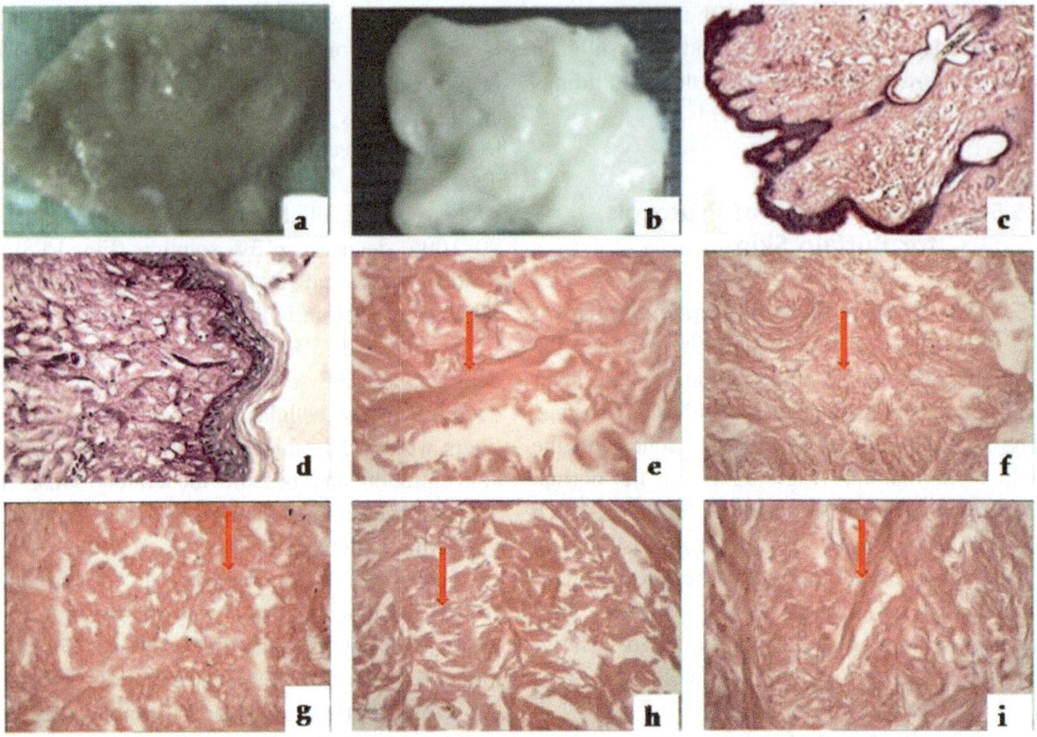

Fig. 15.5 **a–i** Optimization of acellular protocols for buffalo skin

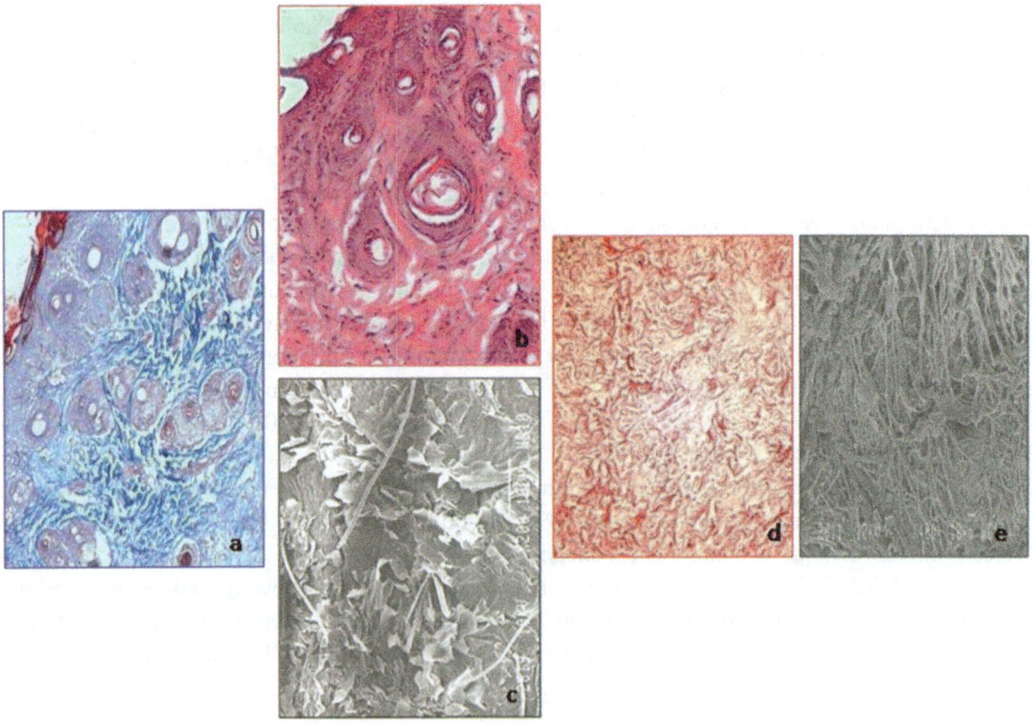

Fig. 15.6 Optimization of acellular protocols for rat skin **a** native rat skin (Masson trichrome stain) **b** native rat skin (H&E stain) **c** native rat skin (SEM picture) **d** acellular rat skin (H&E stain) **e** acellular rat skin (SEM picture)

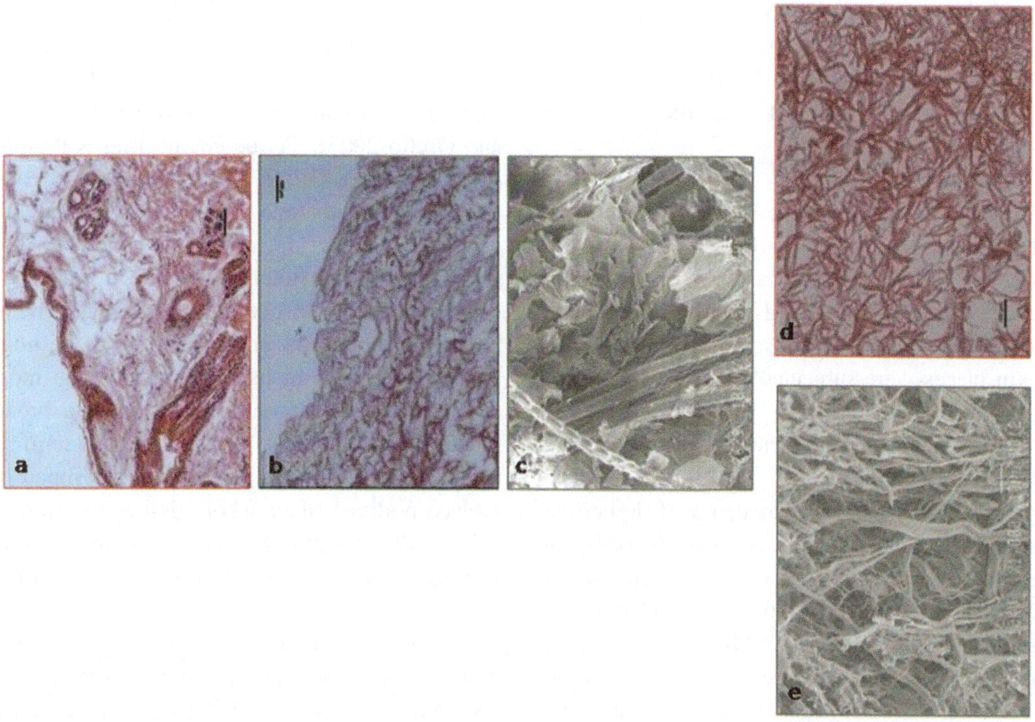

Fig. 15.7 Optimization of acellular protocols for rabbit skin **a** native rabbit skin (H&E stain) **b** de-epithelized rabbit skin (H&E stain) **c** native rabbit skin (SEM picture) **d** acellular rabbit skin (H&E stain) **e** acellular rabbit skin (SEM picture)

15.5.10.2 Protocols for Preparation of Acellular Dermal Matrix

The cellular dermis was then placed in different detergent solutions such as 0.5% sodium dodecyl sulphate (SDS), 1% triton x-100, 0.25% tri (n-butyl) phosphate (TNBP), 0.5% SDS, 0.25% TNBP and 1% triton x-100/0.25% TNBP respectively at 37 °C on an orbital shaker. Samples were qualitatively analyzed histologically with hematoxylin and eosin (H&E) staining for cell nuclei and ECM morphology.

15.5.10.3 Microscopic Observations

Microscopic imaging of rat skin before decellularization showed cellularity, dermal adnexa, and other skin structures. Treatment with hypertonic saline solution for 8 h resulted in complete de-epithelialization of the skin. Most protocols resulted in the retention of a considerable number of whole cells and cell fragments. Treatment with 0.5% SDS and 0.25% TNBP detergent combination resulted in complete decellularization of the cellular dermis at 48 h. Histological analysis of the acellular matrix showed that the dermal stromal cells were completely removed and collagen fibers were arranged in loose meshwork. Double stranded DNA contents were 698.8 ± 11.38, and 42.4 ± 2.87 ng/mg in native and 0.5% SDS and 0.25% TNBP treated rat skin, respectively. Aforementioned decellularization protocol for preparation of the MDM from harvested rat skin using hypertonic solution, SDS and TNBP have been adequate. Treatment resulted in complete decellularization and retention of the distinctive, natural, three-dimensional structures of dermal collagen within the prepared matrix (Fig. 15.6a–e) (Kumar et al. 2016).

15.5.11 Another Protocol for Preparation of Acellular Matrix from Rabbit Dermis

Rabbit skin was collected and continuously agitated for 8 h in hypertonic solution (50 mM Tris HCl, 1 M NaCl, 10 mM EDTA) in an orbital shaker. The epidermis was removed easily, and tissue samples were fixed in 10% formalin for histological evaluation. De-epithelialized rabbit skin pieces were subjected to either 1% sodium deoxycholate and tritron x-100 detergents for 48 h. Samples were collected for histological and SEM examination.

The rabbit skin was treated with hypertonic saline solution for 8 h. At 4 h intervals, the epidermis was separated with slight mechanical assistance. However, at 6 h the epidermis was separated spontaneously. Microscopically, the native skin showed cellularity, dermal adnexa, and other skin structures. The hypertonic saline solution treated skin was completely de-epithelialized at 6 h of. The de-epithelialized skin subjected to sodium deoxycholate and tritron x-100 detergents for 24 h showed acellularity. The cellular debris was seen in between the void spaces of collagen fibers in all the samples. The 48 h treated samples showed complete acellularity with the removal of cellular debris from the tissue. The collagen fibers were thin and mildly loose with moderate porosity in tritron x-100 detergent. In 1% sodium deoxycholate detergent treatment the collagen fibers were thin, mildly loose with high porosity. The SEM examination also revealed mildly loosed thin collagenous fibers and porosity (Fig. 15.7a–e) (Kumar et al. 2014).

15.6 Skin Tissue Engineering

A commonly applied definition of tissue engineering, as stated by Langer and Vacanti (1993), is "an interdisciplinary field that applies the principles of engineering and life sciences toward the development of biological substitutes that restore, maintain, or improve tissue function or a whole organ". Tissue engineering has also been defined as "understanding the principles of tissue growth, and applying this to produce functional replacement tissue for clinical use" (MacArthur and Oreffo 2005). Tissue engineering is the use of a combination of cells, engineering and materials methods, and suitable biochemical and physio-chemical factors to improve or replace biological functions. While it was once categorized as a sub-field of biomaterials, having grown in scope and importance it can be considered as a field in its own right. While most definitions of tissue engineering cover a broad range of applications, in practice the term is closely associated with applications that repair or replace portions of or whole tissues like skin.). Often, the tissues involved require certain mechanical and structural properties for proper functioning.

The collagen-based biomaterials are of the utmost importance for tissue engineering and regenerative medicine. Because of its superior biocompatibility collagen is still the protein of choice for biomaterials preparation. It can be extracted from various tissue sources and assembled in combination with other molecules. There is also use in the laboratory as a decellularized ECM in fundamental studies or as tissue replacement material in medical applications. Tissue engineering has progressed rapidly in recent years and has now emerged as a potential alternative to tissue or organ transplantation (Chapekar 2000). It has provided drastic advances in all fields of surgery like ophthalmology, orthopedic surgery, dental surgery, soft tissue surgery, cardiovascular surgery, and neurosurgery.

Tissue engineered skin substitutes emerged in the 1980s. Development was motivated primarily by the critical need for early coverage of extensive injuries in patients with insufficient sources of autologous skin for grafting. To regenerate new tissues biomedical engineering utilizes three tools; cell, scaffold, and growth factors. These tools are not always simultaneously used. For instance, a dermal tissue can be regenerated simply by placing a porous collagen sheet on a full thickness skin wound without cell seeding.

In such cases fibroblasts are regenerated from the surrounding healthy skin tissue, migrate into pores of the sheet, and secrete proteins and glycosaminoglycans which construct a dermal tissue, the sheet simultaneously absorbed from the body. The most important aspect of tissue engineering is the adhesion and proliferation of cells on biological or synthetic scaffold material. Fibronectin is an essential requirement to achieve focal contact between cell and substratum.

The scaffold provides the appropriate environment for cells and the cell synthesizes matrices of the new tissues. It assists proliferation, differentiation, and biosynthesis of the cells. To fulfill the function of a scaffold in tissue engineering, the scaffold should have interconnected micropores so that numerous cells can be seeded, migrate inside, increase the cell number, and should be supplied by sufficient amount of nutrients. An optional pore size should be in the range between 100 and 500 μm (Ikada 2006). The scaffold does not need to stay longer than one month. If the scaffold remains for a longer time than desired, the remaining material may retard the tissue regeneration rather than promote it. This indicates that the absorption kinetics of scaffold material affects the success rate of tissue engineering.

Tissue engineering is the process of creating functional 3-D tissue combining cells with scaffolds or devices that facilitate cell growth, organization, and differentiation (Huang et al. 2010). Skin substitutes are of great benefit in the treatment of patients with full thickness wounds but there is a need for improvement with respect to wound closure with minimal contraction and early vascularization. The living skin equivalents refer to skin products made mainly of cells, extracellular matrix material, or combination of cells and matrices. The living tissue equivalents (cell seeded collagen and fibrin gels) have several advantages over synthetic alternatives including being a natural cell substrate, allowing cellularity to be achieved directly, and being conducive to cell spreading and ECM formation (Neidert et al. 2002). Large number of natural polymers including fibrin, hyaluronic acid, fibrinogen, and collagen has been used as biologically superior dermal substitutes with collagen showing the most potential (Powell and Boyce 2006). The collagen-based dermal substitutes have favorable physical and biological properties (Ruszczak 2003). The majority of dermal replacement materials currently in use are based on scaffolds of bovine collagen or human donor allodermis. Among the biocompatible wound dressing materials, collagen is the most promising skin substitute or wound dressing biomaterial due to minimal inflammation, cytotoxicity, and its property to promote cellular growth (Kane et al. 1996). The collagen is the predominant extracellular protein in the granulation tissue of healing wounds and a rapid increase in the synthesis of this protein in the wound area occurs soon after an injury (Nwomeh et al. 1999). The collagen also plays a role in hemostasis (Lockhart et al. 2001). The collagen is the most abundant protein in mammals and is crucial in tissue structure. Type-I collagen is the major constituent of skin.

The field of tissue engineering deals with the reorganization of tissues by in vitro developed novel biomaterials. The surgical cases in animals' present a myriad of conditions, wherein, it is desired to adopt surgical interventions to ensure complete regeneration of damaged tissue. The collagen matrices have been used as a substitute for bringing about tissue reorganization. Ideal collagen matrices should be able to provide right biological and physiological environment to ensure homologous cell and extra cellular matrix (ECM) development and it is because of this property of collagen that it has been used extensively for tissue reorganization. Due to their advantageous biological properties, collagen-based scaffolds are still under extensive investigation for use as matrices for tissue engineering. The advantages are that natural matrix materials mimic natural ECM structure and composition; emulate native stimulating effects of ECM on cells and allow the incorporation of growth factors and other matrix proteins to further enhance cell functions.

15.7 Skin Tissue Engineering Work Done in the Division of Surgery

Decellularized rabbit skin was used to develop bioengineered collagen scaffolds for the repair of abdominal wall defects in rats. Bioengineered cell seeded scaffolds were found comparatively better than non-cell seeded scaffolds for the repair of abdominal wall defects in rats. Decellularized rabbit skin was found better and economical substitute for the costly polypropylene synthetic mesh for the repair of abdominal wall defects (Mohsina et al. 2015).

Collagen-based decellularized biomaterials as 3-D scaffold for tissue engineering of skin were developed. Evaluation of attachment, proliferation, and migration of fibroblast cells on acellular matrices was done for bioengineered collagen matrices to be used for reconstructive surgery (Kumar 2016). Mouse embryonic fibroblast cells were cultured *in vitro* and were seeded on decellularized tissues. These decellularized seeded tissues with mouse embryonic fibroblast cells were subjected to histological and scanning electron microscopic examinations. Light microscopic as well as scanning electron microscopic examinations revealed that the mouse embryonic fibroblast cells got attached, proliferated, and migrated over the acellular matrix (Fig. 15.8a, b). These scaffolds can be used for reconstructive surgery (Kumar et al. 2014).

Collagen-based biomaterials were used to develop 3-D bioengineered ECM seeded with mesenchymal stem cells (MSC) for dermal wounds in rats (Mathew et al. 2012). Collection, isolation, and characterization of rat and canine bone marrow derived MSC were performed. Tissue engineering construct of acellular rat and rabbit dermal matrix seeded with rat and canine MSC were developed and efficacy of this bioengineered 3-D matrix was tested for the repair of full thickness skin defects in rat model. The study was conducted on seventy clinically healthy adult rats of either sex. The animals were randomly divided into 7 groups of ten animals each. One 2 × 2 cm^2 size full thickness skin wound was created on the dorsum of each animal for assessing the healing potential of tissue engineered scaffolds. The defects of groups II, III, and IV were repaired with acellular rat dermal matrix, acellular rat dermal matrix seeded with rat MSC and acellular rat dermal matrix seeded with canine MSC, respectively. The defects of groups V, VI and VII were repaired with acellular rabbit dermal matrix, acellular rabbit dermal matrix seeded with rat MSC and acellular rabbit dermal matrix seeded with canine BMSC, respectively. Group I animals were treated with autograft. Macroscopically, all the

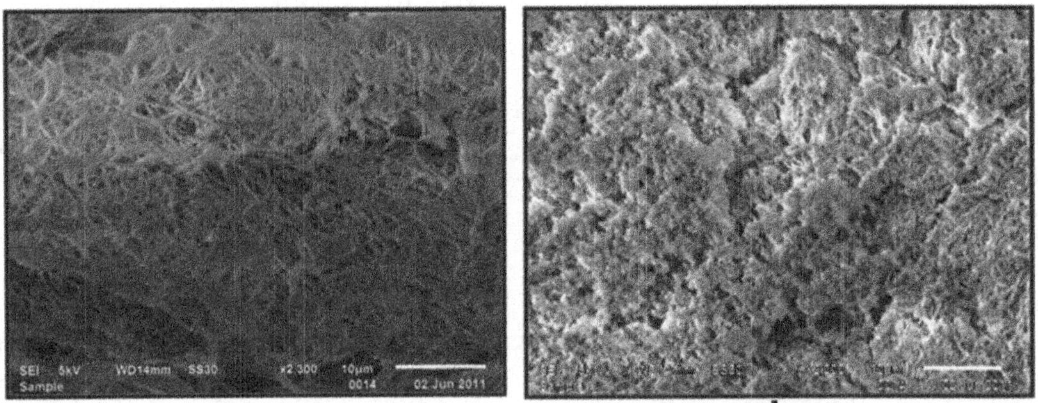

a b

Fig. 15.8 **a** and **b** SEM observation showing attachment and migration of mouse embryonic fibroblast cells over the acellular matrix

wounds implanted with stem cell seeded acellular tissue engineering scaffold revealed rapid reduction in wound size during post-surgery. Acellular rat dermal matrix seeded with rat BMSC has an edge over canine BMSC in full thickness skin wound healing (Mathew 2014).

15.8 Clinical Application of Developed Dermal Matrices in Different Species of Animals

15.8.1 Dermal Matrices Tested in Buffalo

Acellular dermal matrices of bubaline origin were tested for the reconstruction of umbilical/ventral hernias in buffalo calves. The dermal matrices were applied as inlay grafts in 4 buffalo calves having umbilical/ventral hernias. The matrices were anchored in position by 2–0 silk using horizontal mattress sutures in all cases. The details of these cases are presented in a tabular form (Table 15.2).

Present study revealed that acellular dermal grafts of buffalo origin proved to be a good biocompatible biomaterial for the reconstruction of hernia. It is easy to prepare and can be clinically practiced in other species of animals (Fig. 15.9a–f) (Kumar et al. 2012a).

Acellular dermal matrix prepared after decellularization of rabbit skin was used for hernioplasty in a buffalo. The animal was 4 years of age having a large swelling in the ventral abdominal region on the right side over the past 15 days. The case was diagnosed as ventrolateral hernia. The hernial ring was 15 cm in diameter and hernioplasty was performed using an acellular rabbit dermal matrix. The animal recovered from the condition. There were no complications of graft rejection and recurrence. On day 45 postoperatively the abdominal wall was normal in appearance (Fig. 15.10a–e) (Kumar et al. 2014c). Umbilical hernioplasty using decellularized rabbit skin in a buffalo calf was also reported (Mohsina et al. 2013).

15.8.2 Dermal Matrices Tested in Cattle

Acellular dermal matrix (ADM) has been used to close a large defect of ventral abdominal wall due to partial abdominal wall agenesis in a day old cow calf. The abdominal viscera were exposed to the environment and the defect was large enough for the abdominal muscles to get sutured/apposed. The visceral organs were cleaned using normal saline and carefully reduced into the abdominal cavity. The abdominal muscles were

Table 15.2 Summary of 4 clinical cases of hernias repaired with acellular bubaline dermal matrices

S. No.	Species	Breed	Age	Sex	Type of Hernia	Size of hernial ring (Diameter)	Biomaterial applied	Outcome/ Complication if any
1	Buffalo calf	Non-Descript	6 M	M	Umbilical Hernia	8 cm	Acellular bubaline dermis	Cured
2	Buffalo calf	Non-Descript	8 M	M	Umbilical Hernia	6 cm	Acellular bubaline dermis	Cured
3	Buffalo calf	Non-Descript	6 M	F	Umbilical Hernia	8 cm	Acellular bubaline dermis	Cured
1	Buffalo calf	Non-Descript	12 M	F	Ventral Hernia	15 × 12 cm	Acellular bubaline dermis	Cured

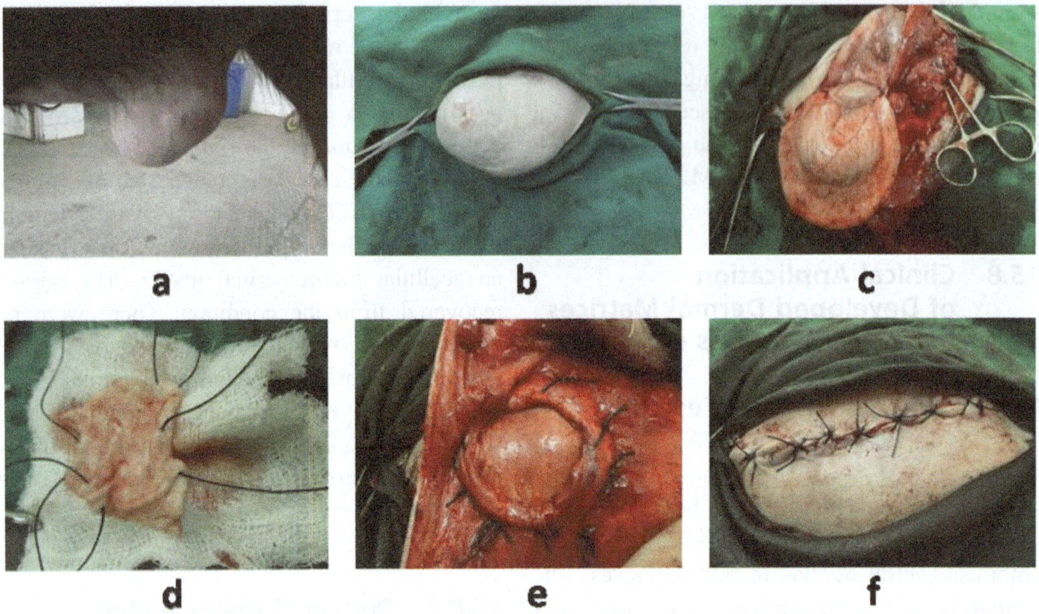

Fig. 15.9 Umbilical hernia in a buffalo calf repair with acellular bubaline dermal matrix. **a** buffalo calf showing large umbilical hernia **b** close-up view of hernia after surgical preparation of site **c** opening of hernial sac showing hernial contents **d** preplacement of silk sutures in acellular dermal matrix **e** complete suturing of graft as inlay graft **f** complete closure of defect after skin suturing

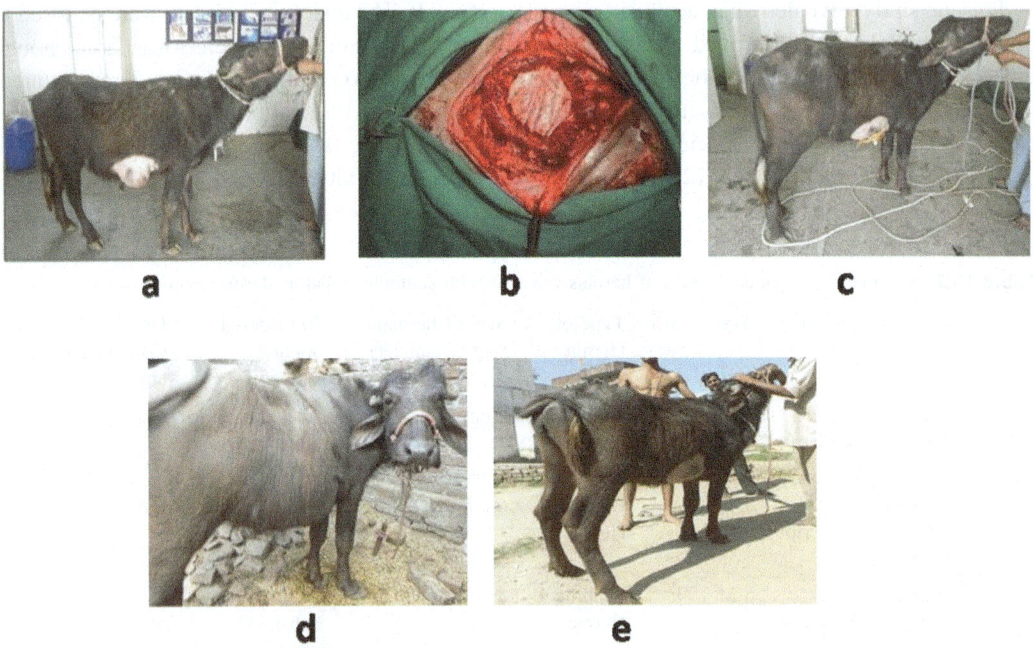

Fig. 15.10 Ventral hernia in a buffalo repair with acellular dermal matrix of rabbit origin. **a** buffalo showing large ventral hernia **b** placing of acellular dermal matrix for repair of hernial ring as inlay graft **c** immediate postoperative view after complete closure of defect **d** postoperative view after day 15 postoperatively **e** postoperative view after day 45 postoperatively

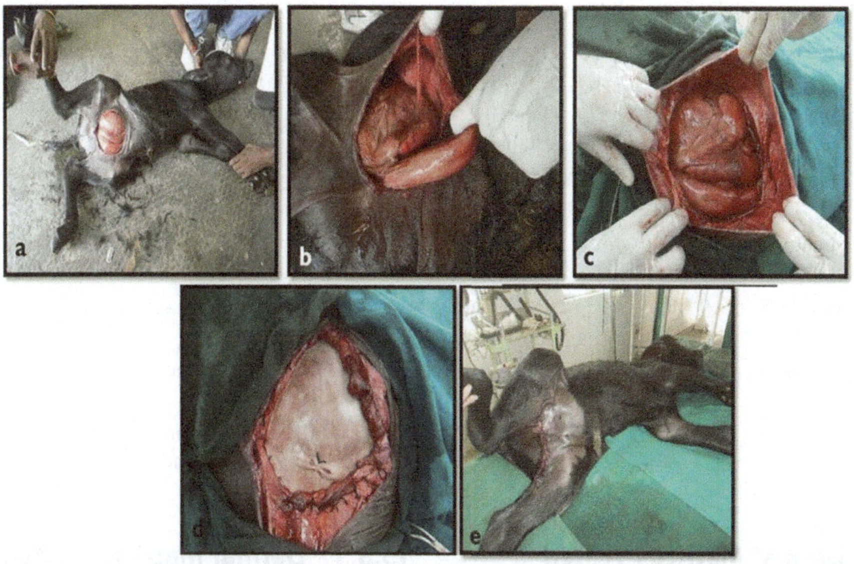

Fig. 15.11 Use of acellular dermal matrix (ADM) to repair large ventral abdominal wall defect due to partial abdominal wall agenesis in a day old cow calf. **a** a day old calf showing large ventral abdominal wall defect due to abdominal wall agenesis **b** cleaning of abdominal organs **c** freshening of abdominal wall muscles **d** placing of ADM for repair of abdominal wall defect as inlay graft **e** postoperative view after complete suturing of skin

freshened and the ADM was placed as a sublay and was sutured using no. 1 surgical silk in an interrupted pattern. The skin was undermined and extended to close the site. The animal was administered with Inj. Ceftriaxone (20 mg/kg) and Inj. Meloxicam (0.2 mg/kg) prior to surgery as well as for 5 days postoperatively. The skin sutures were removed on 15th post-op day (Fig. 15.11a–e) (Mathew et al. 2014).

15.8.3 Dermal Matrices Tested in Goats

Acellular dermal matrix (ADM) of bovine and caprine origin were used for repair of hernias in goats. Hernia in a goat of a size 4 × 3 cm was repaired with acellular dermal graft of buffalo origin. The dermal matrix was made acellular as per the technique standardized in our laboratory. To prevent early degradation, it was treated with 0.05% glutaraldehyde for 18–20 h at room temperature. The graft was fixed after making hernial ring freshened with silk at a continuous pattern. The case was successfully treated (Gangwar et al. 2003). In another case, xenogenic acellular dermal graft without crosslinking with glutaraldehyde was used for the repair of ventral hernia in a non-descript goat (Gangwar et al. 2004). ADM of caprine origin was tested in six goats. The skin of goat was de-epithelialized using 2 M sodium chloride solution and 0.25% trypsin, and further decellularized using 2% sodium deoxycholate, which was then used as graft material. Under xylazine sedation and local infiltration analgesia, the abdominal wall hernias were repaired with ADM graft using inlay graft technique. All animals had an uneventful recovery without clinical signs of wound dehiscence, infection, or recurrence of hernia during 6 months' follow-up periods. The details of these cases are presented in a tabular form (Table 15.3).

The results of the present study demonstrated that ADM of caprine origin serves as a viable option for the repair of abdominal wall hernias in goats (Fig. 15.12a–c) (Kumar et al. 2013a).

Table 15.3 Cases of abdominal wall hernias in goats

S. No.	Breed	Age	Sex	Type of Hernia	Hernial ring (Diameter) (cm)	Hernial contents	Reducible
1	Jamunapari	8 M	M	Ventral Hernia	5	Small intestine	Yes
2	Sirohi	6 M	M	Ventral Hernia	4	Small intestine	Yes
3	Sirohi	17 M	M	Ventral Hernia	9	Omentum	Yes
4	Sirohi	9 M	M	Incisional Ventral hernia	6	Small intestine	Yes
5	Crossbred	36 M	M	Incisional Ventral hernia	12	Fat	Yes
6	Crossbred	4 M	F	Umbilical hernia	4	Small intestine and mesentry	Yes

15.8.4 Dermal Matrices Tested in Horses

Acellular dermal matrix of rat origin was used for surgical repair of ventral hernia in horses. The skin from rats was de-epithelialized using hypertonic solution and further decellularized with 0.5% sodium dodecyl sulphate and 0.25% tri-(n-butyl)-phosphate. Six client-owned adult males, weighing between 300 and 500 kg (mean weight: 366.67 kg) of undefined breed were used in this study. The animals had history of progressive development of swelling at the ventral abdomen and lateral body wall after blunt trauma and gore injury. Palpation at the ventral abdomen and lateral body wall revealed a large, soft, reducible hernia with a discernible ring, which was painless in five horses but painful in the sixth horse. Multiple loops of bowel could be palpated traversing the hernial ring. The bowel appeared to be located within the subcutaneously. The size of the hernial ring ranged from 7 to 12 cm (mean size: 10.83 cm) in diameter. Under general anesthesia, the hernial ring was exposed and repaired with the ADM graft using inlay graft technique. All animals had an uneventful recovery without clinical signs of wound dehiscence, infection, or recurrence of hernias during the 6-month follow-up period (Fig. 15.13a–e) (Kumar et al. 2013b).

15.8.5 Dermal Matrices Tested in Dogs

In the search of biologically functional biomaterial, murine dermal matrix (MDM) was developed and clinically evaluated as a biological mesh in dogs. Fresh skins of murine origin were de-epithelialized and further decellularized with different detergents over a range of periods. Decellularization completeness was confirmed qualitatively and histologically with hematoxylin and eosin as well as a quantitative Pico-Green assay. Under anesthesia, inguinal and perineal defects were repaired with MDM in ten dogs having a body weight of 16.81 ± 3.88 kg and hernial ring size of 21.66 ± 1.60 cm^2 with good results. Complete de-epithelialization was achieved with hypertonic saline solution at 8 h. Histologically the matrix treated with 0.5% sodium dodecyl sulphate (SDS)/0.25% tri (n-butyl) phosphate (TNBP) for 48 h demonstrated complete acellularity and orderly arranged collagen fibers. Defects repaired with MDM healed completely without graft rejection. SDS-PAGE analysis revealed a changed serum protein distribution pattern on post-implantation day 15 as that of day 0 and became normal on post-implantation day 30. Enzyme linked immunosorbent assay confirmed a significantly ($P < 0.05$) increased antibody titer on day 15 as compared to day 0, thereafter it decreased on day 30 as that of day 15

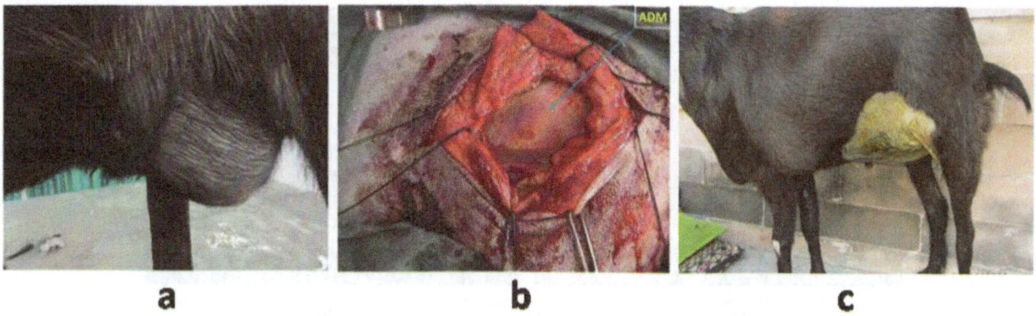

Fig. 15.12 Ventral abdominal hernia in a goat repaired with acellular caprine dermal matrix. **a** image showing unusual large incisional ventral hernia in a male crossbred goat **b** intraoperative image showing placement of ADM to the peritoneum after tying of the sutures on the external sheath of rectus abdominis muscle **c** complete recovery at one month postoperatively

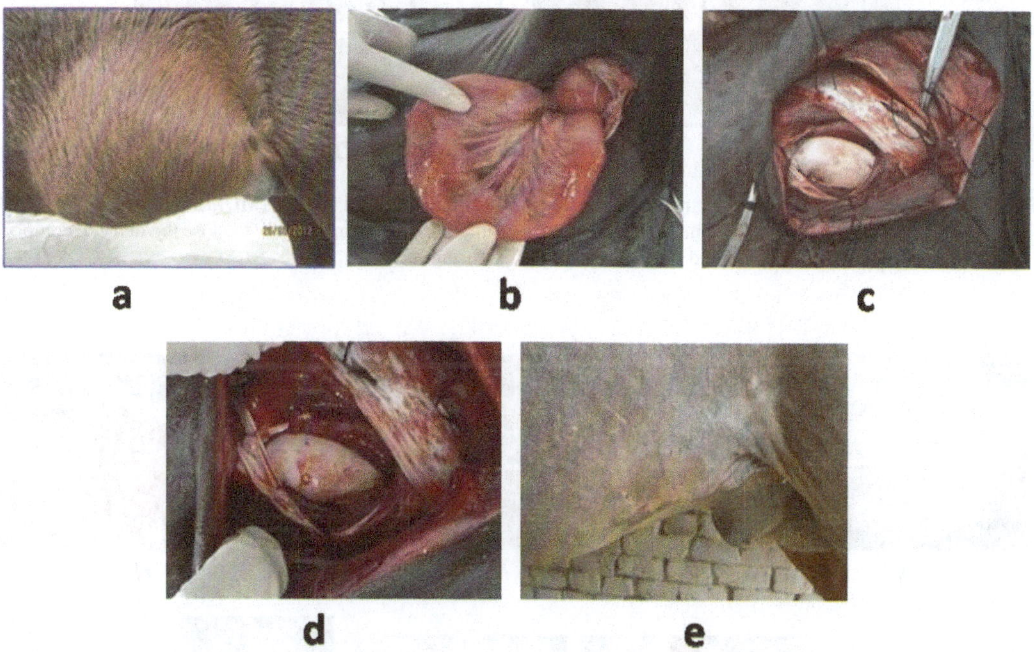

Fig. 15.13 Repair of large ventro-lateral hernia in a horse with acellular dermal matrix of rat origin. **a** horse showing large swelling on the left ventro-lateral abdomen region **b** showing intestines as hernial contents **c** placing of acellular dermal matrix of rat origin for repair of hernial ring **d** biomaterial was placed as inlay graft **d** complete recovery at two months postoperatively

but still remained significantly ($P < 0.05$) higher than day 0. These results suggest that 0.5% SDS/0.25% TNBP combination for 48 h can be used for complete decellularization of rat skin and developed MDM may be safely used as a biological mesh in dogs (Fig. 15.14a–d) (Kumar et al. 2016). Xenogenic crosslinked acellular dermal graft was used for the repair of perineal hernia in a dog (Purohit et al. 2008). Acellular dermal matrix has been used for perineal hernioplasty in a Bhotia dog (Fig. 15.15a–g) (Mathew et al. 2013). Acellular dermal matrix of rabbit origin was used for the repair of hernias in dogs (Kumar et al. 2012b; Remya et al. 2014).

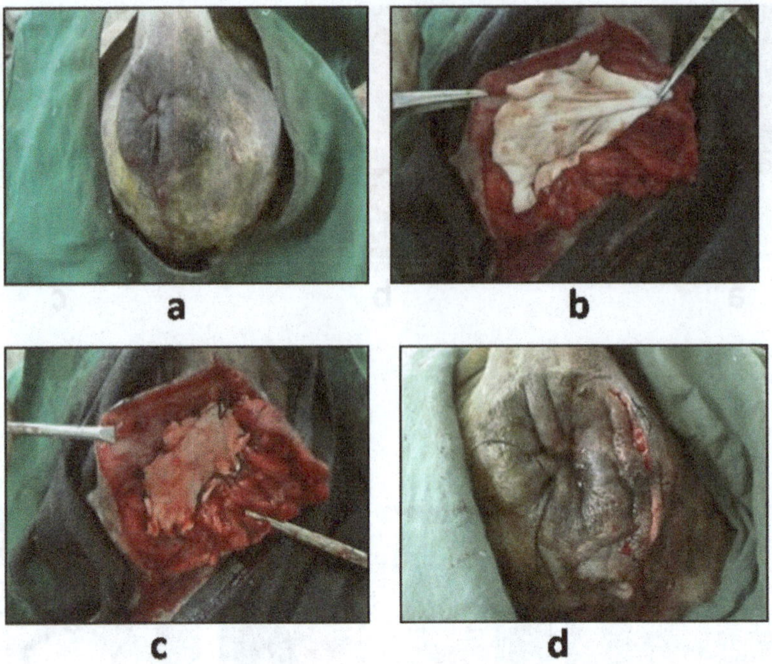

Fig. 15.14 Perineal hernia in a dog repaired with acellular murine dermal matrix (MDM). **a** dog showing large swelling at the perineal region **b** placement of biomaterial-acellular murine dermal matrix to close the defect **c** complete closure of the defect **d** status of wound after complete skin suturing

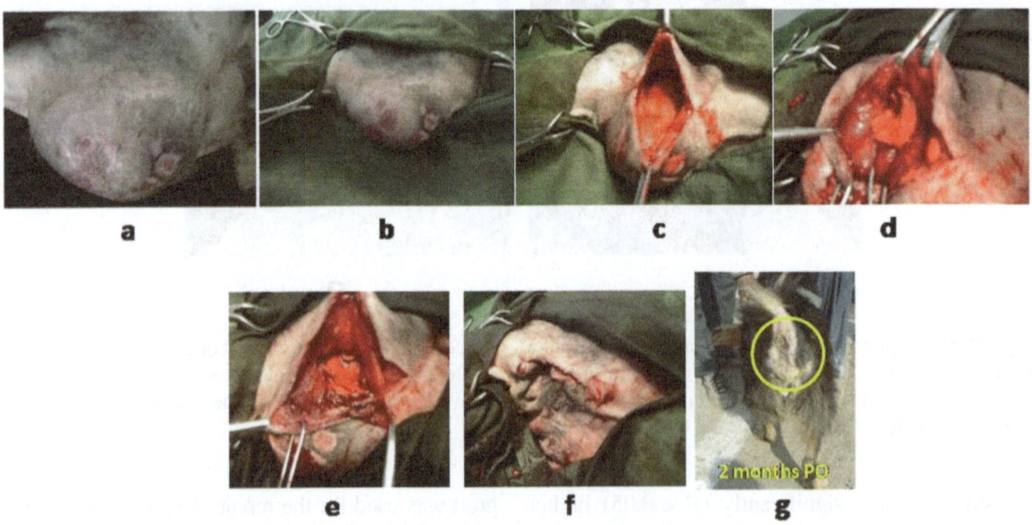

Fig. 15.15 Perineal hernia in a Bhutia dog repaired with acellular dermal matrix of rabbit. **a** nine years old male Bhutia dog showing large swelling at the perineal region **b** preparation of surgical site **c** opening of hernia sac **d** placement of biomaterial-decellularized rabbit skin to close the defect **e** complete suturing of biomaterial to fill the defect **f** skin suturing **d** complete recovery at two months postoperatively

15.9 Conclusion

Acellular dermal matrices are composed of extracellular matrix (ECM) and are typically derived by decellularization of native tissues. Decellularization of a tissue or an organ depends mainly on its density, organization, and geometric properties. Physical, chemical, and enzymatic methods, biological detergents alone or in combination are used to lyse cells, followed by rinsing to remove cell remnants. Preservation of three-dimensional ultrastructure of ECM is highly desirable, which requires proper decisions regarding the agents and techniques involved in the process. In the present study, we have standardized protocols for preparing acellular matrices from the skin of different species of animals. For different skin collected from different species of animals, we standardized the individual protocol. The percentage of biological detergent used, time for making them acellular have been optimized. The decellularization was confirmed by histological examination, DNA content analysis, and SEM analysis. In the end, we succeeded in the preservation of three-dimensional ultrastructure of ECM which is highly desirable and requires proper decisions regarding the agents and techniques involved in the process. These acellular dermal matrices are used in different reconstructive procedures in veterinary patients and can be applicable to human patients also. This chapter describes different protocols to develop acellular dermal matrices from different species of animals for clinical applications.

References

Allaire E, Brunewal P, Mandet C, Becquemin J, Michel J (1997) The immunogenicity of extracellular matrix in arterial xenografts. Surgery 122:73–81

Allman AJ, McPherson TB, Badylak SF, Merrill LC, Kallakury B, Sheehan C, Raeder RH, Folk JE (1980) Transglutaminases. Annu Rev Biochem 49:517–531

Angell DL, Angell WW (1976) Heart valve stent. USA patent 3983581

Badylak SF (2002) The extracellular matrix as a scaffold for tissue reconstruction. Semin Cell Dev Biol 13 (5):377–383

Badylak S, Kokini K, Tullius B, Simmons-Byrd A, Morff R (2002) Morphologic study of small intestinal submucosa as a body wall repair device. J Surg Res 103(2):190–202

Burke JF, Yannas IV Jr, Quinby WC (1981) Successful use of a physiologically acceptable artificial skin in the treatment of extensive burn injury. Ann Surg 194:413–428

Cartmell JS, Dunn MG (2000) Effect of chemical treatment on tendon cellularity and mechanical properties. J Biomed Mater Res 49:134–140

Cebotari S, Tudorache I, Jaekel T, Hilfiker A, Dorfman S, Ternes W (2010) Detergent decellularization of heart valves for tissue engineering: toxicological effects of residual detergents on human endothelial cells. Artif Organs 34(3):206–210

Chapekar MS (2000) Tissue engineering: challenges and opportunities. J Biomed Mater Res Appl Biomater 53:617–620

Coito AJ, Kupiec-Weglinsky JW (1996) Extracellular matrix protein by standers or active participants in the allograft rejection cascade? Ann Transplant 1:14–18

Dahl SL, Koh J, Prabhakar V, Niklason LE (2003) Decellularized native and engineered arterial scaffolds for transplantation. Cell Transplant 12:659–666

Dewanjee MR (1985) Mayo Foundation, assignee. Treatment of Collagenous Tissue with Glutaraldehyde and Aminodiphosphoate Calcification Inhibitor. USA patent 4553974

Dzemeshkevich SL, Konstantinov BA, Gromova GV, Lyudinovskova RA, Kudrina LL (1994) The mitral valve replacement by the new-type bioprostheses (features of design and long-term results). J Cardiovasc Surg 35(6 Suppl 1):189–191

Freytes DO, Badylak SF, Webster TJ, Geddes LA, Rundell AE (2004) Biaxial strength of multilaminated extracellular matrix scaffolds. Biomaterials 25:2353–2361

Gangwar AK, Sharma AK, Kumar N, Kumar N (2003) Surgical management of ventral hernia with glutaraldehyde treated acellular dermal graft in a goat. Vet Pract 4:25–26

Gangwar AK, Sharma AK, Kumar N, Maiti SK, Kumar N (2004) Xenogenic acellular dermal graft for the repair of ventral hernia in a non-descript goat. Indian Vet Med J 28:95–96

Gilbert TW, Sellaroa TL, Badylak SF (2006) Decellularization of tissues and organs. Biomaterials 27:3675–3683

Gilbert TW, Freund J, Badylak SF (2009) Quantification of DNA in biologic scaffold materials. J Surg Res 152 (1):135–139

Grauss RW, Hazekamp MG, Oppenhuizen F, Van Munsteren CJ, Giltenberger-de Groot AC, De Ruiter MC (2005) Histological evaluation of decellularized porcine aortic valves: matrix changes due to different decellularization methods. Eur J Cardiothorac Surg 27:566–571

Gulati AK (1988) Evaluation of acellular and cellular nerve grafts in repair of rat peripheral nerve. J Neurosurg 68(1):117–123

Gulati AK, Cole GP (1990) Nerve graft immunogeneicity as a factor determing axonal regeneration in the rat. J Neurosurg 72:114–122

Gulati AK, Cole GP (1994) Immunogenicity and regenerative potential of acellular nerve allograft to repair peripheral nerve in rats and rabbits. Acta Neurochir Wein 126:158–164

Hardin-Young J, Parenteau NL (2002) Bilayered skin constructs. In: Atala A, Lanza RP (eds) Methods of tissue engineering. Academic Press, California, pp 1177–1188

Hodde M, Hiles A (2002) Virus safety of a porcine-derived medical device: evaluation of a viral inactivation method. Biotechnol Bioeng 79:211–216

Huang S, Xu Y, Wu C, Sha D, Fu X (2010) In vitro constitution and in vivo implantation of engineered skin constructs with sweat glands. Biomaterials 31:5520–5525

Ikada Y (2006) Challenges in tissue engineering. J Res Soc Interface 3:589–601

Kane JB, Tompkins RG, Yarmush ML, Burk JF (1996) Burn Dressing. In: Ratner BD, Hoffman AS, Schoen FJ, Lemens JE (eds) Biomaterial Science. Academic Press, San Diego, pp 360–370

Kim MS, Hong KD, Shin HW, Kim SH, Kim SH, Lee MS, Jang WY, Khang G, Lee HB (2005) Preparation of porcine small intestinal submucosa sponge and their application as a wound dressing in full-thickness skin defect of rat. Int J Biol Macromolles 36:54–60

Klein B, Schiffer R, Hafemann B, Klosterhalfen B, Zwadlo-Klarwasser G (2001) Inflammatory response to a porcine membrane composed of fibrous collagen and elastin as dermal substitute. J Mater Sci Mater Med 12(5):419–424

Konstantinovic ML, Lagae P, Zheng F, Verbeken EK, De Ridder D, Deprest JA (2005) Comparison of host response to polypropylene and non-cross-linked porcine small intestine serosal-derived collagen implants in a rat model. BJOG 112(11):1554–1560

Kropp BP, Cheng EY, Lin HK, Zhang Y (2004) Reliable and reproducible bladder regeneration using unseeded distal small intestinal submucosa. J Urol Supplement 172(4):1710–1771

Kumar N (2016) Development of novel biological soft tissue materials for tissue engineering. In: ISCM-2016, 2nd International conference on soft tissue materials, 12–16, December, 2016, MNIT, Jaipur, Rajasthan, ID 1010, p 14

Kumar V, Kumar N, Gangwar AK, Sharma AK, Singh H, Saxena AC, Negi M, Kaarthick DK (2012a) Acellular dermal grafts for the reconstruction of Umbilical/Ventral hernias in buffalo calves. Vet Pract 13(1):90–92

Kumar V, Mathew DD, Kumar N, Gangwar AK, Saxena AC, Remya V, Mohsina A (2012) Acellular dermal matrix of rabbit origin for the repair of hernias in dogs. In: XXXVI annual congress of ISVS, Anand, Gujrat, during 1–3 November, 2012, Abstract, pp 99–100

Kumar V, Kumar N, Mathew DD, Gangwar AK, Saxena AC, Remya V (2013a) Repair of abdominal wall hernias using acellular dermal matrix in goats. J Appl Anim Res 41(1):117–120. https://doi.org/10.1080/09712119.2012.73822

Kumar V, Gangwar AK, Mathew DD, Ahmed RA, Saxena AC, Kumar N (2013b) Acellular dermal matrix for surgical repair of ventral hernia in horses. J Equine Vet Sci 33(4):238–243. https://doi.org/10.1016/j.jevs.2012.06.017

Kumar N, Gangwar AK, Mathew DD, Shrivastava S, Negi M, Singh H, Sharma AK, Maiti SK, Vellachi R, Sonal, Singh MP (2014) Development of collagen based decellularized biomaterials as 3-D scaffold for tissue engineering of skin. Trends Biomater Artif Organs 28(3):83–91

Kumar N, Mathew DD, Gangwar AK, Remya V, Mohsina A, Maiti SK, Sharma AK (2014) Reconstruction of large ventro-lateral hernia in a buffalo with acellular dermal matrix: a method for treating large hernias in animals–a case report. Veterinarski Arhiv 84(6):691–99

Kumar V, Gangwar AK, Kumar N (2016) Evaluation of the murine dermal matrix as a biological mesh in dogs . Proc Natl Acad Sci, India Sect B Biol Sci 86:953–960. https://doi.org/10.1007/s40011-015-0543-8

Laemmli UK (1970) Cleavage of structural proteins during the assembly of the head of bacteriophage T4. Nature 227:680–685

Langer R, Vacanti JP (1993) Tissue engineering. Science 260(5110):920–926

Liotta DS, Ferrari HM, Pisanu AJ and Donato FO (1978) Low profile gulteraldehyde-fixed porcine aortic prosthetic device. USA patent 4079468

Lockhart LK, Pampolina C, Nickolaychuk BR, McNicol A (2001) Evidence for a role of phospholipase A2 in platelet activation in response to low concentrations of collagen. J Thromb Haemost 85 (5):882–889

MacArthur BD, Oreffo RO (2005) Bridging the gap. Nature 433(7021):19

Malone JM, Brendel K, Duhamil RC, Reinert RL (1984) Detergent- extracted small diameter vascular prostheses. J Vasc Surg 1:181–191

Mathew DD (2014) Development of bioengineered 3-D acellular dermal matrices seeded with mesenchymal stem like cells for dermal wounds in rats, PhD Thesis, Submitted to Deemed University, IVRI, Izatnagar, Uttar Pradesh, India

Mathew DD, Kumar N, Sharma AK, Shrivastava S, Gangwar AK, Negi M, Himani Remya V, Karthick DT, Maiti SK, Kurade NP (2012) Collagen based biomaterials as 3-D scaffold for tissue engineering. In: International conference on design of biomaterials, XXIII annual meeting of SBAOI and VI annual general meting of STERMI, 9–11, December, 2012, IISc, Bangalore, Abstract on Page 239

Mathew DD, Remya V, Sivanarayanan TB, Sangeetha P, Pawde AM, Mahendran K, Mohsina A, Sharma AK, Kumar N (2013a) Perineal hernioplasty in a Bhotia dog: a case report. In: XXXVII annual congress of ISVS, 6 -8 November 2013, Mannuthy, Thrissur, Kerala, Abstract in Small Animal Surgery (Poster), Page 245–246

Mathew DD, Remya V, Madhu DN, Sivanarayanan TB, Kumar N, Sharma AK (2014) Partial abdominal wall agenesis in a calf and its surgical treatment using acellular dermal matrix. In: XXXVIII Annual Congress of ISVS, 15–16 October, 2014, Bikaner, Rajasthan, Abstract in Ruminant Surgery Session-I, p 74

Mohsina A, Gangwar AK, Mathew DD, Remya V, Maiti SK, Sharma AK, Kumar N (2013) Umbilical hernioplasty using decellularised rabbit skin in a buffalo calf. In: XXXVII annual congress of ISVS, 6–8 November, 2013, Mannuthy, Thrissur, Kerala, Abstract in Large Animal Surgery (Poster), Page 235

Mohsina A, Kumar N, Sharma AK, Mishra B, Mathew DD, Remya V, Shrivastava S, Negi M, Kritaniya D, Tamilmahan P, Maiti SK, Shrivastava S, Singh KP (2015) Bioengineered acellular dermal matrices for the repair of abdominal wall defects in rats. Hernia 19(2):219–229. https://doi.org/10.1007/s10029-014-1308-7

Neidert MR, Lee ES, Oegema TR, Tranquillo RT (2002) Enhanced fibrin remodeling *in* vitro with TGF-beta 1, insulin and plasmin for improved tissue equivalents. Biomaterials 23:3717–3731

Nwomeh BC, Liang HX, Cohen IK, Yager DR (1999) MMP-8 is the predominant collagenase in healing wounds and non-healing ulcers. J Surgical Res 81 (21):189–195

O'Neill P, Booth AE (1984) Use of porcine dermis as a dural substitute in 72 patients. J Neurosurg 61(2):351–354

Patino MG, Neiders ME, Andreana S, Noble B, Cohen RE (2003) Cellular inflammatory response to porcine collagen membranes. J Periodontal Res 38 (5):458–464

Powell HM, Boyce ST (2006) EDC crosslinking improves skin substitute strength and stability. Biomaterials 27:5821–5827

Purohit S, Kumar N, Sharma AK, Dewangan R, Maiti SK (2008) Xenogenic crosslinked acellular dermal graft for repair of perineal hernia in a dog. In: XXXII annual congress of ISVS, Namakkal, Tamil Nadu, 6–8 November, 2008, Abstract 3.27

Ratner BD (2006) Biomaterials tutorial: an introduction to biomaterials. University of Washington Engineering Biomaterials. http://www.uweb.engr.washington.edu/research/tutorials/introbiomat.html. Accessed on 5th June 2007

Ratner BD, Hoffman AS, Schoen FJ, Lemons JE (1996) Biomaterials science: an introduction to materials in medicine. Academic Press, New York, pp 1–10, 84–94

Remya V, Saxena AC, Ninu AR, Kumar R, Mohan D, Kumar N (2014) Hernioplasty for correction of ventral hernia in a dog using acellular dermal matrix. In: XXXVIII annual congress of ISVS, 15–16 October, 2014, Bikaner, Rajasthan, Abstract in Small Animal Surgery Session-II, p 233

Ruszczak Z (2003) Effect of collagen matrices on dermal wound healing. Adv Drug Deliv Rev 55:1595–1611

Schechter I (1975) Prolonged retention of glutaraldehyde-treated skin allografts and xenografts: immunological and histological studies. Ann Surg 182(6):699–704

Schmidt CE, Baier JM (2000) Acellular vascular tissues: natural biomaterials for tissue repair and tissue engineering. Biomaterials 21(22):2215–2231

Takami Y, Matsuda T, Yoshitake M, Hanumadass M, Walter RJ (1996) Dispase/detergent treated dermal matrix as a dermal substitute. Burns 22:182

van der Rest M, Garrone R (1991) Collagen family of proteins. Faseb J 5(13):2814–2823

Weadock K, Olson RM, Silver FH (1984) Evaluation of collagen crosslinking techniques. Biomater Med Devices Artif Organs 11(4):293–318

Woods T, Gratzer PF (2005) Effectiveness of three extraction techniques in the development of a decellularized bone–anterior cruciate ligament–bone graft. Biomaterials 26:7339–734

Yannas IV (1996b) Natural materials. In: Ratner BD, Hoffman AS, Schoen FJ, Lemon JE (eds) Biomaterial Science. Academic Press, San Diego, pp 84–94

Yannas IV, Ratner BD, Hoffman AS, Schoen FJ (1996) Natural materials. In. Biomaterials science. Academic Press, San Diego, pp 84–94

Yoganathan AP (1995) Cardiac valve prosthesis. In: Bronzino JD (ed) The biomedical engineering hand book. CRC Press, Boca Raton, FL, pp 1847–1870

Zhang MH, Chen J, Kirilak Y, Willers C, Xu J, Wood D (2005) Porcine small intestine submucosa (SIS)is not an acellular collagenous matrix and contains porcine DNA: possible implications in human implantation. J Biomed Mater Res B Appl Biomater 73(1):61–67

Tissue Engineering in Skin Substitute 16

Wojciech Łabuś, Diana Kitala, Michał Szapski, Agnieszka Klama-Baryła, Małgorzata Kraut, and Wojciech Smętek

Abstract

Thermal injuries may cause significant damage to large areas of the skin. Extensive and deep burn wounds require specialized therapy. The optimal method in the strategy of treating extensive, full thickness burns (III°) is the use of autologous split thickness skin grafts STSG (Busuioc et al. Rom J Morphol Embryol 4:1061–1067, 2012; Kitala D, Kawecki M, Klama-Baryła A, Łabuś W, Kraut M, Glik J, Ryszkiel I, Kawecki MP, Nowak M. Allogeneic vs. Autologous Skin Grafts in the Therapy of Patients with Burn Injuries: A Restrospective, Open-label Clinical Study with Pair Matching. Adv Clin Exp Med. 2016 Sep-Oct;25(5):923–929.; Glik J, Kawecki M, Kitala D, Klama-Baryła A, Łabuś W, Grabowski M, Durdzińska A, Nowak M, Misiuga M, Kasperczyk A. A new option for definitive burn wound closure - pair matching type of retrospective case–control study of hand burns in the hospitalized patients group in the Dr Stanislaw Sakiel Center for Burn Treatment between 2009 and 2015. Int Wound J. 2017 Feb 21. https://doi.org/10.1111/iwj.12720. [Epub ahead of print]; Prim et al. May 24Wound Repair Regen., 2017; Grossova et al. Mar 31Ann Burns Fire Disasters 30:5–8, 2017). The main limitation of that method is the inadequate amount of healthy, undamaged skin (donor sites), which could be harvested and used as a graft. Moreover, donor sites are an additional wounds that require analgesic therapy, leave scars during the healing process and they are highly susceptible to infection (1–6). It must be emphasized that in terms of the treatment of severe, deep and extensive burns, and there should be no doubt that the search for a biocompatible skin substitute that would be able to replace autologous STSG is an absolute priority. The above-mentioned necessitates the search for new treatment methods of severe burn wounds. Such methods could consider the preparation and application of bioengineered, natural skin substitutes. At present, as the clinical standard considered by the physicians may be use of available biological skin substitutes, e.g., human allogeneic skin, in vitro cultured skin cells, acellular dermal matrix ADM and revitalized ADMs, etc. (Busuioc et al. Rom J Morphol Embryol 4:1061–1067, 2012; Kitala D, Kawecki M, Klama-Baryła A, Łabuś W, Kraut M, Glik J, Ryszkiel I, Kawecki MP,

W. Łabuś (✉) · D. Kitala · M. Szapski · A. Klama-Baryła · M. Kraut · W. Smętek
Dr Stanislaw Sakiel Center for Burn Treatment, Siemianowice Śląskie, Poland

M. Szapski
Gyncentrum, Laboratory of Molecular Biology and Virology, Katowice, Poland

W. Smętek
Warsaw University of Technology, Warsaw, Poland

A. Kajbafzadeh (ed.), *Decellularization Methods of Tissue and Whole Organ in Tissue Engineering*, Advances in Experimental Medicine and Biology 1345,
https://doi.org/10.1007/978-3-030-82735-9_16

Nowak M. Allogeneic vs. Autologous Skin Grafts in the Therapy of Patients with Burn Injuries: A Restrospective, Open-label Clinical Study with Pair Matching. Adv Clin Exp Med. 2016 Sep-Oct;25(5):923–929.; Glik J, Kawecki M, Kitala D, Klama-Baryła A, Łabuś W, Grabowski M, Durdzińska A, Nowak M, Misiuga M, Kasperczyk A. A new option for definitive burn wound closure - pair matching type of retrospective case–control study of hand burns in the hospitalised patients group in the Dr Stanislaw Sakiel Center for Burn Treatment between 2009 and 2015. Int Wound J. 2017 Feb 21. https://doi.org/10.1111/iwj.12720. [Epub ahead of print]; Prim et al. May 24Wound Repair Regen., 2017; Grossova et al. Mar 31Ann Burns Fire Disasters 30:5–8, 2017; Łabuś et al. FebJ Biomed Mater Res B Appl Biomater 106:726–733, 2018).

16.1 History

Historically, the first use of human allogeneic skin dates back to World War II. Natural skin substitutes obtained by the means of tissue engineering, and it is possible to decrease the area of the donor sites and shorten the time of hospitalization among burned patients (Kitala et al. 2016). It has been shown that the use of biological skin substitutes (such as allogeneic skin grafts obtained from human cadaver donors) is beneficial even when a single clinical application did not lead to complete wound closure, which in turn forced the implementation of further procedures (e.g., the need for autologous STSG application). However, even than the optimal preparation of the wound for autologous STSG has been achieved (e.g., proper cleansing and vascularization of the wound and final determination of the degree of skin damage, etc.). In addition, the area required for harvesting of autologous skin have been decreased (Kitala et al. 2016; Glik et al. 2017; Prim et al. 2017). It has been shown that the above-mentioned phenomena may significantly reduce the time of hospitalization (Kitala et al. 2016). However, it should be remarked that the use of "traditional" allogeneic biostatic skin grafts is only a temporary solution, because such biological dressings are characterized by high risk of graft rejection (Busuioc et al. 2012; Kitala et al. 2016; Prim et al. 2017; Łabuś et al. 2018; Kawecki et al. 2018). Therefore, the benefits of using acellular dermal matrix ADM (a natural, bioengineered, cell-free skin substitutes) inspire to continue the work aimed at improving the properties of these biomaterials (Łabuś et al. 2018; Kawecki et al. 2018). Despite the clinical application of tissue-engineered acellular dermal matrices (ADM) is a relatively new approach in the history of wound healing the use of elements of natural, dermal cell-free ECM is an increasingly common strategy for regenerative medicine and tissue engineering, especially in the manner of burn wound or chronic wound management (Kitala et al. 2016; Grossova et al. 2017; Łabuś et al. 2018; Kawecki et al. 2018; Barrio et al. 2016; Barkan et al. 2010; Cortiella et al. 2010; Sellaro et al. 2010; Cheng et al. 2009; Ross et al. 2009; Stern et al. 2009; Wainwright and Bury 2011; Bing et al. 2019; Nafisi et al. 2017; Begum et al. 2016; Gowda et al. 2016; Guo et al. 2014).

Skin is a collagen-rich tissue and is a rich source of biomaterials used in tissue engineering (Barrio et al. 2016; Barkan et al. 2010). As a result of cell removal from allogeneic human dermis, one obtains an acellular, collagenous, non-immunogenic lattice that can be de novo revitalized by autologous cells. Such an acellular dermal matrix (ADM) or acellular dermal graft (ADG) might be a stimulus for natural regenerating and reconstructing mechanisms (Busuioc et al. 2012; Kitala et al. 2016; Glik et al. 2017; Prim et al. 2017; Grossova et al. 2017; Łabuś et al. 2018; Kawecki et al. 2018; Barrio et al. 2016; Barkan et al. 2010). The immune response is mainly directed against proteins and lipids of a cell membrane, thus cell extraction from tissue is a promising method to avoid a creation of post-transplant immune response in the body of the recipient (Garcia and Scott 2013; Butterfield 2013; Ott et al. 2008; Evans and Kon 2015). It was found that collagen, with only small inter-species differences among the collagen type I family, is characterized by low immunogenicity clinically (Evans and Kon 2015). Low immunogenicity of collagen type I stems from the insignificant differences in amino acid composition (Valerio et al. 2015). On the other hand, studies have shown that other molecular components of the extracellular matrix (ECM), such as fibronectin, may play a fundamental role in the induction of an immune response in the body of the recipient (Evans and Kon 2015).

Extracellular matrix (ECM) of the skin is especially composed of collagen type I, III, IV and VII and elastin fibers. Additionally, the composition of the skin includes glycosoaminoglycans (heparin, hyaluronic acid, chondroitin), glycoproteins (fibronectin, laminin, entactin) and proteoglycans (heparan sulfate proteoglycan, decorin) (Łabuś et al. 2018; Kawecki et al. 2018; Butterfield 2013; Ott et al. 2008).

What is more it should emphasized that natural ECM, thanks to its characteristic properties (e.g., collagenous lattice, reservoir of growth factors, ECM-cell anchoring areas, optimal pH, CO_2), ensures an optimal microenvironment for homeostatic and regenerative cell development (Butterfield 2013; Ott et al. 2008; Evans and Kon 2015; Schneider et al. 2016). The aim of the clinical usage of cell-free ECM scaffolds is the enhancement of tissue regeneration with possible minimization of an adverse host reaction to allogeneic or xenogeneic biomaterial (Schneider et al. 2016; Roolker and Patt 2000). Thus, the objective of decellularization is to obtain acellular grafts characterized by optimal biological properties, such as lack of remaining cellular elements (e.g., cell membrane phospholipids and proteins, nucleic acids, mitochondria, etc.), lack of immunogenicity, lack of calcification promotion, and lack of cytotoxicity (e.g., unrinsed detergents) (Łabuś et al. 2018). Furthermore, cell-free ECM scaffolds should present the optimal mechanical and structural properties that may ensure the biocompatibility of the graft. The maintenance of the ultrastructure composition of the extracellular matrix (ECM) is one of the most important goals of the methods of decellularization (Schneider et al. 2016; Roolker and Patt 2000). It has been shown that the ECM influences a process of mitogenesis and chemotaxis of cells (Schneider et al. 2016) as well as targeted cell differentiation (Barrio et al. 2016; Barkan et al. 2010; Cortiella et al. 2010; Sellaro et al. 2010; Cheng et al. 2009; Ross et al. 2009; Stern et al. 2009). Therefore, it was proved that the ECM induces and promotes constructive remodeling of host tissue (Wainwright and Bury 2011; Bing et al. 2019; Nafisi et al. 2017; Begum et al. 2016; Gowda et al. 2016). These might be understood as the natural ECM displays the presence of bioactive molecules that drive tissue homeostasis and stimulate regeneration (Nafisi et al. 2017; Nahabedian 2009). It should be also mentioned that ECM components such as growth factors and bioactive peptides, e.g., fibronectin that forms integrin base focal adhesion play important roles in defining the microenvironmental niche within which cells function in both normal homeostasis and in response to injury. These gears may enable cell-free ECM dermal scaffolds to generate a niche that is sufficient enough to recruit stem/progenitor cells and support their differentiation into functional skin tissues (Brown and Badylak 2014; Bhat and Bissell 2014; Tharp and Stahl 2015; Brownfield et al. 2013; Coyer et al. 2012; Lee et al. 2016; Fülber et al. 2016; Crapo et al. 2011; Turner and Badylak 2015). These characteristics of the ECM could be pointed as essential, major reasons for clinical application of cell-free ECM dermal scaffolds in regenerative medicine.

As one of the proposed mechanisms by which cell-free ECM scaffolds may support and promote structural and functional remodeling of injured tissues in vivo represents the thesis in which site opportunely retained and organized and at least functional tissue is deposited at the site of scaffold lattice. The outcome of this scenario is a scar-free tissue and return of its functionality (Turner and Badylak 2015). To avoid the default tissue healing response, which would be characterized as antitype of constructive regeneration, each anatomic site or clinical application has distinctive characteristics (e.g., thick skin on feet and hands) and the use of biologic scaffolds in these different locations must be appropriate (Turner and Badylak 2015).

To conclude, as a result of removing cells from human allogeneic dermis, one can obtain a biological scaffold consisting of extracellular matrix (ECM) elements, which on one hand favorably promote the regenerative processes in the recipient organism after transplantation, and on the other hand provide the optimal microenvironment for de novo proliferating autologous cells (Łabuś et al. 2018; Kawecki et al. 2018; Barrio and I, Kirby J, Ali S. 2016; Barkan et al. 2010; Cortiella et al. 2010; Sellaro et al. 2010; Cheng et al. 2009; Ross et al. 2009; Stern et al. 2009; Wainwright and Bury 2011; Bing et al. 2019; Nafisi et al. 2017; Crapo et al. 2011).

In this aspect, decellurization procedures of allogeneic dermis arouse particular interest (Łabuś et al. 2018). There is a wide range of commercially available cell-free skin substitutes of both human and animal origin (e.g., human ADM—Alloderm, porcine ADM—Strattice, Permacol, or decellurized bovine pericardium—Veritas) used in modern surgery. These biomaterials are characterized by sufficient biocompatibility and stability for long-term substitution of damaged tissues (Łabuś et al. 2018; Kawecki et al. 2018). Furthermore, there are a number of known patent claims regarding methods of human and animal production of ADM. Among others, there is a solution known from patent application no. CN102218162 (A)—Preparation method of homologous acellular dermal matrix for the production of cell-free dermal matrices by means of hyperosmotic separation of the epidermis from the dermis and subsequent incubation of the dermis in sodium dodecyl sulfate solution aided by ultrasound treatment for additional breakdown of cells. The material was then rinsed. In the next step, the preparations were trypsinized and washed again. The ADMs thus obtained were sterilized (CN102218162 (A)—Preparation method of homologous acellular dermal matrix). Another example of the method of producing ADM is the solution known from the patent claim No. KR20150143311 (A)—Manufacturing method of bio-graft or bio-implant encompassing cross-linked acellular dermal matrix", which relates to a multi-step method of removing cells based on epidermal removal, then removing cells from the dermis, lyophilization of the dermis. In the next stage, the dermis was cross-linked and further lyophilized (KR20150143311 (A)—Manufacturing method of bio-graft or bio-implant compositions comprising cross-linked acellular dermal matrix). Another example of a patented method of producing ADM may be a solution known from the "CN105126170 (A)—Acellular dermal matrix and preparing method of acellular dermal matrix" covering two stages: enzymatic method (proteolytic enzyme) in the first stage and chemical method (surfactant) in the second stage. The ADMs thus prepared were lyophilized (CN105126170 (A)—Acellular dermal matrix and preparing method of acellular dermal matrix). Another example of the method of producing ADM is the solution known from the patent no. CN105079880 (A)—Preparing the method of heterogeneous acellular matrix substrate with good biocompatibility (CN105079880 (A)—Preparing method of heterogeneous acellular dermal matrix substrate with good biocompatibility). The invention relates to a method for the preparation of heterogeneous ADM. The method was based on a multistage method including, inter alia, removal of skin grafts of appropriate thickness, then separation of the epidermis and dermis, removal of cells from the dermis, subsequent cross-linking, virus inactivation, refining and final radiation sterilization (Co-60) (CN105079880 (A)—Preparing method of heterogeneous acellular dermal matrix substrate with good biocompatibility). In contrast to the aforementioned methods, the method described in the patent application "CN102293690 (A)—Preparation method of freeze-thawing xenogeneic laser microporous irradiated acoustic molecular dermal matrix" uses only the physical method—cyclic freezing and thawing of xenogeneous dermis. The preparations were subject to final radiation sterilization (CN102293690 (A)—Preparation method of freeze-thawing xenogeneic laser microporous irradiated acellular dermal matrix and product thereof). In the next proposed solution with the number CN102225218 (A) - "Method for preparing acellular dermal matrix by utilizing ultrasonic wave", a solution was proposed involving the incubation of the examined dermis in a hyperosmotic solution for 0.5 to 12 h. The operation was repeated 1–6 times. In the next stage, the material was subjected to ultrasound treatment and then rinsed six times in PBS. The next stage involved the sterilization of the material (CN102225218 (A)—Method for preparing acellular dermal matrix by utilizing ultrasonic wave). When analyzing the methods of production of cell-free matrices suggested in the abovementioned patent claims, it should be noted that almost all of them have been based on a long-term and/or multistage/complex methodology. It should be emphasized that methods based on the use of chemical agents, such as SDS (CN102218162 (A)—Preparation method of homologous acellular dermal matrix; CN105126170 (A)—Acellular dermal matrix and preparing method of acellular dermal matrix) will require additional time necessary for efficient washing out of the cytotoxic detergent (e.g., SDS). That time can be prolonged up to several days (Xu et al. 2014). In addition, it should be remembered that most of the produced ADMs require the final sterilization. Before distribution and allocation, all obtained ADMs must be microbiologically tested. Continuing, microbiological tests take another 7–14 days (Łabuś et al. 2018). Therefore, it should not be doubted that the time and effort required to perform the abovementioned procedures may be much too long. This might be the basis for further research and improvement of the decellurization methods. Especially, when a severely burned patient requires a immediate specialized treatment.

55The selection of the optimal method of decellurization should be rationally planned and scientifically justified (Busuioc et al. 2012; Kitala et al. 2016; Kawecki et al. 2018; Crapo et al. 2011). The process of decellurization consists of chemical, physical and enzymatic techniques of elimination of cellular components (Schmidt and Baier 2215). It must be stressed that the preserving the natural nanostructure and composition properties of obtained ADM is highly desirable (Uygun et al. 2010; Petersen et al. 2010; Nakayama et al. 2010; Allen et al. 2010).

It should be stressed that the decellurization methods do not allow for the removal of the total content of cellular material (Kitala et al. 2016; Kawecki et al. 2018; Crapo et al. 2011). Going further, it has been shown that remaining and unremoved allogeneic or xenogeneic cellular debris may adversely affect the promotion of the regenerative processes of biological scaffolds in vivo after transplantation (Brown and Badylak 2014; Crapo et al. 2011). Therefore, the use of tissue engineering techniques including procedures for tissue and organ decellularization can be a critical element of clinical success (Brown and Badylak 2014; Bhat and Bissell 2014; Tharp and Stahl 2015; Brownfield et al. 2013; Coyer et al. 2012; Lee et al. 2016; Fülber et al. 2016; Crapo et al. 2011; Turner and Badylak 2015). What is remarkable a research for an optimal decellurization method is still an immense field for research?

Continuing it has been also shown that remaining cellular and intracellular debris (e.g., cell membrane components, DNA, mitochondria, etc.) that may remain after bioengineering of allogeneic or xenogeneic biomaterial may lead to undesired host reactions (e.g., adverse immune response or calcification) consequently resulting in degradation of the transplant (Zhang et al. 2010). Since the immune response is primarily directed against proteins and lipids of the cell membrane, thus tissue decellurization is a promising method to prevent the induction of an immune response in a body of the patient after the transplantation (Nakayama et al. 2010). The removal of cellular components should minimize immunologically induced inflammation, which may undermine the process of biodegradation of transplanted bioprosthesis. It should be once again emphasized that the immune response cannot be completely eliminated by the process of removing cells from tissues (Busuioc et al. 2012; Kitala et al. 2016). It has been shown that not only an adverse immune response might be the cause of the clinical failure and degradation of the graft. It was found that unremoved nucleic acids after implantation may play a role as a calcification agents. Therefore, an efforts should be taken for residual DNA and RNA extraction (Allen et al. 2010; Zhang et al. 2010).

In addition, it has been demonstrated that the use of in vitro cultured skin cells, such as autologous keratinocytes CEA and fibroblasts can significantly reduce the risk of death in the specific group of burned patients (Kitala et al. 2018). The use of autologous keratinocytes (CEA) cultured in vitro under good manufacturing practice GMP conditions is an alternative clinical standard in the therapy of extensive full thickness burns in patients with the lack of donor sites, already available for over three decades (4, 5). This method results in the acceleration of integument regeneration process, however in the long run, hypertrophied scars unacceptable for functional and esthetic reasons may occur. We can hypothesize that a rational reason for that undesirable reaction may be the lack of ECM lattice in the full thickness skin wounds. The absence of ECM does not support newly colonized regenerative cells with adequate homeostasis for their growth and proliferation. In order to provide the required and favorable conditions, the newly growing fibroblasts may hyperproduce the ECM elements, e.g., collagen. What in the long-term observation can lead to the formation of hypertrophic scars? The considered reasonable solution for that serious clinical problem could be ADM revitalization with in vitro cultured fibroblasts and keratinocytes. It may be a promising strategy in the production of biovital, biocompatible and "semi-autologous" skin graft. The essence of that method is the use of cell-free extracellular matrix of allogeneic dermis, serving as a scaffold for in vitro cultured skin cells. The main benefits offered by this technique include the introduction of a scaffold for the newly forming tissue, ensuring mechanical protection that is able to withstand the forces occurring in vivo until the developed skin substitute becomes an integral part of the body.

The mentioned above motivate to further work on improving the production of cell-free ECM scaffolds, such as ADMs which could be synergistically supported by live, cultured skin cells—fibroblasts and/or keratinocytes.

16.2 Materials and Methods

The allogeneic human skin is generally harvested during cadaver multi-organ/multi-tissue donations or autopsy/coroner donations. All actions must be carried out with strict aseptic hygienic regime. Prior to collection, it must be ascertained whether the donor fulfilled the legal criteria for qualification, thus not consenting to the collection of cells, tissues and organs from his corpses for transplantation. Multi-organ donors of allogenic skin and live skin donors of small skin sample for in vitro cell culture must be qualified for the following procedures in accordance with the EU Commission Directive 2004 23 EC, the Commission Directive 2006 17 EC and the Commission Directive 2006 86 EC.

All activities related to work with cellular and tissue material should be carried out in a clean room environment [e.g., GMP A class (ISO 5) in the area of GMP B class (ISO 7)].

As mainly mentioned, decellurization methods can be distinguished by chemical methods (detergents, acids, bases, chelating agents, solutions hypertonic and hypotonic), physical (freezing, sonication, agitation, mechanical removal) and biological (enzymes) and mixed that synergistically combine the efficiency of the above these methods (Brown and Badylak 2014; Keane et al. 2015; Łabuś et al. 2015). All of them got their advantages and disadvantages.

ADMs are mainly derived in a multistage process of applying proteolytic enzymes onto cadaveric, allogeneic human skin (Lataillade et al. 2010; Barker 2011). However, the process of decellularization may consist of chemical (e.g., alcohols, acids, basis, detergents, hypo- and hypertonic solutions), enzymatic (e.g., nucleases, trypsin, collagenase, lipase, dispase, thermolysin and α-galactosidase), as well as physical (e.g., direct mechanical force, ultrasounds and freeze–thaw) techniques of cellular components eradication (Barker 2011; Liem et al. 2013).

Enzymes are the most commonly used factors in the methodology of cell removal (Zhang et al. 2009; Kulig et al. 2013). One of the most essential enzymes used in the protocols of tissue-engineered human skin is dispase. Dispase is a relatively delicate peptidase that reaches its optimal activity in the physiological pH and temperature. Dispase hydrolyzes bonds of N-terminal peptides of non-polar amino acids. Such a sequence is highly frequent in collagen (Kulig et al. 2013). This mild protease generally does not damage the integrity of the cell membrane, thus is widely used in the processes of isolation and passaging of primary cell lines (Kulig et al. 2013). Moreover, the use of dispase leads to the separation of dermis and epidermis, which is crucial for establishing an in vitro culture of keratinocytes and fibroblasts separately (Zhang et al. 2009; Kulig et al. 2013) or for production of ADM (Busuioc et al. 2012; Łabuś et al. 2015). It has been shown that the procedure of mixed use of dispase and then trypsin showed a better decellurization effect in a comparison to a procedure that implies the use of trypsin alone. The observations were confirmed also in relation to the tissue of thicker composition (Prasertsung et al. 2008).

Another example of a widely used proteolytic enzyme is trypsin. Trypsin is a serine endopeptidase that is formed from trypsinogen in the pancreas. Trypsin's active form is converted in the small intestine by the enzyme enteropeptidase. This enzyme catalyzes the hydrolysis of the carboxyl group both in lysine and arginine (Kulig et al. 2013). In regards to tissue engineering of human skin, trypsin is used to isolate keratinocytes and fibroblasts from a small skin graft in order to prosecute in vitro cell culture (Schmidt and Baier 2215). Furthermore, trypsin is a widely proposed substance for decellularization (Łabuś et al. 2018; Kawecki et al. 2018; Crapo et al. 2011). Additionally, trypsin removes non-helical collagen peptides that are responsible for the host immune response (Crapo et al. 2011).

Another proposed group of enzymes used in the decellurization procedures is a group of collagenases. Use of collagenase in the procedures of cell removing may be justified only in cases where the use of clinical decellurized bioprostheses do not require to maintain the integrity of collagen structure (Flynn 2010).

It was mentioned above that unremoved nucleic acids after implantation may play a role as a calcification agents. Therefore, an efforts should be taken for residual DNA and RNA extraction (Booth et al. 2002). Nucleases (e.g., DNases and RNases) cleave nucleic acid sequences, and thus can facilitate the removal of nucleotides after decellurization procedures (Petersen et al. 2010; Mallis et al. 2014).

Sodium lauryl sulfate or sodium dedecyl sulfate (SDS) is the most commonly used anionic surfactant in decellularization procedures. SDS lowers the surface tension of aqueous solutions. In addition, SDS destroys non-covalent interactions of the native protein (Ballal et al. 2013). SDS is composed mainly of a mixture of sodium alkyl sulfates of molecular formula $C_{12}H_{25}OSO_3Na$. Since its introduction in 1972, SDS has found its application in laboratory practice as an efficient reagent to dissolve cell membranes and to elute proteins (Macheiner et al. 2014; Stanton and Billmire 2002). The range of SDS concentrations currently used in tissue engineering ranges from 0.03 to 1% in an environment of physiological saline (0.9% NaCl) or 9.0% NaCl solution. Suggested time for this process is from 24 to 72 h (Brown and Badylak 2014). The main advantage of SDS usage results from the ability of this detergent to extract fat from the surface of biomaterials (Xu et al. 2014; Choi et al. 2013).

Another example of a detergent widely proposed among the decellularization procedures is Triton X-100, which can effectively expel cell debris from thicker tissue, such as dermis, wherein the enzymatic and osmotic methods are insufficient. Usage of this detergent is accompanied by a change in the structure and composition in the ECM. But it was found that as a result of this seemingly adverse whether a reduction of adverse immune responses in vivo was noticed (Meyer et al. 2006).

SDS appears to be a more effective agent than Triton X-100 in the procedures for removal of cell nuclei and nucleic acids from thick tissues with maintaining the mechanical characteristics of the natural tissue (Lumpkins et al. 2008).

Another anionic detergent proposed for decellurization is 3-[(3-cholamidopropyl) dimethylammonio]-1-propanesulfonate (CHAPS). As mentioned above, CHAPS is an effective decellurization agent (Gilpin et al. 2014; Faulk et al. 2015).

The use of hypertonic and hypotonic solutions is also an example of chemical methods of cell extraction from the tissues. These methods act as decellurization factors by means of osmotic lysis with minimal impact on the structure of the ECM (Booth et al. 2002). Methods of this type can be used as synergistic support of other cell removing factors. What might be confirmed by the results obtained by Booth et al. (2002)?

Another example of chemical substances routinely used in the cell removing processes is the use of chelating agents (e.g., Ethylenediaminetetraacetic acid EDTA). Chelating agents form one or two coordination bond with the central metal ion. The heterocyclic rings comprise a central metal atom as an integral part of the ring. There are biological systems, which form chelates with metals such iron-binding porphyrine group of hemoglobin or magnesium-binding chlorophyll. In terms of chemical chelators are used to remove ions from solutions, while their use, in the medical context, is directed against microorganisms but also are used in the treatment of poisoning by metal and chemotherapeutic protocols. The rationale for the use of EDTA is that as a chelating agent, it separates the multivalent ions from solution, to form a substantially permanent bond. So that the metalloproteinase activity is inhibited (Ballal et al. 2013). Matrix metalloproteinases enable the mammary cells to invade by remodeling their surrounding ECM. The mechanism of the signaling interaction between ECM and matrix metalloproteinases is relevant to morphogenetic and physiological contexts. It has been shown that metalloproteinase expression and activity are required for mammary epithelial invasion/branching within dense collagen I lattices. (Gomes et al. 2015). In this manner, chelating agents are used to dissociate cells from ECM by binding metal cations at cell-ECM adhesion sites.

The rationale for ethylenediaminetetraacetic acid (EDTA) usage is that it is a chelating agent that separates the multivalent ions from the solution by forming solid bonds with ions so that the metalloproteinase activity is inhibited (Cheng et al. 2009).

In addition, a number of chemicals that do not lead to remove cells but are used to crosslink the collagen-rich material are proposed (Ishida et al. 2014). Among them, a dominant role may play: glutaraldehyde, formaldehyde, polyepoxy compounds and polyurethane, asylum azide, caroboimid compounds, hexamethyl diisocyanate, tannic acid, tartaric acid, vitamin B2-induced UVA crosslinking and genipine (Ishida et al. 2014; Genovese et al. 2014; Remlinger et al. 2012; Smith et al. 2012; Tam et al. 2015; Koch et al. 2015; Zhai et al. 1190; Jiang et al. 2013; Hey et al. 1990; Charron et al. 2016). As often described the procedure for processing, the dermis can cite two-step method based on enzymatic digestion of cellular elements and then crosslinked structure of the extracellular matrix with glutaraldehyde in order to mask residual cellular structures. However, the use of glutaraldehyde may lead to damage of the integrity of the collagen, which affects adversely the characteristics of the transplant and increases its toxicity (Jiang et al. 2013; Hey et al. 1990). In a number of original works, many authors, during the preparation of cell-free skin substitutes, resigned from step of crosslinkers stabilizing such as, for example, glutaraldehyde (Ishida et al. 2014; Genovese et al. 2014; Remlinger et al. 2012; Smith et al. 2012; Tam et al. 2015; Koch et al. 2015; Zhai et al. 1190; Jiang et al. 2013; Hey et al. 1990; Charron et al. 2016; Jang et al. 2016).

The main advantage of the tissue decellularization by means of physical processes over the chemical methods is the lack of the introduction of potentially toxic chemicals to the tissue. But it may not prevent the occurrence of specific adverse side effects (Khor 1997). It should also be noted that the use of physical methods as a cell removing agents often requires prolonged exposure of tissues to the agent, which can cause degradation of the material (Moore et al. 1996).

Processing of tissue by cyclic freeze/thaw method is effective for lysis of cells in tissues and organs, but leaves unresolved fragments of cell membranes and intracellular elements. Thus, it is required to repeat the subsequent cycles. A single cycle of freezing and thawing may, however, reduce adverse immune reactions, including leukocyte infiltration, within the ECM (Lehr et al. 2011). What is more a number of immediately repeated consecutive cycles of freezing and thawing may be used to remove cells from tissues or organs (Cortiella et al. 2010; Gardin et al. 2015b; Pulver et al. 2014), but on the other hand these may increase a certain degree of loss of protein and other adverse structural changes and quality of the ECM. In order to reduce this negative phenomenon, it is proposed to use the cryoprotectants as, for example, 5% trehalose (Pulver et al. 2014). Since the processing of tissues using the cyclic freezing and thawing can bring a minor changes in the ultrastructure of the ECM (Pulver et al. 2014) therefore, such a procedure should be applied only when these effects are acceptable in the final ECM product (Gardin et al. 2015b; Pulver et al. 2014).

A mechanical removal of the cells is an another example of physical decellurization methods from the tissues and organs. The surface-inhabiting cells of the tissue or organ (e.g., the urinary bladder, small intestine, amnion) can be effectively removed by mechanical abrasion in combination with the use of enzymes (Schenke-Layland et al. 2003), hypertonic saline, or chelating agents in order to support and facilitate the separation process of cells from the underlying basement membrane. However, such methods invariably lead to damage of the ultrastructure and the integrity of the ECM scaffold (Schenke-Layland et al. 2003; Pchelintsev et al. 2016). Some serious doubts connected with the effectiveness of this decellurizating method for tissues and organs with the cells inhabited deep within the ECM.

Among the physical decellularization methods, a freeze drying is also mentioned (Li et al. 2015; Wang et al. 2015). Freeze drying is undoubtedly a recognized and effective method for long-term storage of biological materials (Silva et al. 2016; Abbah et al. 2015; Chen et al. 2016) but in terms of the effectiveness of cell removing from tissues and organs a similar controversy as cyclic freezing and thawing can raise. Removal of defaulting cell membranes, nucleic acids and other cellular organelles can remain at unsatisfactory levels.

A thick and dense tissue, e.g., dermis requires a complex protocol of cell removing and prolonged exposure to the decellurization agents. The complexity and duration of cell removing processes depend in this case on the size and thickness of the tissue and the desired properties required of the processed biomaterials (e.g., macrostructure and nanostructure of ECM, proteins of the basement membrane, growth factors, etc.) (Kawecki et al. 2018; Crapo et al. 2011). Disorders of the ECM may include the deletion defined structural components (collagens, glycosaminoglycans, GAGs, laminin, fibronectin and growth factors, etc.) from the decellurized biological material. The logical consequence of the loss of certain, often key components of ECM nanostructure, must be a change in mechanical and biological properties of biological materials and the final product. The result of these changes can be reduced ECM's ability to support the growth and differentiation of cells in vitro and in vivo, mechanical failure or, e.g., induction of the host immune response after transplantation (Crapo et al. 2011). It should be once more stressed that the decellurization methods do not allow for the removal of total cellular material (Kawecki et al. 2018; Crapo et al. 2011).

As mentioned above, the widely proposed dermis decellurization methodology is time and labor-consuming. The work of Łabuś et al. (Łabuś et al. 2015) has developed a fast and effective method of producing ADM. The essence of the method was the use of initially sterile (electron beam, 35 kGy), biostatic allogeneic human skin grafts decellurized and revitalized under GMP A class conditions (ISO 5) in the environment of GMP B (ISO 7). The assumption was proposed that former radiation sterilization (electron beam, 35 kGy) could have affected dermal ECM and its cells, so the decellularization process became much easier to be conducted. On the other hand, the decellularization of unsterilized human dermis, proposed under the study conditions (short time, single-step, low concentration of the decellularization factors), came out to be highly inefficient and infectious. The method presented by Łabuś et al. (2015) is a quick and effective method of removing cells from human dermis, especially in the comparison to other authors (CN102218162 (A)—Preparation method of homologous acellular dermal matrix; KR20150143311 (A)—Manufacturing method of bio-graft or bio-implant compositions comprising cross-linked acellular dermal matrix; CN105126170 (A)—Acellular dermal matrix and preparing method of acellular dermal matrix; CN105079880 (A)—Preparing method of heterogeneous acellular dermal matrix substrate with good biocompatibility; CN102293690 (A)—Preparation method of freeze-thawing xenogeneic laser microporous irradiated acellular dermal matrix and product thereof.; CN102225218 (A)—Method for preparing acellular dermal matrix by utilizing ultrasonic wave). According to GMP guidelines, the production of sterile medicinal products that was conducted aseptically at all stages in special clean areas does not require the final sterilization. However, strict and precise quality control protocols for the obtained ADMs remain to be conducted (including, e.g., microbiological testing). If the sample is contaminated, then a final sterilization is required.

For reasons of biological security, a decellurized ADMs prior it clinical application do need a final sterilization (Hodde and Hiles 2002; Marsit et al. 2014; Singh et al. 2007; Djefal et al. 2007). The use of sterilization procedures such as, e.g., exposure of the material to ethylene oxide, gamma or electron beam radiation may significantly influence the structure of the ECM and may negatively affect a mechanical properties of the biomaterial (Sun and Leung 2008).

16.2.1 Cell Seeding

An alternative for autologous STSGs may be available for over three decades transplantation of in vitro cultured autologous keratinocytes (cultured epithelial autograft CEA) (Łabuś et al. 2015; Dearth et al. 2016). The main benefit of that method is acceleration of regenerative process. Whereas in long-term observation a functional disorders and unacceptable cosmetic effect of hypertrophic scars may occur (Dearth et al. 2016). The possible mechanism of that undesired effect was mentioned above.

In vitro cell culture on the tissue-engineered cell-free ECM dermal scaffolds is a promising strategy in the production of "new tissues" - a biovital and stable skin graft (Łabuś et al. 2018). The essence of that method is the use of ADM as a scaffold for in vitro cultured skin cells. The main benefits of that technique are a promotion of wound remodeling and regeneration and mechanical protection. Such a graft would be able to become an integral part of the body (Kitala et al. 2018; Badylak et al. 1998; Bhrany et al. 2006; Bolland et al. 2007; Hopper et al. 2003; Callcut et al. 2006; Wainwright 1995; Gardin et al. 2015a; Vacanti et al. 1998; Seland et al. 2011; Bannasch et al. 2008; Kim and Mooney 1998).

16.2.2 Clinical Applications

The clinical use of skin substitutes is intended to protect the microenvironment of the wound from the adverse influence of the external environment and to enable a rapid closure of the wound. Skin substitutes are expected to show good adhesion to the wound, optimal permeability, adequate mechanical properties (strength and flexibility), to be a barrier to microbes, be safe in use. Additionally, they must not be infectious, immunogenic and toxic. They should support the reconstruction of natural tissue by biodegradation/remodeling (Kawecki et al. 2018; Crapo et al. 2011).

A widely described, commercially available cell-free skin substitute is Alloderm.

Alloderm is a biological dressing made of human skin, available in lyophilized form. It is a matrix deprived of the epidermis and cells obtained from the dermis. Alloderm has preserved basal membrane (Callcut et al. 2006; Wainwright 1995; Shakespear 2005), and it easily gets incorporated into the wound environment without rejection and does not cause an inflammatory reaction (Shakespear 2005). The overriding goal of Alloderm is to minimize tendon shrinkage and scar tissue scarring and to maintain the physiological functionality of the body areas within which tendons are found, e.g., neck, shoulders, elbow, knee, ankle, wrist and hand (Callcut et al. 2006; Wainwright 1995; Shakespear 2005).

16.2.3 Limitations

Today's research should focus on developing a fast, effective and safe method of production of acellular dermal matrix (ADM) and/or acellular dermal graft (ADG) transplants and their revitalization in vitro by the recipient's autologous cells. As a result of such a procedure, one could theoretically get a full-blown alive, stable skin graft (Łabuś et al. 2018). In an ideal theoretical situation, this type of transplant should be characterized by all the features of natural skin (Table I), including dermis and epidermis inhabited by naturally occurring cells and skin appendages. However, this type of procedure has a number of significant technical limitations and difficulties (Łabuś et al. 2018). First of all, among the difficulties one can point to obtaining a truly completely decellurized ECM structure that did not acquire cytotoxicity traits during the preparation (Łabuś et al. 2018). The cytotoxicity of the ECM scaffold, followed by the loss of biocompatibility may occur as a result of the use of some substances that remove cells from tissues (e.g., sodium dodecyl sulfate SDS) (Łabuś et al. 2018; Kawecki et al. 2018; Crapo et al. 2011). According to that a diametrically opposite results were obtained in the work of Xu et al. (2014). In their study, the effectiveness of a solution of 0.05% trypsin EDTA, 0.5% SDS and 3% triton X-100 were compared as cell removing factors from porcine annulus. The obtained scaffolds were in vitro revitalized by fibroblasts. The results showed that only the matrix obtained by incubation in trypsin occurred the fibroblasts growth, while the scaffolds obtained by chemical methods (SDS and Triton X-100) were not rehabitated by cells (Xu et al. 2014). What clearly proves that chemical agents such SDS and Triton X-100 could cause a high-cytotoxic effect.

Incomplete decellurization and cytotoxicity may lead to adverse reactions in the body of the recipient after the transplant (Kawecki et al. 2018). As a further difficulty in the course of cell removal preparations, it can be pointed out that during the process of removing cells from tissues, too many valuable ECM elements may be undesirable, resulting in drastic deterioration of mechanical properties (e.g. flexibility, tensile strength, etc.) (Kawecki et al. 2018). In addition, it may also have a negative impact on the biological properties of the matrix (Kawecki et al. 2018; Cortiella et al. 2010)—in this aspect it should be emphasized that the behavior in the ECM structure, e.g., collagen IV (component of the basal membrane ADM) is extremely important as the basement membrane (e.g., desmosomes, collagen IV) is the place of attachment/anchorage for the epidermis, including for example for developing keratinocytes (Begum et al. 2016; Gowda et al. 2016). That is extremely important in the manner of production of a biovital skin substitute by revitalizing ADM (Łabuś et al. 2018; Begum et al. 2016; Gowda et al. 2016). Therefore, achieving a kind of "golden mean" between the efficiency of removing cells from tissues and the maximum preservation of an unharmed ECM structure, which will ensure adequate biological, structural and mechanical properties can be considered as the basic task of removing cells from tissues (Kawecki et al. 2018).

Although the issue of revitalization of cell-free dermal matrixes (ADM) is not a new area of research, the lack of inward inhabitation of fibroblasts into ADM is still one of the main limitations of this procedure (Łabuś et al. 2018; Guo et al. 2014; Garcia and Scott 2013). It was shown that the source of this problem is the compact and dense structure of collagen fibers building the ECM extracellular matrix of the dermis (Butterfield 2013; Ott et al. 2008; Evans and Kon 2015). This problem can be overcome by the use of collagenases, which will lead to the relaxation of the collagen structure of ADM. However, one should remember about the high risk of matrix structure degradation (Kawecki et al. 2018).

Briefly, it could be proposed that measurement criteria for biological properties of ECM scaffolds could asses:

- Lack of visible nuclear material in a histological section stained with hematoxilin and eozine (H&E) or 4',6-diamidino-2-phenylindole (DAPI),
- ‹ 50 ng ds DNA per 1 mg of dry weight of ECM,
- ‹ 200 bp DNA fragment length (Kitala et al. 2016),
- Lack of intracellular membrane compartments (e.g., mitochondria),
- Lack of cell membrane elements,
- A presence of unremoved and undamaged ECM elements (collagen, GAG, fibronectin, etc.)
- Lack of cytotoxicity of obtained ECM scaffold.

16.3 Conclusions

Production of natural cell-free ECM dermal scaffolds for the safe clinical use requires the implementation of procedures for removing cells from tissues and organs. Choosing the right cell, removing technique should be based on a substantive knowledge and understanding of the mechanism of acting of used cell removing agents. The optimal technique should assume the proper use of cell removing agent (or sequentially appropriate agents) and the optimum conditions for the maximum operation efficiency.

References

Abbah SA, Delgado LM, Azeem A, Fuller K, Shologu N, Keeney M, Biggs MJ, Pandit A, Zeugolis DI (2015) Harnessing hierarchical nano- and micro-fabrication technologies for musculoskeletal tissue engineering. Adv Healthc Mater 4(16):2488–2499

Allen RA, Seltz LM, Jiang H, Kasick RT, Sellaro TL, Badylak SF et al (2010) Adrenal extracellular matrix scaffolds support adrenocortical cell proliferation and function in vitro. Tissue Eng Part A 16:3363

Badylak SF, Record R, Lindberg K (1998) Small intestinal submucosa: a substrate for in vitro cell growth. J Biomater Sci Polym Ed 9:863–878

Ballal NV, Tweeny A, Baumgartner JC, Ginjupalli K, Saraswathin V (2013) Effect of maleic acid and ethylenediaminetetraacetic acid on the shear bond strength of RealSeal SE sealer to root canal dentin. Eur J Prosthodont Restor Dent 21:152

Bannasch H, Stark GB, Knam F, Horch RE, Föhn M (2008) Decellularized dermis in combination with cultivated keratinocytes in a short- and long-term animal experimental investigation. JEADV 22:41–49

Barkan D, Green JE, Chambers AF (2010) Extracellular matrix: a gatekeeper in the transition from dormancy to metastatic growth. Eur J Cancer 46:1181–1188

Barker TH (2011) The role of ECM proteins and protein fragments in guiding cell behavior in regenerative medicine. Biomaterials 18:4211–4214

Begum T, Farrelly PJ, Craigie RJ (2016) Non-cross-linked porcine acellular dermal matrix (Strattice Tissue Matrix) in pediatric reconstructive surgery. J Pediatr Surg 51(3):461–464

Bhat R, Bissell MJ (2014) Of plasticity and specificity: dialectics of the microenvironment and macroenvironment and the organ phenotype. Wiley Interdiscip Rev Dev Biol 3(2):147–63

Bhrany AD, Beckstead BL, Lang TC i wsp (2006) Development of an esophagus acellular matrix tissue scaffold. Tissue Eng 12:319–330

Bing Z, Feng L, Wu C-S, Du J-T, Liu Y-F, Liu S-X (2019) Acellular dermal matrix contributes to epithelialization in patients with chronic sinusitis. J Biomater Appl 088532821882263

Bolland F, Korossis S, Wilshaw SP i wsp (2007) Development and characterisation of a full-thickness acellular porcine bladder matrix for tissue engineering, Biomaterials 28:1061–1070

Booth C, Korossis SA, Wilcox HE, Watterson KG, Kearney JN, Fisher J, Ingham E (2002) Tissue engineering of cardiac valve protheses I: development and histological characterisation of an acellular porcine scaffold. J Heart Valve Dis 11:457

Brown BN, Badylak SB (2014) Extracellular matrix as an inductive scaffold for functional tissue reconstruction. Transl Res 163:268

Brownfield DG[1], Venugopalan G, Lo A, Mori H, Tanner K, Fletcher DA, Bissell MJ (2013) Patterned collagen fibers orient branching mammary epithelium through distinct signaling modules. Curr Biol 23 (8):703–709

Busuioc CJ, Popescu FC, Mogoşanu GD, Pârvănescu H, Streba L, Mogoantă L (2012) Histological and immunohistochemical study of cutaneous angiogenesis process in experimental third-degree skin burns treated with allograft. Rom J Morphol Embryol 4:1061–1067

Butterfield JL (2013) 440 consecutive immediate, implant-based, single-surgeon breast reconstructions in 281 patients: a comparison of early outcomes and costs between surgimend fetal bovine and alloderm human cadaveric acellular dermal matrices. Plast Reconstr Surg 5:940

Callcut RA, Sshurr MJ, Sloan M i wsp (2006) Clinical experience with Alloderm: a one-staged composite dermal/epidermal replacement utilizing processed cadevar dermis and thing autograft. Burns 32:583–588

Charron PN, Fenn SL, Poniz A, Oldinski RA (2016) Mechanical properties and failure analysis of visible light crosslinked alginate-based tissue sealants. J Mech Behav Biomed Mater 10(59):314–321

Chen S, Zhang Q, Nakamoto T, Kawazoe N, Chen G (2016) Gelatin scaffolds with controlled pore structure and mechanical property for cartilage tissue engineering. Tissue Eng Part C Meth 22(3):189–198

Cheng NC, Estes BT, Awad HA, Guilak F (2009) Chondrogenic differentiation of adipose-derived adult-stem cells by a porous scaffold derived from native articular cartilage extracellular matrix. Tissue Eng Part A 15:231–241

Choi S, Cheong Y, Shin JH, Jin KH, Park HK (2013) Inflammatory effect of monopolar radiofrequency treatment on collagen fibrils in rabbit skins. J Biomed Nanotechnol 9:1403

CN102218162 (A)—Preparation method of homologous acellular dermal matrix (2008)

CN102225218 (A)—Method for preparing acellular dermal matrix by utilizing ultrasonic wave (2011)

CN102293690 (A)—Preparation method of freeze-thawing xenogenic laser microporous irradiated acellular dermal matrix and product thereof (2011)

CN105079880 (A)—Preparing method of heterogeneous acellular dermal matrix substrate with good biocompatibility (2015)

CN105126170 (A)—Acellular dermal matrix and preparing method of acellular dermal matrix (2015)

Cortiella J, Niles J, Cantu A, Brettler A, Pham A, Vargas G et al (2010) Influence of acellular natural lung matrix on murine embryonic stem cell differentiation and tissue formation. Tissue Eng Part A 16:2565–2580

Coyer SR, Singh A, Dumbauld DW, Calderwood DA, Craig SW, Delamarche E, Garcı´a AJ (2012) Nanopatterning reveals an ECM area threshold for focal adhesion assembly and force transmission that is regulated by integrin activation and cytoskeleton tension. J Cell Sci 125:5110–5123. Accepted 15 July 2012

Crapo PM, Gilbert TW, Badylak SF (2011) An overview of tissue and whole organ decellularization processes. Biomaterials 32:3233

Dearth CL, Keane TJ, Carruthers CA, Reing JE, Huleihel L, Ranallo CA, Kollar EW, Badylak SF (2016) The effect of terminal sterilization on the material properties and in vivo remodeling of a porcine dermal biologic scaffold. Acta Biomater 15(33):78–87

Del Barrio IM, Kirby J, Ali S (2016) The role of chemokine and glycosaminoglycan interaction in chemokine-mediated migration in vitro and in vivo. Methods Enzymol. 570:309–333.

Djefal A, Tahtat D, Khodja AN, Bouzid SS, Remane N (2007) Validation and substantiation of 25 kGy as sterilization dose for lyophilized human amnion membrane. Cell Tissue Bank 8:9

Evans AK, Kon ND (2015) Extracellular matrix for repair of type IV laryngotracheo-esophageal cleft. Int J Pediatr Otorhinolaryngol 79(12):2484–2486

Faulk DM, Wildemann JD, Badylak SF (2015) Decellularization and cell seeding of whole liver biologic scaffolds composed of extracellular matrix. J Clin Exp Hepatol 5(1):69–80

Flynn LE (2010) The use of decellularized adipose tissue to provide an inductive microenvironment for the adipogenic differentiation of human adipose-derived stem cells. Biomaterials 31:4715

Fülber J, Maria DA, Silva LC, Massoco CO, Agreste F, Baccarin RY (2016) Comparative study of equine mesenchymal stem cells from healthy and injured synovial tissues: an in vitro assessment. Stem Cell Res Ther 7(1):35

Garcia O Jr, Scott JR (2013) Analysis of acellular dermal matrix integration and revascularization following tissue expander breast reconstruction in a clinically relevant large-animal model. Plast Reconstr Surg 5:741

Gardin C, Ricci S, Ferroni L, Guazzo R, Sbricoli L, De Benedictis G, Finotti L, Isola M, Bressan E, Zavan B (2015b) Decellularization and delipidation protocols of bovine bone and pericardium for bone grafting and guided bone regeneration procedures. PLoS ONE 10:1371

Gardin C, Ricci S, Ferroni L, Guazzo R, Sbricoli L, De Benedictis G, Finotti L, Isola M, Bressan E, Zavan B (2015a) Decellularization and delipidation protocols of bovine bone and pericardium for bone grafting and guided bone regeneration procedures. PLoS One 10 (7):e0132344, 20 Jul 2015

Genovese L, Zawada L, Tosoni A, Ferri A, Zerbi P, Allevi R, Nebuloni M, Alfano M (2014) Cellular localization, invasion, and turnover are differently influenced by healthy and tumor-derivedextracellular matrix. Tissue Eng Part A 20(13–14):2005–2018

Gilpin SE, Guyette JP, Gonzalez G, Ren X, Asara JM, Mathisen DJ, Vacanti JP, Ott HC (2014) Perfusion decellularization of human and porcine lungs: bringing the matrix to clinical scale. J Heart Lung Transplant 33(3):298–308

Glik J, Kawecki M, Kitala D, Klama-Baryła A, Łabuś W, Grabowski M, Durdzińska A, Nowak M, Misiuga M, Kasperczyk A (2017) A new option for definitive burn wound closure - pair matching type of retrospective case-control study of hand burns in the hospitalised patients group in the Dr Stanislaw Sakiel Centre for Burn Treatment between 2009 and 2015. Int Wound J

14(5):849–855. https://doi.org/10.1111/iwj.12720. Epub 2017 Feb 21. PMID: 28220597; PMCID: PMC7949498

Gomes AM, Bhat R, Correia AL, Mott JD, Ilan N, Vlodavsky I, Pavão MS, Bissell M (2015) Mammary branching morphogenesis requires reciprocal signaling by heparanase and MMP-14. J Cell Biochem 116 (8):1668–1679

Gowda AU, McNichols CH, Asokan I, Matthews JA, Buckingham EB, Sabino J, Maddox JS, Slezak S, Rasko Y, Singh DP (2016) Porcine acellular dermal matrix for hernia repair in transplant patients. Ann Plast Surg 77(6):674–677

Grossova I, Zajicek R, Kubok R, Smula MC (2017) The treatment of palmar contact burns in children: a five-year review. Ann Burns Fire Disasters 30(1):5–8

Guo R, Teng J, Xu S, Ma L, Huang A, Gao C (2014) Comparison studies of the in vivo treatment of full-thickness excisional wounds and burns by an artificial bilayer dermal equivalent and J-1 acellular dermal matrix. Wound Repair Regen 22(3):390–398

Hey KB, Lachs MC, Worthy MJ, Wood EJ (1990) Crosslinked fibrous collagen for use as a dermal implant: control of the cytotoxic effects of glutaraldehyde and dimethylsuberimidate. Biotech Appl Biochem 12:85

Hodde JP, Hiles M (2002) Virus safety of a porcine-derived medical device: evaluation of a viral inactivation method. Biotechnol Bioeng 79:211

Hopper RA, Woodhouse K, Semple JL (2003) Acellularization of human placenta with preservation of the basement membrane: a potential matrix for tissue engineering. Ann Plast Surg 51:598–602

Ishida Y, Sakakibara S, Terashi H, Hashikawa K, Yamaoka T (2014) Development of a novel method for decellularizing a nerve graft using a hypertonic sodium chloridesolution. Int J Artif Organs 37 (11):854–860

Jang J, Kim TG, Kim BS, Kim SW, Kwon SM, Cho DW (2016) Tailoring mechanical properties of decellularized extracellular matrix bioink by vitamin B2-induced photo-crosslinking. Acta Biomater 15 (33):88–95

Jiang T, Ren XJ, Tang JL, Yin H, Wang KJ, Zhou CL (2013) Preparation and characterization of genipin-crosslinked rat acellular spinal cord scaffolds. Mater Sci Eng C Mater Biol Appl 1:3514

Kawecki M, Łabuś W, Klama-Baryla A, Kitala D, Kraut M, Glik J, Misiuga M, Nowak M, Bielecki T, Kasperczyk A (2018) A review of decellurization methods caused by an urgent need for quality control of cell-free extracellular matrix' scaffolds and their role in regenerative medicine. J Biomed Mater Res B Appl Biomater 106(2):909–923. https://doi.org/10.1002/jbm.b.33865. Epub 2017 Feb 14. PMID: 28194860

Keane TJ, Swinehart IT, Badylak SF (2015) Methods of tissue decellularization used for preparation of biologic scaffolds and in vivo relevance. Methods 84:25

Khor E (1997) Methods for the treatment of collagenous tissue for bioprotheses. Biomaterials 18:95

Kim BS, Mooney DJ (1998) Development of biocompatible synthetic extracellular matrices for tissue engineering. Trends Biotechnol 16:224–230

Kitala D, Kawecki M, Klama-Baryła A, Łabuś W, Kraut M, Glik J, Ryszkiel I, Kawecki MP, Nowak M (2016) Allogeneic versus autologous skin grafts in the therapy of patients with burn injuries: a restrospective, open-label clinical study with pair matching. Adv Clin Exp Med 25(5):923–929

Kitala D, Klama-Baryła A, Kraut M, Łowicki D, Łabuś W, Glik J, Misiuga M, Nowak M, Lisowska G, Olszewski R, Kawecki M (2018) Does the transplantation of keratinocytes really reduce the risk of death? Survival analysis of patients hospitalized at the Dr Stanisław Sakiel Centre for Burns Treatment in 2008–2015. Adv Clin Exp Med. 23 Nov 2018. https://doi.org/10.17219/acem/76327. [Epub ahead of print]

Koch H, Hammer N, Ossmann S, Schierle K, Sack U, Hofmann J, Wecks M, Boldt A (2015) Tissue engineering of ureteral grafts: preparation of biocompatible crosslinked ureteral scaffolds of porcine origin. Front Bioeng Biotechnol 23:3

KR20150143311 (A)—Manufacturing method of bio-graft or bio-implant compositions comprising cross-linked acellular dermal matrix (2015)

Kulig KM, Luo X, Finkelstein EB, Liu XH, Goldman SM, Sundback CA, Vacanti JP, Neville CM (2013) Biologic properties of surgical scaffold materials derived from dermal ECM. Biomaterials 23:5776–5784

Łabuś W, Kawecki M, Glik J, Maj M, Kitala D, Misiuga M, Klama-Baryła A, Kraut M, Nowak M (2015) Own experience from the use of a substitute of an allogeneic acellular dermal matrix revitalized with in vitro cultured skin cells in clinical practice. Pol Przegl Chir 10:929

Łabuś W, Glik J, Klama-Baryła A, Kitala D, Kraut M, Maj M, Nowak M, Misiuga M, Marcinkowski A, Trzebicka B, Poloczek R, Kawecki M (2018) Atomic force microscopy in the production of a biovital skin graft based on human acellular dermal matrix produced in-house and in vitro cultured human fibroblasts. J Biomed Mater Res B Appl Biomater 106 (2):726–733

Lataillade JJ, Bey E, Thepenier C, Prat M, Leclerc T, Bargues L (2010) Skin engineering for burns treatment. Bull Acad Natl Med 7:1339–1351

Lee DE, Ayoub N, Agrawal DK (2016) Mesenchymal stem cells and cutaneous wound healing: novel methods to increase cell delivery and therapeutic efficacy. Stem Cell Res Ther 7(1):37

Lehr EJ, Rayat GR, Chiu B, Churchill T, McGann LE, Coe JY et al (2011) Decellularization reduces immunogenicity of sheep pulmonary artery vascular patches. J Thorac Cardiovasc Surg 141:1056

Li W, Ma G, Brazile B, Li N, Dai W, Butler JR, Claude AA, Wertheim JA, Liao J, Wang B (2015) Investigating the potential of amnion-based scaffolds

as a barrier membrane for guided bone regeneration. Langmuir 11:8642
Liem PH, Morimoto N, Ito R, Kawai K, Suzuki S (2013) Autologous skin reconstruction by combining epidermis and acellular dermal matrix tissue derived from the skin of giant congenital melanocytic nevi. J Artif Organs 16:332–342
Lumpkins SB, Pierre N, McFetridge PS (2008) A mechanical evaluation of three decellularization methods in the design of a xenogeneic scaffold for tissue engineering the temporomandibular joint disc. Acta Biomater 4:808
Macheiner T, Kuess A, Dye J, Saxena AK (2014) A novel method for isolation of epithelial cells from ovine esophagus for tissue engineering. Biomed Mater Eng 24:1457–1468
Mallis P, Gontika I, Poulogiannopoulos T, Zoidakis J, Vlahou A, Michalopoulos E, Chatzistamatiou T, Papassavas A, Stavropoulos-Giokas C (2014) Evaluation of decellularization in umbilical cord artery. Transplant Proc 46:3232
Marsit N, Dwejen S, Saad I, Abdalla S, Shaab A, Salem S, Khanfas E, Hasan A, Mansur M, Abdul SM (2014) Substantiation of 25 kGy radiation sterilization dose for banked air dried amniotic membrane and evaluation of personnel skill in influencing finished product bioburden. Cell Tissue Bank 15:603
Meyer SR, Chiu B, Churchill TA, Zhu L, Lakey JR, Ross DB (2006) Comparison of aortic valve allograft decellularization techniques in the rat. J Biomed Mater Res A 79:254
Moore MA, Chen WM, Philips RE, Bohachevsky JK, Mollroy BK (1996) Schrinkage temperature protein extraction assay as a measure of stabilisation of photooxidized tissue. J Biomed Mat Res 32:209
Nafisi N, Akbari ME, Mahjoub F, Mohseni MJ, Sabetkish S, Khorramirouz R, Tehrani M, Kajbafzadeh AM (2017) Application of human acellular breast dermal matrix (ABDM) in implant-based breast reconstruction: An experimental study. Aesthetic Plast Surg 41 (6):1435–1444
Nahabedian MY (2009) AlloDerm performance in the setting of prosthetic breast surgery, infection, and irradiation. Plast Reconstr Surg 6:1743
Nakayama KH, Batchelder CA, Lee CI, Tarantal AF (2010) Decellularized rhesus monkey kidney as a three-dimensional scaffold for renal tissue engineering. Tissue Eng Part A 16:2207
Ott HC, Matthiesen TS, Goh SK, Black LD, Kren SM, Netoff TI et al (2008) Perfusiondecellularized matrix: using nature's platform to engineer a bioartificial heart. Nat Med 14:213
Pchelintsev NA, Adams PD, Nelson DM (2016) Critical parameters for efficient sonication and improved chromatin immunoprecipitation of high molecular weight proteins. PLoS One 11(1):e0148023
Petersen TH, Calle EA, Zhao L, Lee EJ, Gui L, Raredon MB et al (2010) Tissue-engineered lungs for in vivo implantation. Science 329:538
Prasertsung I, Kanokpanont S, Bunaprasert T, Thanakit V, Damrongsakkul S (2008) Development of acellular dermis from porcine skin using periodic pressurized technique. J Biomed Mater Res B Appl Biomater 85:210
Prim PM, Kim HS, Shapiro LE, Lee JS, Kaan JH, Jackson JD, Yoo JJ, Atala A, Lee SJ (2017) In vitro skin expansion: wound healing assessment. Wound Repair Regen. https://doi.org/10.1111/wrr.12550
Pulver, Shevtsov A, Leybovich B, Artyuhov I, Maleev Y, Peregudov A (2014) Production of organ extracellular matrix using a freeze-thaw cycle employing extracellular cryoprotectants. Cryo Lett 35:400
Remlinger NT, Wearden PD, Gilbert TW (2012) Procedure for decellularization of porcine heart by retrograde coronary perfusion. J Vis Exp (70):e50059
Roolker W, Patt TW, van Dijk CN i wsp (2000) The Gore-tex prosthetic ligament as a salvage procedure in deficient knees. Knee Surg Sports Traumatol Arthrosc 8, 20:2000.
Ross EA, Williams MJ, Hamazaki T, Terada N, Clapp WL, Adin C, Ellison GW, Jorgensen M, Batich CD (2009) Embryonic stem cellsproliferate and differentiate when seeded into kidney scaffolds. J Am Soc Nephrol 20:2338–2347
Schenke-Layland K, Vasilevski O, Opitz F, Konig K, Riemann I, Halbhuber KJ et al (2003) Impact of decellularization of xenogeneic tissue on extracellular matrix integrity for tissue engineering of heart valves. J Struct Biol 143:201
Schmidt Ch, Baier JM (2000) Acellular vascular tissues: natural biomaterials for tissue engineering. Biomaterials 21:2215
Schneider KH, Aigner P, Holnthoner W, Monforte X, Nürnberger S, Rünzler D, Redl H, Teuschl AH (2016) Decellularized human placenta chorion matrix as a favorable source of small-diameter vascular grafts. Acta Biomater 29:125–34 Acta Biomater.1 Jan 2016
Seland H, Gustafson CJ, Johnson H i wsp (2011) Transplantation of acellular dermis and keratinocytes cultured on porous biodegradable microcarriers into full-thickness skin injuries on athymic rats. Burns 37:99–108
Sellaro TL, Ranade A, Faulk DM, McCabe GP, Dorko K, Badylak SF, Strom SC (2010) Maintenance of human hepatocyte function in vitro by liver-derived extracellular matrix gels. Tissue Eng Part A 16:1075–1082
Shakespear PG (2005) The role of skin substitutes in the treatment of burn injuries. Clin Dermatol 23:413–418
Silva SS, Oliveira NM, Oliveira MB, da Costa DP, Naskar D, Mano JF, Kundu SC, Reis RL (2016) Fabrication and characterization of Eri silk fibers-based sponges for biomedical application. Acta Biomater 1(32):178–189
Singh R, Purohit S, Chacharkar MP, Bhandari PS, Bath AS (2007) Microbiological safety and clinical efficacy of radiation sterilized amniotic membranes for treatment of second-degree burns. Burns 33:505

Smith JG, Smith AJ, Shelton RM, Cooper PR (2012) Antibacterial activity of dentine and pulp extracellular matrix extracts. Int Endod J 45:749

Stanton RA, Billmire DA (2002) Skin resurfacing for the burned patent. Clin Plast Surg 29:29–51

Stern MM, Myers RL, Hammam N, Stern KA, Eberli D, Kritchevsky SB, Soker S, Van Dyke M (2009) The influence of extracellular matrix derived from skeletal muscle tissue on the proliferation and differentiation of myogenic progenitor cells ex vivo. Biomaterials 30:2393–2399

Sun WQ, Leung P (2008) Calorimetric study of extracellular tissue matrix degradation and instability after gamma irradiation. Acta Biomater 4:817

Tam H, Zhang W, Feaver KR, Parchment N, Sacks MS, Vyavahare N (2015) A novel crosslinking method for improved tear resistance and biocompatibility of tissue based biomaterials. Biomaterials 66:83

Tharp KM, Stahl A (2015) Bioengineering beige adipose tissue therapeutics. Front Endocrinol (lausanne). 20 (6):164

Turner NJ, Badylak SF (2015) The use of biologic scaffolds in the treatment of chronic nonhealing wounds. Adv Wound Care 4:490

Uygun BE, Soto-Gutierrez A, Yagi H, Izamis ML, Guzzardi MA, Shulman C et al (2010) Organ reengineering through development of a transplantable recellularized liver graft using decellularized liver matrix. Nat Med 16:814

Vacanti JP, Langer R, Upton J, Marler JJ (1998) Transplantation of cells in matrices for tissue regeneration. Adv Drug Deliv Rev 33:165–182

Valerio IL, Campbell P, Sabino J, Dearth CL, Fleming M (2015) The use of urinary bladder matrix in the treatment of trauma and combat casualty wound care. Regen Med 10(5):611–622

Wainwright DJ (1995) Use of an acellular allograft dermal matrix (AlloDerm) in the management of full-thickness burns. Burns 21:243–248

Wainwright DJ, Bury SB (2011) Acellular dermal matrix in the management of the burn patient. Aesthetic Surg J 31(7 Supplement):13S-23S

Wang S, Oldenhof H, Goecke T, Ramm R, Harder M, Haverich A, Hilfiker A, Wolkers WF (2015) Sucrose diffusion in decellularized heart valves for freeze-drying. Tissue Eng Part C Meth 21(9):922–931

Xu H, Xu B, Yang Q, Li X, Ma X, Xia Q, Zhang Y, Zhang C, Wu Y, Zhang Y (2014) Comparison of decellularization protocols for preparing a decellularized porcine annulus fibrosus scaffold. PLoS ONE 9:1–13

Zhai W, Zhang H, Wu C, Zhang J, Sun X, Zhang H, Zhu Z, Chang J (2014) Crosslinking of saphenous vein ECM by procyanidins for small diameter blood vessel replacement. J Biomed Mater Res B Appl Biomater 102:1190

Zhang X, Deng Z, Wang H, Yang Z, Guo W, Li Y, Ma D, Yu C, Zhang Y, Jin Y (2009) Expansion and delivery of human fibroblasts on micronized acellular dermal matrix for skin regeneration. Biomaterials 30:2666–2674

Zhang Q, Raoof M, Chen Y, Sumi Y, Sursal T, Junger W et al (2010) Circulating mitochondrial DAMPs cause inflammatory responses to injury. Nature 464:104

Extracellular Matrix Scaffold Using Decellularized Cartilage for Hyaline Cartilage Regeneration

17

Seyed Mostafa Monzavi,
Abdol-Mohammad Kajbafzadeh,
Shabnam Sabetkish, and Alexander Seifalian

Abstract

The repair of osteochondral defects is among the top ten medical needs of humans in the 21st centuries with many countries facing rapidly aging population involved with osteoarthritis as a major contributor to global disease burden. Tissue engineering methods have offered new windows of hope to treat such disorders and disabilities. Regenerative approaches to cartilage injuries require careful replication of the complex microenvironment of the native tissue. The decellularized hyaline cartilage derived from human allografts or xenografts is potentially an ideal scaffold, simulating the mechanical and biochemical properties, as well as biological microarchitecture of the hyaline cartilage. There have been many attempts to regenerate clinically viable hyaline cartilage tissue using decellularized cartilage-derived extracellular matrix with stem cell technology. This chapter describes the reproducible methods for hyaline cartilage decellularization and recellularization. In addition, quality control and characterization requirements of the product at each step, as well as the clinical applications of final product have been discussed.

Keywords

Decellularization · Hyaline Cartilage · Recellularization · Tissue Scaffolds · Tissue Engineering · Regenerative Medicine

S. M. Monzavi · A.-M. Kajbafzadeh · S. Sabetkish
Pediatric Urology and Regenerative Medicine Research Center, Children's Medical Center, Tehran University of Medical Sciences, Tehran, Iran

S. M. Monzavi · A.-M. Kajbafzadeh
Department of Tissue Engineering and Applied Cell Sciences, School of Advanced Technologies in Medicine, Tehran University of Medical Sciences, Tehran, Iran

A. Seifalian (✉)
Nanotechnology and Regenerative Medicine Commercialization Centre (NanoRegMed Ltd.), The London BioScience Innovation Centre, London, UK

A. Seifalian
UCL Centre for Nanotechnology and Regenerative Medicine, Division of Surgery and Interventional Science, University College London, London, UK

17.1 History

Hyaline cartilage is a translucent highly specialized connective tissue found in a number of organs in the human body. These include: (1) articulating surfaces of bones in synovial joints, also known as articular cartilage, (2) costal cartilages at the ventral terminals of ribs that link the upper ten pairs of ribs to the sternum, (3) supporting rings within the elastic walls of the trachea and the larger bronchial tubes, (4) part of the supporting framework of the larynx, (5) the external flexible part of the nose

A. Kajbafzadeh (ed.), *Decellularization Methods of Tissue and Whole Organ in Tissue Engineering*, Advances in Experimental Medicine and Biology 1345,
https://doi.org/10.1007/978-3-030-82735-9_17

forming the nostrils and (6) mechanical support for developing bones (Watkins and Mathieson 2009; Tissues et al. 2017).

Hyaline cartilage is mainly composed of extracellular matrix (ECM) with a low density of cells, which for example a mature articular cartilage consists of 70–80% water, 15–20% collagen fibrils, 10–15% proteoglycans, non-collagenous proteins and glycoproteins, and 2–5% cells as per the tissue wet weight (Buckwalter and Mankin 1998; Palukuru et al. 2014; Poole et al. 2001; Sophia Fox et al. 2009). The collagenous part of hyaline cartilage is primarily (>90%) made of collagen type II, although types I and IV-XI are also present in and contribute to the structure (Sophia Fox et al. 2009). In addition to limited cellularity, adult hyaline cartilage is generally devoid of blood vessels, nerves and lymphatics, (Buckwalter and Mankin 1998; Palukuru et al. 2014) as such the healing capacity of the tissue is majorly compromised. As an example, in osteoarthritis, in which biochemical breakdown of the articular cartilage following derangement of ECM, disruption of collagen fibers and degradation of proteoglycans occur, the patient experiences progressive inflammation and degeneration of the joint due to limited regenerative potential of the cartilage which will ultimately result in pain, deformity and reduced mobility, especially in the absence of appropriate medical treatments (Ghasemi et al. 2017; Goldring and Goldring 2007; Xia et al. 2014; Jivan et al. 2015). The repair of chondral defects is among the top ten medical needs and currently considered as an unmet clinical need with many countries facing a rapidly aging population involved with osteoarthritis as a major contributor to the global disease burden (Safiri et al. 2020; Murray and Lopez 2013). Cell-based tissue engineering strategies have offered new windows of hope to treat such disorders and disabilities (Huang et al. 2016; Raghunath et al. 2007). In addition to osteoarthritis, patients with nasal agenesis, nasal burns and defects, as well as those with injury, obstruction or tumors in the laryngotracheal tree would take benefit from such regenerative technologies (Grevemeyer et al. 2014; Hoshi et al. 2017; Crowley et al. 2015; Moradi et al. 2017; Ghorbani et al. 2017a).

Healing of a hyaline cartilage differs from the fibrous cartilage formation following injuries, which culminate in a stiff tissue at the injury site and long-term performance complications. Hence, the best practices on hyaline cartilage regeneration should focus on careful replication of the complex microenvironment of the native tissue. To produce a chondral unit, a combination of a suitable scaffold, apposite cells and stimulatory/growth factors is needed (Huang et al. 2016; Bautista and Bilgen 2018). The decellularized hyaline cartilage derived from human allografts or xenografts is potentially an ideal scaffold, simulating the mechanical and biochemical properties, as well as biological microarchitecture of the hyaline cartilage. There have been many attempts to regenerate clinically viable hyaline cartilage tissue using decellularized cartilage-derived ECM with stem cell technology. It has been shown in a preclinical study that the implantation of decellularized rat trachea between the paravertebral muscles of nude mice results in histologically mimicking cartilage tissue (Kajbafzadeh et al. 2015). This research provided evidence for utilizing the body as a bioreactor for total graft recellularization, thereby positioning the idea of homologous transplantation for the treatment of segmental trachea defects (Kajbafzadeh et al. 2015). However, studies in rodents should be repeated in larger animal models before moving to clinical study. A similar protocol using a decellularized porcine articular tissue implanted orthotopically to chondral defects of pigs resulted in cartilaginous neo-tissue (Nie et al. 2020). Recently, a suitable decellularization protocol has been recommended for the fabrication of 3D tissue constructs using decellularized cartilage seeded with nasal septal chondrocytes (Galuzzi et al. 2018). There have been many other similar protocols for decellularization and recellularization of hyaline cartilage regeneration, which gradually move toward the development of a standard protocol; however, still much work is required to come into an international standard protocol for the clinical setting (Bautista and Bilgen 2018; Benders et al. 2019; Thakkar et al. 2015). The aim of this chapter is to provide critical evaluation and compilation of well-tailored protocols for the fabrication of hyaline cartilage using decellularized cartilage-derived bio-scaffolds with full characterization and quality control for clinical application.

17.2 Materials

All the consumables and equipment required for the decellularization of hyaline cartilage and subsequent development of recellularized hyaline cartilage using the corresponding decellularized ECM are listed in Table 17.1. The listed materials are for the three steps of tissue harvesting, decellularization and recellularization. For clinical application, the entire procedure should be carried out under good laboratory practice (GLP). All media, buffers and solutions should be certified for human use and manufactured under GLP. The materials should be used at room temperature unless otherwise stated.

17.3 Methods

17.3.1 General Considerations

The fabrication process of a hyaline cartilage neo-tissue using a decellularized cartilage-derived ECM seeded with autologous or allogenic cells comprises three main steps, as mentioned earlier (i.e., tissue harvesting, decellularization and recellularization). In all steps, aseptic techniques (or GLP standards for clinical application) must be followed to ensure the sterility of the final product (Benders et al. 2019; Thakkar et al. 2015; JoVE Science Education Database 2020):

(1) All materials, consumables and manual equipment must be sterilized.

(2) All procedures should be performed inside a validated class II biosafety cabinet preferably within a biosafety level 3 laboratory. Before entering lab and wearing the lab clothing, wash hands with soap or hand-rub with 70–75% ethanol for at least 30 s.

(3) The operator should wear protective clothing and equipment including a medical mask, spunbond bouffant cap, sterile surgical gown covering complete torso and upper extremities particularly the wrists, sleeves, safety goggles/spectacles and sterile surgical gloves.

17.3.2 Tissue Harvesting

In order to procure suitable tissue for decellularization, the cartilage and surrounding tissue can be excised either from human cadaver or discarded human tissue following surgery or the corpse of slaughtered animals. The intact articular joint (i.e., skin around the joint is still intact) shall be washed with copious water and common detergents and then shall be packaged appropriately and transferred into the lab. Note that other than articular cartilage, the hyaline cartilage can be obtained from nasoseptal or tracheal (from the human/animal cadaver or the discarded tissues after surgery) cartilages (Kajbafzadeh et al. 2015; Griffin et al. 2016; Galliger and Panoskaltsis-Mortari 2018; Graham et al. 2016). The next steps of articular cartilage harvesting are as follows (Benders et al. 2019; Thakkar et al. 2015; Mohammadie et al. 2017):

1. After opening the package in the lab, the intact joint should be soaked with alcohol and deconex™ for at least a minute each.
2. Inside the biosafety cabinet, the skin, fat and muscle covering the joint area of interest are removed by using a scalpel and surgical tweezers without imposing damage to the underlying joint.
3. Prior to arthrotomy (separation of the joint), the appropriate articular surface of the joint should be designated by performing extension and flexion of the joint.
4. The remaining overlying layers of fat, muscles and tendons are removed with a scalpel and surgical tweezers to expose the joint area.
5. An incision is made to reach the joint cavity cautiously, while being aware of synovial fluid, which can drip from the cavity or

Table 17.1 Materials and equipment required for tissue harvesting, decellularization and recellularization of hyaline cartilage

Step	Materials	Details and features
1 and 2. Tissue harvesting and decellularization	*Consumables*	
	Washing solution	1% (v/v) penicillin–streptomycin + 1% (v/v) amphotericin B in PBS
	Phosphate-buffered saline	0.01 M, pH 7.4
	Hypotonic buffer	10 mM Tris–HCl, pH 8.0
	Protease inhibitor cocktail	2.95 μl/ml protease inhibitor cocktail
	Chondroitinase ABC	0.125 U/ml chondroitinase ABC + 50 mM Tris–HCl, 60 mM sodium acetate + 2 mM EDTA, pH 8.0
	Anionic detergent	0.1% (w/v) sodium dodecyl sulfate + 0.1% (w/v) EDTA in hypotonic buffer
	Nuclease solution	100 U/ml DNase (Thermo Fisher) + 1 U/ml RNase + 10 mM magnesium chloride + 50 μg/ml BSA + 50 mM Trizma base
	Peracetic acid	0.1% (v/v) peracetic acid in PBS
	24-well plates	non-treated, suitable for cell culture
	Petri dishes	100 and 150 mm
	Conical tubes	15 and 50 mL
	Equipment	
		Orbital shaker, surgical tweezers, scalpel, pipette set
3. Recellularization	*Consumables*	
	Phosphate-buffered saline	0.01 M, pH: 7.4
	Basic medium	α-MEM or DMEM or DMEM-F12 + 10% FBS + 0.2 mM L-ascorbic acid 2-phosphate + 2 mM L-glutamine + 1% (v/v) penicillin–streptomycin
	Proliferation medium	α-MEM or DMEM or DMEM-F12 + 10% FBS + 0.2 mM L-ascorbic acid 2-phosphate + 2 mM L-glutamine + 1% (v/v) penicillin–streptomycin + and 1 ng/mL bFGF-2
	Chondrocyte culture medium	DMEM or DMEM-F12 + 10% FBS + 1 × non-essential amino acids + 10 mM HEPES buffer + 0.2 mM L-ascorbic acid 2-phosphate + 0.4 mM proline + 1% (v/v) penicillin–streptomycin
	Chondrogenic medium	DMEM or DMEM-F12 + 1% GlutaMAX + 0.2 mM L-ascorbic acid 2-phosphate + 1% (v/v) penicillin–streptomycin + 0.4 mM proline + 100 μg/mL sodium pyruvate + 50 μg/mL insulin transferrin selenium-premix
	Lymphocyte separation medium (Density gradient medium)	Neutral, high-mass, hydrophilic polysaccharide with density adjusted to 1.077 g/ml
	Collagenase type II solution	0.15% collagenase type II in DMEM
	Trypsin solution	0.05% trypsin–EDTA
	Supplementation	10 ng/mL TGF-β3 + 100 nM dexamethasone
	Ethanol	70–100% (v/v)
	Culture flasks	Surface treated and nontreated
	24-well plates	Surface treated and nontreated, suitable for cell culture
	Petri dishes	100 and 150 mm
	O-rings	Different sizes (e.g., 14 × 1.78 mm or 13.9 × 2.62 mm)
	Equipment	
		Cell scraper, cell counter, centrifuge, inverted microscope, pipette set

be splashed on spectacles, clothing and periphery.

6. The joint is opened up completely by cutting through the structural tendons supporting the joint and interconnecting adnexal tissues.
7. The cartilage should be inspected for any macroscopic damage, and the tissue may be discarded in the absence of glossy and smooth appearance or in case of evident blistering, clefts or defects.
8. Using a sterile scalpel, the cartilage is resected from the subchondral bone. The incision must approach completely down to the subchondral bone to remove the deep zone of the cartilage. Alternatively, for research purposes, tissue pieces (usually 2 × 2 mm pieces) can be excised from the cartilage with a metal puncher without the need for removal of entire cartilage and then the tissue pieces should be washed carefully (Mohammadie et al. 2017).

Note 1. Nasoseptal cartilages can be obtained from routine surgical procedures (i.e., septoplasties and septorhinoplasties) or from human/animal cadaver (Graham et al. 2016). The nasal tissue harvested from cadaver should undergo washing and asepsis. For harvesting the nasoseptal cartilages, in addition to overlying skin and soft tissues, removal of perichondrium to generate an isolated intact cartilage tissue has been recommended (Graham et al. 2016).

Note 2. To obtain tracheal cartilage, the trachea should be opened by making a full-length incision on the dorsal side. The internal epithelial layers and smooth muscle should be removed by tweezers and each cartilage ring should be separated from others by cutting (Kajbafzadeh et al. 2015; Galliger and Panoskaltsis-Mortari 2018).

After harvesting the cartilage plug/slices (from any source mentioned above, i.e., articular, nasoseptal or tracheal cartilage), they should be washed in suitable petri-dishes for several times, and then should be transferred into 50 mL conical tubes containing the washing solution. Once the collection is completed, the cartilage slices are snap-freezed in liquid nitrogen for 5 min. Then the samples are thawed and the solution is removed and replaced with new washing solution. This can be repeated several times.

17.3.2.1 Decellularization

Decellularization relates to the procedure of treating a tissue/organ with a combination of physical stress and chemical/enzymatic agents to eliminate cellular and nuclear constituents while preserving the ECM structure that can be used for research/therapeutic applications. In a decellularization process, the principal challenge is complete cell removal while limiting damage to the ECM. Chemical decellularization is a common technique that is performed through the use of detergents. In this technique, nonionic detergents such as triton X-100 and ionic detergents such as sodium dodecyl sulfate (SDS) are used. A decellularized hyaline cartilage should mimic the ECM composition, organization and bioactivity of the native tissue while providing very limited immunogenicity (Ye et al. 2013). Some scientists have preferred precluding the enzymatic digestion to create potential advantages, i.e., minimizing the damage and derangement of ECM architecture and integrity (Kajbafzadeh et al. 2014). In clinic, patients may benefit from this approach, as in the absence of these enzymes, safety snags, i.e., toxicity, inflammation, irritation and teratogenicity, due to enzyme residues that may adhere to ECM even after profuse washing (Crapo et al. 2011), can be prevented. Nonetheless, many protocols include enzymatic digestion as a necessary step to ensure complete removal of cellular and genetic components of the ECM. Careful washing and precise quality control assays after enzymatic digestion may mitigate such safety concerns.

Decellularization Procedure

Despite some discrepancies, the current protocols for decellularizing the hyaline cartilage share some common steps (Bautista and Bilgen 2018; Kajbafzadeh et al. 2015; Benders et al. 2019; Thakkar et al. 2015; Mohammadie et al. 2017). The protocol developed by Bautista and Bilgen (Bautista and Bilgen 2018) provides reproducible results and is described with minor modifications (considering the other quoted protocols) as follows:

1. Under sterile condition, cartilage plug/slice (s) are transferred to a single cryovial.
2. Using sterile scissors, multiple strips of filter paper are prepared. Then, using sterile tweezers, each strip is immersed in PBS and is placed in each cryovial.
3. Cryovials are placed in 20 °C freezer for at least 3 h (or until crystal formation), followed by thawing on bench for 2–4 h. Of note, this step may be repeated.
4. Using sterile tweezers, filter paper from each cryovial is removed.
5. One ml hypotonic buffer to each cryovial is added, while ensuring the entire cartilage specimen is submerged in the solution.
6. Cryovials are placed in 20 °C freezer for at least 3 h (or until crystal formation), followed by thawing on bench for 2–4 h.
7. Cryovial content is emptied into one well of a 24-well plate and old solution is removed.
8. One ml chondroitinase ABC solution with 2.95 μl/ml protease inhibitor cocktail is added to each well.
9. The plate is incubated at 37 °C for 24 h with agitation on orbital shaker set at 180 RPM. The lid of plate may be secured with one/two strips of lab tape all around the plate and its lid. The edges of the plates may be sealed with parafilm to prevent solution evaporation.
10. In the next day, old solution is aspirated, followed by addition of 1 ml hypotonic buffer with protease inhibitor cocktail to each well. The plate undergoes 24 h of incubation at 45 °C with agitation on orbital shaker set at 180 RPM.
11. In the next day, old solution is aspirated, followed by addition of 1 ml anionic detergent (0.1%) with protease inhibitor cocktail to each well. The plate is incubated at 45 °C for 24 h with agitation on orbital shaker set at 180 RPM. If SDS is used at higher concentrations such as 1% or 2%, the duration required for SDS application might be shorter. As an instance, the use of 2% SDS for 6 or 8 h has resulted in complete histological decellularization of bovine cartilage (Elder et al. 2009).
12. Steps 10 and 11 can be repeated twice for a total of three detergent cycles.
13. Old solution is aspirated, followed by addition of 1 ml PBS with protease inhibitor cocktail to each well. The plate is agitated on an orbital shaker at 180 RPM at room temperature for 30 min. This step can be repeated once more.
14. Old solution is aspirated, followed by addition of 1 ml PBS with protease inhibitor cocktail to each well. The plate undergoes 24 h of incubation at 45 °C with agitation on orbital shaker set at 180 RPM.
15. Old solution is aspirated, followed by addition of 1 ml nuclease solution to each well. The plate is incubated at 37 °C for 3 h with agitation on orbital shaker set at 180 RPM.
16. Old solution is aspirated, followed by addition of 1 ml PBS with protease inhibitor cocktail to each well. The plate is agitated on an orbital shaker at 180 RPM at room temperature for 1 h.
17. Old solution is aspirated, followed by addition of 1 ml peracetic acid solution to each well. The plate is agitated on an orbital shaker at 180 RPM at room temperature for 3 h.
18. Old solution is aspirated, followed by addition of 1 ml washing solution. The plate is agitated on an orbital shaker at 180 RPM at room temperature for 30 min. This step can be repeated once more.
19. Old solution is aspirated, followed by addition of 1 ml washing solution. The plate is agitated on an orbital shaker at 180 RPM at room temperature for 24 h.
20. Old solution is aspirated, followed by addition of 1 ml washing solution. The samples are ready for characterization.

Characterization of the Decellularized Hyaline Cartilage

The decellularized hyaline cartilage should be characterized using sections of the paraffin-embedded samples to be analyzed with H&E

staining for morphology (Fig. 17.1), DAPI staining for DNA content, Alician blue staining for glycosaminoglycans (GAG) content and Safranin-O staining for proteoglycan content. No detectable nuclear material should be detected in DAPI or H&E stainings. The amount of double-stranded DNA (dsDNA) in decellularized ECM should be less than 50 ng of dsDNA per mg of the dry ECM. For collagen typing, immunohistochemical staining of type I and II collagen can be performed. Although alternatively, Picrosirius red staining is able to visualize collagen I fibers. Additionally, Russell-Movat's (pentachrome) staining can also be used for histological demonstration of the ECM to differentiate collagen, elastin, mucin and fibrin in tissue sections. Masson's trichome staining can also be used to visualize collagen fibers.

It is also important to examine the microscopic alterations in the ECM structure and the biochemical composition via imaging techniques such as scanning/transmission electron microscopy and microcomputed tomography to ensure the preservation of ECM composition (Fig. 17.1). Further, the amount of sulfated GAG and collagen can be evaluated by dimethylmethylene blue and hydroxyproline assays, respectively. To estimate the maximum elongation and tensile strength of the decellularized ECM, tensile test should be considered.

17.3.2.2 Recellularization

To prepare a hyaline cartilage neo-tissue, the decellularized hyaline cartilage should be seeded with the cell of interest via the dynamic process of recellularization. The most common cells used for recellularization of hyaline cartilage scaffolds include chondrocytes and mesenchymal stromal/stem cells (MSCs), (Bautista and Bilgen 2018; Thakkar et al. 2015; Ghorbani et al. 2017b; New et al. 2017) albeit the ultimate goal should be set to identification of the most reliable source to produce a mechanically stable and biologically compatible cartilage.

Isolation and Expansion of Chondrocytes

To produce a bioengineered hyaline cartilage, chondrocytes are logically the most convenient cell of choice for recellularization, as they secrete collagen and GAGs, thereby remodeling and maintaining the cartilage baseline matrix. To isolate and expand chondrocytes either from the recipient or a donor or from a xenogeneic source, the following steps should be considered (Thakkar et al. 2015; Lau et al. 2015):

1. The harvested cartilage is transferred into a 100-mm petri-dish containing basic medium.
2. The cartilage plug/slices are minced into small pieces (2 × 2 mm).
3. The cartilage pieces are transferred into a flask containing medium and collagenase type II solution and undergo 24 h incubation at 37 °C. Plugs turn into single cell suspension, otherwise, the remaining chondrocyte aggregates can become single cell via mechanical force (pipetting the solution with narrow tip).
4. To remove the remaining aggregates, the cell suspension is filtered through a 100-μm cell strainer.
5. The cells are washed twice with PBS supplemented with antibiotics.
6. The cells are centrifuged. After removal of the supernatant, the cell pellet is resuspended in basic medium.
7. The cells are seeded at 5000 cells/cm^2 (Passage 0) in T175 cm^2 culture flasks filled with 15 mL of chondrocyte culture medium and the flasks undergo a 3-week incubation at 37 °C in a humidified atmosphere containing 5% CO_2. The medium is refreshed every 3 days.
8. When confluence reaches 80%, the cells are trypsinized with 1 mL of trypsin solution covering the bottom surface of the flask for 3 min at 37 °C incubation. Next, basic medium is added to inactivate the enzyme. Alternatively, PBS supplemented with 0.05% EDTA for 5 min at 4 °C in refrigerator may be used for cell detachment. The flask may be gently tapped to ensure complete detachment of cells.
9. The cell suspension is transferred to a tube and is centrifuged at 200 g for 10 min.

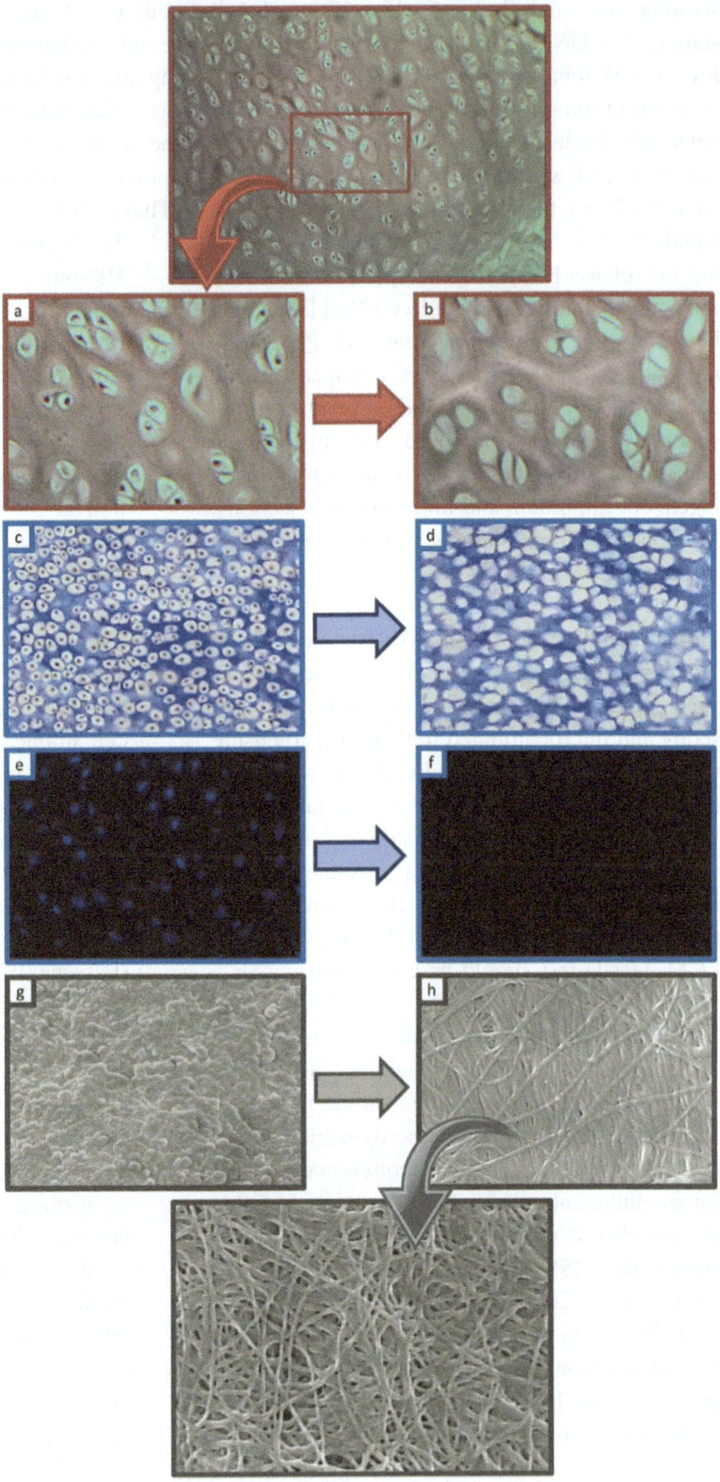

Fig. 17.1 Native (**a**, **c**, **e**, **g**) versus decellularized hyaline cartilage (**b**, **d**, **f**, **h**): **a** and **b**, H&E staining. **c** and **d**, Trichrome staining. **e** and **f**, DAPI staining. **g** and **h**, Scanning electron microscope micrographs

10. The supernatant is discarded and the cell pellet is suspended in basic medium.
11. The cells are ready for characterization. The chondrocytes can be characterized using safranin O or toluidine blue staining as well as COL1, COL2, ACAN immunohistochemical staining. Alcian blue staining for localization of GAG content and flowcytometry of the cells for CD44 and CD54 may also be helpful (Oseni et al. 2013a; Isyar et al. 2016).

Isolation and Expansion of Mesenchymal Stem Cells

MSCs derived from different sources such as bone marrow, umbilical cord blood or adipose tissue may also be used for recellularization and development of a hyaline cartilage neo-tissue, although they require an additional step for differentiation into chondroblasts/chondrocytes. The isolation and characterization of MSCs from different sources have been described in details elsewhere (Moradi et al. 2017; Mushahary et al. 2018), and here the isolation process from bone marrow aspirate is briefly provided:

1. The bone marrow aspirate is suspended in basic medium.
2. The cells are filtered through a 70-µm cell strainer to remove bone spicules, muscle fragments and cell clumps.
3. The cells undergo density gradient centrifugation using the lymphocyte separation medium to isolate mononuclear cells (MNCs). At this step, to avoid coculture of MSCs and monocytes that are both adherent to the plastic surface, monocyte component of the MNCs can be depleted through a magnetic-activated cell sorting using a GMP-grade monocyte isolation kit.
4. The MNCs are seeded at a density of 5×10^5 cells per flask in T175 cm^2 culture flasks filled with 15 mL of an FBS-free basic medium (Thakkar et al. 2015).
5. After 3–4 h, the non-adherent cells that are overlayered on the adherent cells should be removed from the flask after gentle pipetting, followed by replacing with fresh proliferation medium.
6. The flasks undergo incubation at 37 °C in a humidified atmosphere containing 5% CO_2 for two weeks (or until reaching confluence) while the medium is refreshed twice a week. Of note, the aggregates of adherent spindle-shaped cells appear on the days 3–4 after seeding.
7. After harvesting the adherent cells at the end of expansion period, the MSCs are characterized with expression of CD105, CD73 and CD90, and lack expression of CD45, CD34, CD14 or CD11b, CD79α or CD19 and HLA-DR surface markers (Moradi et al. 2017; Mushahary et al. 2018; Dominici et al. 2006; Kamguyan et al. 2017).
8. To differentiate the MSCs to chondrocytes before cell seeding (optional), the harvested MSCs are resuspended in chondrogenic medium, supplemented with 10^{-7} M dexamethasone and 10 ng/mL TGF-β3 (Thakkar et al. 2015). The cells undergo incubation at 37 °C in a humidified atmosphere containing 5% CO_2 for 2–3 weeks while refreshing the medium every 3 days.

Cell Seeding

1. The decellularized hyaline cartilage pieces are soaked in 100% ethanol for 30 min followed by 15 min for drying.
2. The scaffolds are washed twice with PBS.
3. Secured with O-rings, the scaffold pieces are transferred to a suitable plate filled with chondrogenic culture medium and are incubated for 4 h (Thakkar et al. 2015).
4. The scaffold pieces are seeded with 20×10^3 cells/mm^3 and are incubated for 3–4 h to allow the cells to attach to the scaffold.
5. The seeded scaffolds undergo 3-week incubation at 37 °C in a humidified atmosphere containing 5% CO_2 in chondrocyte culture medium, while changing the medium every 3 days (Fig. 17.2).
6. The recellularized hyaline cartilage should be characterized with H&E staining for morphology, Almar Blue staining for cell viability, Alcian blue staining for localization of GAG content, Safranin O for demonstrating the proteoglycan content, DAPI staining for

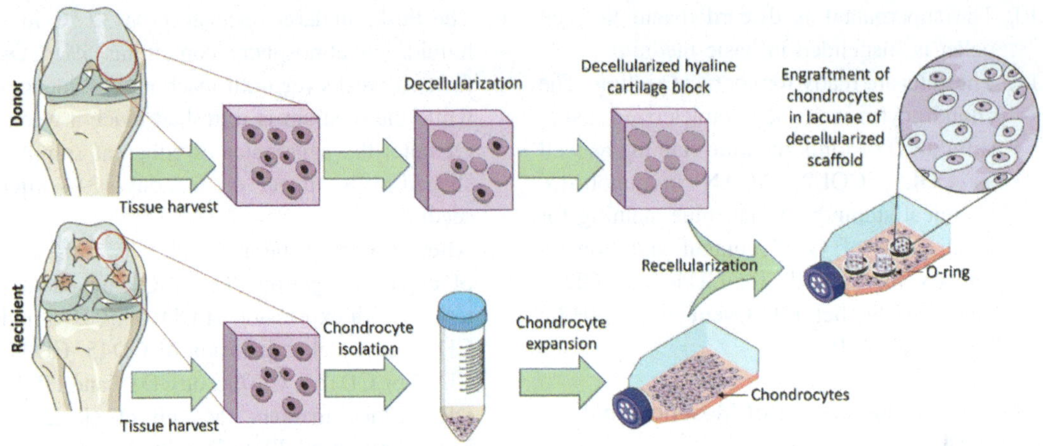

Fig. 17.2 Schematic diagram of fabrication processes of hyaline cartilage neo-tissue using decellularized hyaline cartilage ECM seeded with chondrocytes. The decellularized hyaline cartilage of a donor tissue is cocultured with chondrocytes isolated and expanded from the recipient's articular cartilage biopsy specimen. After 3–4 weeks, living hyaline cartilaginous graft establishes. (Some of the graphical elements used in this figure were generated via *mindthegraph.com*, which are made availablefreely by the website under a free culture Creative Commons license (CCBY-SA 4.0))

assessment of cell viability and the relative distribution of cells over the scaffold. The presence of collagen within engineered hyaline cartilage and structure of collagen network scan be assessed with scanning electron microscope and Picrosirius red staining.

17.4 Research and Clinical Applications

Use of decellularized hyaline cartilage for repair of chondral defects has been investigated in a number of studies. The matrix was shown to support the growth of primary chondrocytes and MSC-derived chondrocytes along with reconstitution of collagen and proteoglycans in the process of in vitro culture and in the host (Ghorbani et al. 2017a; Cheng et al. 2011; Yang et al. 2010; Dai et al. 2019; Oseni et al. 2013b). A tailor-made study showed MSC-loaded decellularized cartilage scaffolds could repair osteochondral defect in rabbit model by filling with hyaline cartilage. The repaired tissue was revealed to possess comparable mechanical and biochemical properties with native cartilage (Kang et al. 2014). However, it is still unclear whether MSCs can undergo in situ differentiation to chondrocytes after engraftment or it is more reasonable to differentiate MSCs to chondrocytes prior to transplantation. Perhaps, use of stem cells versus chondrocytes for recellularization of decellularized ECM should be more carefully investigated.

To date, four products based on decellularized ECM for treatment of hyaline cartilage defects have been authorized to enter into the medical market (Table 17.2). Chondrofix® (Zimmer Biomet Inc., USA) is an off-the-shelf osteochondral allograft that is composed of donated human decellularized hyaline cartilage and the underlying cancellous bone. It has been produced with the purpose of homologous repair of osteochondral lesions in diarthrodial joints either due to osteoarthritis or sport injuries. Despite producing a relatively high failure rate, the use of this product has shown to result in improved articular performance and daily function in a fraction of the recipients and complete bony incorporation and cartilage integrity (Beer et al. 2019; Long et al. 2016; Reynolds and Bishai 2014). The other off-the-shelf product containing the decellularized hyaline cartilage, known as BioCartilage® (Arthrex, Inc., USA), is composed

Table 17.2 Marketed products using a decellularized ECM for hyaline cartilage defects

Product	Company	Composition		Trial registry codes	Applications
		ECM	Cellular component		
Chondrofix®	Zimmer Biomet, Inc	Decellularized osteochondral human articular cartilage	None	*NCT01410136*	Chondral and osteochondral lesions of grade III-IV
BioCartilage®	Arthrex, Inc	Micronized osteochondral human articular cartilage	None	*NCT02203071* *NCT02309957* *NCT01183637*	Augmentation of microfracture procedures (Serves as a scaffold over a microfractured defect)
MACI®	Vericel Corp	Decellularized porcine peritoneum	Expanded autologous chondrocytes	*NCT00719576* *NCT03588975* *NCT01251588*	Single or multiple full-thickness cartilage defects of the knee with or without bone involvement in adults
NOVOCART®	TETEC AG	Decellularized bovine pericardium	Expanded autologous chondrocytes	*NCT01656902* *NCT01957722*	Chondral and osteochondral lesions of grade III-IV, Focal traumatic defects, Osteochondrosis dissecans

of micronized, dehydrated, allogenic cartilage ECM that can form a paste when mixed with platelet-rich plasma (PRP) or bone marrow aspirate concentrate. The resultant paste can be used as an augmentative material in microfracture procedures. This product has shown favorable regenerative effects on chondral lesions, though still no long-term evaluation of the product on humans is available (Sutherland et al. 2015; Hirahara and Mueller 2015).

While the decellularized ECM itself possess the chondrocyte anchoring sites providing a receptive niche for autologous residing chondrocytes in the diseased joint, and so potentially having the chondroinductive capacity to generate hyaline cartilage in vivo without ex vivo seeding of cells (Kajbafzadeh et al. 2015), the more clinically translational approach might be ex vivo cell-loading on the scaffold prior to transplantation (Jacobi et al. 2011). The autologous cultured chondrocytes seeded on decellularized porcine collagen membrane, commercially available as MACI® (Vericel Corp., USA), has been in the US market since 1999 and finally achieved FDA approval for clinical use in December 2016 (Jacobi et al. 2011). NOVOCART® 3D (TETEC AG, Germany), a product similar to MACI®, composed of decellularized bovine collagen membrane seeded with autologous expanded chondrocytes, has been in clinical use in Europe since 2003. Both products have provided very good success rate in cartilage repair and improvement of functionality in patients with full thickness cartilage defects of the knee joint with promising long-term efficacy (Huang et al. 2016; Saris et al. 2014; Zak et al. 2014; Erickson et al. 2018; Kreuz et al. 2019).

17.5 Limitations

Structural integrity, functional maturation and scale-up in manufacturing are the main challenges and impediments to cartilage bioengineering. Methods that enable maintaining the physicochemical properties of decellularized ECM, while harnessing its natural regenerative capabilities needs further investigation. Despite having biological properties similar to native tissue, the decellularized hyaline cartilage may be compromised with loss of biomechanical strength and stability during the process of decellularization owing to derangement of ECM as evidenced in some studies (Ye et al. 2013; Liao et al. 2008). Moreover, the ECM may lose its integrated microarchitecture during the chemical washing steps which therefore fail to represent the innate microenvironment for chondrocyte homing and recellularization (Ye et al. 2013; Liao et al. 2008). Balancing between the complete elimination of cellular and nuclear components, in order to avoid an immune response after transplantation, while preserving ECM composition, is a challenge that needs further explorations. Moreover, tissue sources and storage condition before decellularization technique also influence the quality of ECM that require meticulous scrutiny.

In tracheobronchial disorders, the use of bioengineered cartilage may result in lumen obstruction or stenosis (Crowley et al. 2015). To overcome these challenges, use of a stable non-biodegradable polymeric support may provide a solution. In this context, cylindrical mesh, nanocomposites and hybrid bio-scaffolds (3D printed nanoparticles incorporated to decellularized ECM) epithelialized with chondrocytes or MSCs may provide a solution and produce an organ-shaped and mechanically stable trachea (Raghunath et al. 2007; Crowley et al. 2015; Ghorbani et al. 2017b). Moreover, trachea replacement using tissue engineered tracheas may be associated with complications such as infection, inflammation, calcification, ischemia and subsequent rejection of the transplanted tissue, (Crowley et al. 2015; Bader and Macchiarini 2010; Etienne et al. 2018) the problems that should be carefully addressed in preclinical studies and require further investigations.

Although, the chondrocytes are appeared to be the ideal source of cells for recellularization, their use is hampered owing to their limited number in cartilage tissue, which therefore requires extensive in vitro expansion (Ye et al. 2013; Chung and Burdick 2008). This expansion may additionally cause the dedifferentiation of mature chondrocytes to earlier lineages, thereby causing lower synthesis of collagen type II or predominantly synthesis of collagen type I (Ye et al. 2013; Goessler et al. 2004, 2006). MSCs are abundantly and readily available cells from different tissue sources. However, they require defined medium and costly growth factors to differentiate into chondrocytes, making them a less economically-rational choice for recellularization. In addition, bone marrow-derived MSCs may express collagen type X and MMP-13 inducing hypertrophic chondrogenesis with the potential of mineralization when exposed to osteogenic stimuli (Ye et al. 2013; Muraglia et al. 1998). The adipose-derived MSCs, on the other hand, have shown to produce inferior matrix with attenuated mechanical integrity, as compared with chondrocytes (Mauck et al. 2006). To improve cell viability and to enhance cartilaginous ECM formation, one potential approach would be the use of PRP that is rich in chondrogenic growth factors (TGF- β, IGF-1) (Haleem and Chu 2010; Jivan et al. 2021).

After transplantation of tissue-engineered hyaline cartilage, local injection of autologous PRP, which is shown to be safe and non-immunogenic, can activate the seeded cells on the bio-scaffold and maintain their viability, thereby substantiating the collagen formation, increasing the ECM integrity, facilitating the neovascularization and ultimately enhancing the engraftment rate (Haleem and Chu 2010; Jivan et al. 2021; Barbon et al. 2019).

17.6 Conclusions

Tissue grafts and cellular engineering may offer a clinical solution for the critically damaged hyaline cartilages. To manufacture a 3D decellularized biological scaffold, removal of the allogeneic or xenogeneic cellular antigens and elimination of immunogenicity are essential. However, the decellularization process of hyaline cartilage should be optimized to preserve the composition, bioactivity and structural integrity of the ECM. Moreover, to generate a viable functional cell-loaded biocompatible hyaline cartilage, suitable cells for recellularization should be expanded in sufficient numbers in well-defined cell culture conditions within in vivo or ex vivo physiological bioreactors. The methods described in this chapter consider these aspects to achieve products suitable for clinical translation. The fabricated tissues should be well characterized and quality controlled with the simulation of biomechanical properties for target organs. Altogether, suitable and reproducible methods must be further optimized for decellularization and recellularizations based on the potential clinical applications.

References

Bader A, Macchiarini P (2010) Moving towards in situ tracheal regeneration: the bionic tissue engineered transplantation approach. J Cell Mol Med 14(7):1877–1889

Barbon S, Stocco E, Macchi V, Contran M, Grandi F, Borean A et al (2019) Platelet-rich fibrin scaffolds for cartilage and tendon regenerative medicine: from bench to bedside. Int J Mol Sci 20(7):1701

Bautista CA, Bilgen B (2018) Decellularization and recellularization of cartilage. Methods Mol Biol 1577:139–146

Beer AJ, Tauro TM, Redondo ML, Christian DR, Cole BJ, Frank RM (2019) Use of allografts in orthopaedic surgery: safety, procurement, storage, and outcomes. Orthop J Sports Med 7(12): 2325967119891435

Benders KEM, Terpstra ML, Levato R, Malda J (2019) Fabrication of decellularized cartilage-derived matrix scaffolds. J Vis Exp (143): e58656

Buckwalter JA, Mankin HJ (1998) Articular cartilage: tissue design and chondrocyte-matrix interactions. Instr Course Lect 47:477–486

Cheng NC, Estes BT, Young TH, Guilak F (2011) Engineered cartilage using primary chondrocytes cultured in a porous cartilage-derived matrix. Regen Med 6(1):81–93

Chung C, Burdick JA (2008) Engineering cartilage tissue. Adv Drug Deliv Rev 60(2):243–262

Crapo PM, Gilbert TW, Badylak SF (2011) An overview of tissue and whole organ decellularization processes. Biomaterials 32(12):3233–3243

Crowley C, Birchall M, Seifalian AM (2015) Trachea transplantation: from laboratory to patient. J Tissue Eng Regen Med 9(4):357–367

Dai L, He Z, Jiang Y, Zhang X, Ren S, Zhu J et al (2019) One-step strategy for cartilage repair using acellular bone matrix scaffold based in situ tissue engineering technique in a preclinical minipig model. Am J Transl Res 11(10):6650–6659

Dominici M, Le Blanc K, Mueller I, Slaper-Cortenbach I, Marini F, Krause D et al (2006) Minimal criteria for defining multipotent mesenchymal stromal cells. The international society for cellular therapy position statement. Cytotherapy 8(4):315–317

Elder BD, Eleswarapu SV, Athanasiou KA (2009) Extraction techniques for the decellularization of tissue engineered articular cartilage constructs. Biomaterials 30(22):3749–3756

Erickson BJ, Strickland SM, Gomoll AH (2018) Indications, techniques, outcomes for matrix-induced autologous chondrocyte implantation (MACI). Oper Tech Sports Med 26(3):175–182

Etienne H, Fabre D, Gomez Caro A, Kolb F, Mussot S, Mercier O et al (2018) Tracheal replacement. Eur Respir J 51(2):1702211

Galliger Z, Panoskaltsis-Mortari A (2018) Tracheal cartilage isolation and decellularization. Methods Mol Biol 1577:155–160

Galuzzi M, Perteghella S, Antonioli B, Tosca MC, Bari E, Tripodo G et al (2018) Human engineered cartilage and decellularized matrix as an alternative to animal osteoarthritis model. Polymers (Basel) 10(7):738

Ghasemi A, Yaghoubi M, Shariati-Sarabi Z, Zargaran B, Monzavi SM, Tavakol AJ (2017) Clinical Improvement of patients with knee osteoarthritis after intra-articular injection of autologous bone marrow-derived mesenchymal stem cells: a case series. Iran J Biotechnol 15:237–238

Ghorbani F, Feizabadi M, Farzanegan R, Vaziri E, Samani S, Lajevardi S et al (2017a) An investigation of topics and trends of tracheal replacement studies using co-occurrence analysis. Tissue Eng Part B Rev 23(2):118–127

Ghorbani F, Moradi L, Shadmehr MB, Bonakdar S, Droodinia A, Safshekan F (2017b) In-vivo characterization of a 3D hybrid scaffold based on PCL/decellularized aorta for tracheal tissue engineering. Mater Sci Eng C Mater Biol Appl 81:74–83

Goessler UR, Bugert P, Bieback K, Baisch A, Sadick H, Verse T et al (2004) Expression of collagen and fiber-associated proteins in human septal cartilage during in vitro dedifferentiation. Int J Mol Med 14(6):1015–1022

Goessler UR, Bugert P, Bieback K, Sadick H, Baisch A, Hormann K et al (2006) In vitro analysis of differential expression of collagens, integrins, and growth factors in cultured human chondrocytes. Otolaryngol Head Neck Surg 134(3):510–515

Goldring MB, Goldring SR (2007) Osteoarthritis. J Cell Physiol 213(3):626–634

Graham ME, Gratzer PF, Bezuhly M, Hong P (2016) Development and characterization of decellularized human nasoseptal cartilage matrix for use in tissue engineering. Laryngoscope 126(10):2226–2231

Grevemeyer B, Bogdanovic L, Canton S, St Jean G, Cercone M, Ducharme NG et al (2014) Regenerative

medicine approach to reconstruction of the equine upper airway. Tissue Eng Part A 20(7–8):1213–1221
Griffin MF, Premakumar Y, Seifalian AM, Szarko M, Butler PE (2016) Biomechanical characterisation of the human nasal cartilages; implications for tissue engineering. J Mater Sci Mater Med 27(1):11
Haleem AM, Chu CR (2010) Advances in tissue engineering techniques for articular cartilage repair. Oper Tech Orthop 20(2):76–89
Hirahara AM, Mueller KW Jr (2015) BioCartilage: a new biomaterial to treat chondral lesions. Sports Med Arthrosc Rev 23(3):143–148
Hoshi K, Fujihara Y, Saijo H, Kurabayashi K, Suenaga H, Asawa Y et al (2017) Three-dimensional changes of noses after transplantation of implant-type tissue-engineered cartilage for secondary correction of cleft lip-nose patients. Regen Ther 7:72–79
Huang BJ, Hu JC, Athanasiou KA (2016) Cell-based tissue engineering strategies used in the clinical repair of articular cartilage. Biomaterials 98:1–22
Isyar M, Yilmaz I, Yasar Sirin D, Yalcin S, Guler O, Mahirogullari M (2016) A practical way to prepare primer human chondrocyte culture. J Orthop 13(3):162–167
Jacobi M, Villa V, Magnussen RA, Neyret P (2011) MACI—a new era? Sports Med Arthrosc Rehabil Ther Technol 3(1):10
Jalali Jivan S, Shariati-Sarabi Z, Tavakol Afshari J, Asadi Sakhmaresi T, Monzavi SM, Hamidi Alamdari D et al (2015) Effectiveness of intra-articular injection of platelet-rich plasma on clinical improvement of patients with knee osteoarthritis. Annal Rheum Dis 74(Suppl 2):1194
Jalali Jivan S, Monzavi SM, Zargaran B, Hamidi Alamdari D, Tavakol Afshari J, Etemad-Rezaie A et al (2021) Comparative analysis of the effectiveness of intra-articular injection of platelet-rich plasma versus hyaluronic acid for knee osteoarthritis: results of an open-label trial. Arch Bone Jt Surg 9(5):500
JoVE Science Education Database (2020) Lab animal research: sterile tissue harvest. Cambridge, MA. J Vis Exp (JoVE). Available from: https://www.jove.com/science-education/10298/sterile-tissue-harvest
Kajbafzadeh AM, Sabetkish S, Heidari R, Ebadi M (2014) Tissue-engineered cholecyst-derived extracellular matrix: a biomaterial for in vivo autologous bladder muscular wall regeneration. Pediatr Surg Int 30(4):371–380
Kajbafzadeh AM, Sabetkish S, Sabetkish N, Muhammadnejad S, Akbarzadeh A, Tavangar SM et al (2015) In-vivo trachea regeneration: fabrication of a tissue-engineered trachea in nude mice using the body as a natural bioreactor. Surg Today 45(8):1040–1048
Kamguyan K, Katbab AA, Mahmoudi M, Thormann E, Zajforoushan Moghaddam S, Moradi L et al (2017) An engineered cell-imprinted substrate directs osteogenic differentiation in stem cells. Biomater Sci 6(1):189–199
Kang H, Peng J, Lu S, Liu S, Zhang L, Huang J et al (2014) In vivo cartilage repair using adipose-derived stem cell-loaded decellularized cartilage ECM scaffolds. J Tissue Eng Regen Med 8(6):442–453
Kreuz PC, Kalkreuth RH, Niemeyer P, Uhl M, Erggelet C (2019) Long-term clinical and MRI results of matrix-assisted autologous chondrocyte implantation for articular cartilage defects of the knee. Cartilage 10(3):305–313
Lau TT, Leong W, Peck Y, Su K, Wang DA (2015) Use of Interim scaffolding and neotissue development to produce a scaffold-free living hyaline cartilage graft. Methods Mol Biol 1340:153–160
Liao J, Joyce EM, Sacks MS (2008) Effects of decellularization on the mechanical and structural properties of the porcine aortic valve leaflet. Biomaterials 29(8):1065–1074
Long WJ, Greene JW, Cushner FD (2016) Early clinical outcomes associated with a novel osteochondral allograft transplantation system in the knee. Adv Orthop Surg. 2016:1–6
Mauck RL, Yuan X, Tuan RS (2006) Chondrogenic differentiation and functional maturation of bovine mesenchymal stem cells in long-term agarose culture. Osteoarthritis Cartilage 14(2):179–189
Mohammadie ZM, Parivar K, Mahdavi Shahri N, Fereidoni M, Hayati-Roodbari N (2017) Technology. Decellularized bovine articular cartilage matrix reinforced by Carboxylated-SWCNT for tissue engineering application. Braz Arch Biol Technol 60: e17160083
Moradi L, Vasei M, Dehghan MM, Majidi M, Farzad Mohajeri S, Bonakdar S (2017) Regeneration of meniscus tissue using adipose mesenchymal stem cells-chondrocytes co-culture on a hybrid scaffold: In vivo study. Biomaterials 126:18–30
Muraglia A, Martin I, Cancedda R, Quarto R (1998) A nude mouse model for human bone formation in unloaded conditions. Bone 22(5 Suppl):131S-S134
Murray CJ, Lopez AD (2013) Measuring the global burden of disease. N Engl J Med 369(5):448–457
Mushahary D, Spittler A, Kasper C, Weber V, Charwat V (2018) Isolation, cultivation, and characterization of human mesenchymal stem cells. Cytometry A 93(1):19–31
New SEP, Ibrahim A, Guasti L, Zucchelli E, Birchall M, Bulstrode NW et al (2017) Towards reconstruction of epithelialized cartilages from autologous adipose tissue-derived stem cells. J Tissue Eng Regen Med 11(11):3078–3089
Nie X, Chuah YJ, Zhu W, He P, Peck Y, Wang DA (2020) Decellularized tissue engineered hyaline cartilage graft for articular cartilage repair. Biomaterials 235:119821
Oseni AO, Butler PE, Seifalian AM (2013a) Optimization of chondrocyte isolation and characterization for large-scale cartilage tissue engineering. J Surg Res 181(1):41–48

Oseni AO, Butler PE, Seifalian AM (2013b) Nasal reconstruction using tissue engineered constructs: an update. Ann Plast Surg 71(2):238–244

Palukuru UP, McGoverin CM, Pleshko N (2014) Assessment of hyaline cartilage matrix composition using near infrared spectroscopy. Matrix Biol 38:3–11

Poole AR, Kojima T, Yasuda T, Mwale F, Kobayashi M, Laverty S (2001) Composition and structure of articular cartilage: a template for tissue repair. Clin Orthop Relat Res (391 Suppl):S26–33

Raghunath J, Rollo J, Sales KM, Butler PE, Seifalian AM (2007) Biomaterials and scaffold design: key to tissue-engineering cartilage. Biotechnol Appl Biochem 46(Pt 2):73–84

Reynolds KL, Bishai SK (2014) In situ evaluation of chondrofix® osteochondral allograft 25 months following implantation: a case report. Osteoarthritis Cartilage 22:S155–S156

Safiri S, Kolahi AA, Smith E, Hill C, Bettampadi D, Mansournia MA et al (2020) Global, regional and national burden of osteoarthritis 1990–2017: a systematic analysis of the Global Burden of Disease Study 2017. Ann Rheum Dis 79(6):819–828

Saris D, Price A, Widuchowski W, Bertrand-Marchand M, Caron J, Drogset JO et al (2014) Matrix-applied characterized autologous cultured chondrocytes versus microfracture: two-year follow-up of a prospective randomized trial. Am J Sports Med 42(6):1384–1394

Sophia Fox AJ, Bedi A, Rodeo SA (2009) The basic science of articular cartilage: structure, composition, and function. Sports Health. 1(6):461–468

Sutherland AJ, Converse GL, Hopkins RA, Detamore MS (2015) The bioactivity of cartilage extracellular matrix in articular cartilage regeneration. Adv Healthc Mater 4(1):29–39

Thakkar S, Fernandes H, Moroni L (2015) Decellularized extracellular matrix scaffolds for cartilage regeneration. Methods Mol Biol 1340:133–151

Tissues C (2017). In: Pollard TD, Earnshaw WC, Lippincott-Schwartz J, Johnson GT (eds) Cell biology. Elsevier, Philadelphia, USA, pp 555–570

Watkins J, Mathieson I (2009) Connective tissues. In: Watkins J, Mathieson I (eds) The pocket podiatry guide: functional anatomy. Churchill Livingstone, Elsevier, London, UK, pp 107–156

Xia B, Di C, Zhang J, Hu S, Jin H, Tong P (2014) Osteoarthritis pathogenesis: a review of molecular mechanisms. Calcif Tissue Int 95(6):495–505

Yang Z, Shi Y, Wei X, He J, Yang S, Dickson G et al (2010) Fabrication and repair of cartilage defects with a novel acellular cartilage matrix scaffold. Tissue Eng Part C Methods 16(5):865–876

Ye K, Felimban R, Moulton SE, Wallace GG, Di Bella C, Traianedes K et al (2013) Bioengineering of articular cartilage: past, present and future. Regen Med 8(3): 333–349

Zak L, Albrecht C, Wondrasch B, Widhalm H, Vekszler G, Trattnig S et al (2014) Results 2 years after matrix-associated autologous chondrocyte transplantation using the novocart 3D scaffold: an analysis of clinical and radiological data. Am J Sports Med 42(7): 1618–1627

Decellularization of Bone Tissue

18

Ha Le Bao Tran, Vu Nguyen Doan,
Quan Minh To, My Thi Ngoc Nguyen,
Van Ngoc Le Trinh,
and Tuyet Thi Vi Le

Abstract

Decellularization technology is a process that uses different methods such as physical, chemical or enzymatic methods in order to eliminate cellular remnants from original tissues or organs while minimizing any adverse effect on the structural properties, biological activity, and mechanical integrity of the remaining ECM. Regenerative medicine uses the most promising therapies to replace or regenerate tissues and organs in human, restore or establish normal functions lost due to disease or injury. By the combination between new biomaterials and cells, one of the goals of regenerative medicine is to create autologous grafts for transplantation therapies in the future.

Various decellularization methods have been developed include chemical treatment, biological treatment and physical treatment. The aim of this chapter is to evaluate the decellularization method and all available materials that preserves the matrix without structural disruption.

Keywords

Extracellular matrix · Scaffolds · Decellularization · Bone

18.1 History

In regenerative medicine, a potential material is the extracellular matrix (ECM). They are the acellular components of tissue, which plays role in supporting structural stability and providing biochemical signals to determine cell's fate. Each tissue type has a specific ECM structure and composition. In recent years, donor tissues and organs such as the kidney (Orlando et al. 2013), lungs (Gilpin et al. 2014), heart (Sánchez et al. 2015) and liver (Mazza et al. 2015), have been used to make the extracellular matrix scaffolds by decellularization technology.

The establishment of decellularized ECM (dECM) is successful if (1) DNA content needs to be less than 50 ng/mg dry ECM, the debris of cell is removed from the tissue without destruction of ECM, and (Badylak 2004), (3) the dECM can be recellularized (Crapo et al. 2011; Yin et al. 2013). Successful decellularization is critical controlled

H. L. B. Tran (✉) · V. N. Doan · Q. M. To · M. T. N. Nguyen · V. N. L. Trinh · T. T. V. Le
Laboratory of Tissue Engineering and Biomedical Materials, Department of Physiology and Animal Biotechnology, Faculty of Biology – Biotechnology, University of Science, Vietnam National University, Ho Chi Minh City (VNU-HCM), Ho Chi Minh City, Vietnam
e-mail: tlbha@hcmus.edu.vn

A. Kajbafzadeh (ed.), *Decellularization Methods of Tissue and Whole Organ in Tissue Engineering*, Advances in Experimental Medicine and Biology 1345,
https://doi.org/10.1007/978-3-030-82735-9_18

by the target tissue in term of structural, biochemical, and physiological characteristics.

Decellularization technology of tissues was published in 2006 (Fu et al. 2014). Many later improvements created scaffolds which are capable of retaining the characteristics of the structural and acellular compositions. Decellularized tissues exhibit good biocompatibility and the removal of the donor cells minimizes the risk of transplant rejection. Moreover, the elimination of cells in tissue structures increases their potential for use in allograft and xenograft (Barakat et al. 2012; Luo et al. 2016; Perniconi et al. 2011; Rieder et al. 2004). The decellularized ECM can be repopulated with patient's cells to produce a personalized tissue. In addition, decellularized ECM is also used in research of drug treatments. Human decellularized adipose tissue ECM was established for the investigation of breast cancer growth and drug treatments (Dunne et al. 2014) or the liver recellularization of rat model was established for assessing human drug metabolism and liver biology (Robertson et al. 2018).

All the methods of tissue derived cell removal will affect the ECM composition and cause a certain degree of ultrastructure disruption. In addition, many factors affect the cell removal efficiency, such as the tissue thickness and the matrix density. Therefore, it is very important to determine which method or combination of them is the most suitable for particular tissue types (Table 18.1).

Decellularization of bone tissue.

Bone tissue is a complex internal and external structure and is a heterogeneous composite material (Boskey 2013). Bones are not just the part of skeleton system that support the body structurally, but also protect the internal organs. They serve as reservoir for bone marrow where new blood cells are produced, and mineral storage for minerals, particularly calcium (Datta et al. 2008; Downey and Siegel 2006). The bone transplantation is a surgical procedure that is necessary in the tissue reparation and regeneration caused by trauma, tumor, infections, as well as from diseases, such as osteoporosis or arthritis and periodontitis in dentistry (Dimitriou et al. 2011; Hashimoto et al. 2011). Among all the transplants made, bone grafts are the second most commonly transplanted tissue, and command a tremendous global market, standing at 3.02 billion US dollars in 2014 with projections reaching 3.48 billion dollars by the end of 2023 (Research G.V. 2018). The loss of bone due to damage is a very challenging problem for the orthopaedic surgeons especially in those cases where loss of bone is massive. Autogenous bone grafts, also known as autografts, are made from your own bone, taken from somewhere else in the body. Autogenous bone grafts are ubiquitously referred to as the gold standard as they do not evoke any immunological response in the patient. Moreover, they contain living cellular elements that enhance bone growth. Although autogenic bone grafting exhibits a distinct therapeutic effect, its application is greatly limited due to the lack of sufficient bone and the secondary injury to donor sites caused by surgery. The allograft and xenograft in this situation can be thought of as better option in terms of availability and structural similarity with

Table 18.1 Decellularization methods and suitable tissue

Types of tissue	Methods	References
Native bone	Combination of chemical and biological treatments	Chen et al. (2015), Sun et al. (2011)
Cortical bone	High-hydrostatic pressurization (HHP) method (avoids the use of any chemical agents)	Hashimoto et al. (2011)
Different tissue	Physical method includes freezing, direct pressure, sonication, agitation, and vacuum-assisted decellularization, high-hydrostatic pressurization (HHP) method	Butler et al. (2017), Gilbert et al. (2006), Hashimoto et al. (2011), Nakamura et al. (2016)

the host. Especially, xenogenic bone grafts are preferred due to its unlimited supply especially when collected from a larger species like bovine and porcine (Bansal et al. 2009; Bracey et al. 2018; Calvo-Guirado et al. 2012; Dimitriou et al. 2011; Elliot and Richards 2011; Feng et al. 2012; Kim et al. 2004; Ledford et al. 2013; Quan et al. 2014; Sladkova et al. 2018). However, the most limiting factor in the usage of allografts or xenografts is the presence of foreign antigens and rejection of the graft. It therefore becomes necessary to remove these antigens to minimize or avoid the immune response. Decellularization is widely adopted as valid strategy to eliminate the antigenicity of xenograft bones in the development of bone substitutes. Decellularization also results in the creation of a natural scaffold, so called extracellular matrix, which serves as ideal material for cell growth, cell differentiation, and tissue regeneration (Chen and Lv 2017; Crapo et al. 2011; Guruswamy Damodaran and Vermette 2018).

In bone tissue, the extracellular matrix (ECM) plays an important role in maintenance and regeneration of tissues. Bone ECM has two phases, including organic and inorganic phase. The organic phase, primarily type I collagen, provides flexibility of tissue. While the inorganic phase, mostly calcium phosphate and hydroxyapatite (HA) provides bone strength (Alford et al. 2015). It is difficult to create the bone ECM in vitro. Therefore, bone decellularization has become popular in bone tissue engineering researches. Similar to other tissues and organs, decellularized bone is a scaffold forming by removal cellular components and preserving the specific micro-architecture and composites of the extracellular matrix. It has unaltered biologic activity and mechanical integrity (Badylak 2004; Gock et al. 2004).

Various bone decellularization methods have been developed include chemical treatment, biological treatment, and physical treatment (Table 18.2).

Table 18.2 History of bone decellularization

Year	References	Methods	Testing model
1988	Jackson (1988)	Snap freezing	Goat knees
2005	Woods and Gratzer (2005)	10 mM tris, EDTA, 1% triton X-100	Porcine bone
2008	Lumpkins et al. (2008)	1% SDS, 1% triton X-100	Porcine disc
		Or 25% acetone/75% ethanol	
2011	Hashimoto et al. (2011)	Hydrostatic pressure	Porcine Femur
	Kheir et al. (2011)	Freezing and thawing 0.1% SDS, 0.1% peracetic acid	Porcine cartilage bone
2012	Marcos-Campos et al. (2012)	0.1% EDTA, 10 mM tris, 0.5% SDS	Bone of bovine wrists
	Pathak et al. (2012)	Freezing and thawing	Buffalo bone
		Acetone: ethanol	
		SDS	
2013	Tamilmahan (2013)	Freezing and thawing	Rabbit bone
	Sawkins et al. (2013)	0.05% trypsin, 0.02% EDTA	Bovine bone
	Chan et al. (2013)	0.1% SDS, EDTA-free 1x	Bovine intervertebral disc
2016	Lee et al. (2016)	0.5% SDS, 0.1% NH4OH	Rat calvaria
2017	Rashmi and Amarpal (2017)	Freezing and thawing	Bovine femur
2018	Bracey et al. (2018)	0.05% trypsin–EDTA, 1.5% peracetic acid, 2% triton X-100	Porcine femur

The aim of this chapter is to describe the decellularization methods, as well as characterization of decellularized bone scaffold. Additionally, cell seeding processes for bone scaffold are also described.

18.2 Materials and Methods

Bone tissue is ranked into hard-dense tissue with porous structure, which addresses much effect on optimization for decellularizing solution to penetrate the tissue. Furthermore, rather than applying each separate decellularize method, either physical, chemical, or enzymatic category, it is required the combination of varied decellularizing solution and delivery methods order to achieve a complete removal of cellular organelles and genetic materials (Crapo et al. 2011; Gilbert et al. 2006; Liu et al. 2013; Woods and Gratzer 2005).

Physical methods have been indicated as agitation in solution, vascular perfusion, thermal shock, ultrasonics, and manual disruption. Numerous studies have used thermal shock, which involves one or more freeze–thaw cycles, for cellular disruption and lysis due to the formation of intracellular ice crystals. It has been shown that freeze–thaw cycles do not significantly alter the mechanical properties of the ECM (Jung et al. 2011; Nonaka et al. 2014). It was shown that bone grafts could undergo up to 8 freeze–thaw cycles without compromising its biomechanical properties (Jung et al. 2011). Although freeze–thaw cycles could produce minor disruptions in the tissue ultrastructure, this technique can minimize the amounts of chemical agents required for effective decellularization. Thus, it was suggested to adopt freeze–thaw cycles at the beginning of the decellularization protocol to enhance the efficacy of future efforts to transport the cellular materials from the tissue (Gilbert 2012). Cartilage bone blocks from the patella were undergone two cycles of −20 °C until crystal formation and room temperature for 2–4 h, followed by two cycles of freeze–thaw in a hypotonic buffer (10 mM Tris–HCl, pH 8.0) before entering the further decellularization process with Sodium dodecyl sulfate (SDS) (Kheir et al. 2011).

In the manner of detergent-based decellularization, non-ionic detergents have been used extensively in decellularization protocols because of their relatively mild effects upon tissue structure. Non-ionic detergents disrupt lipid–lipid and lipid–protein interactions, but leave protein–protein interactions intact so that proteins within a tissue or organ following non-ionic detergent treatment should be left in a functional conformation. Triton X-100 is the most widely studied non-ionic detergent for decellularization protocols. Triton X-100 was used to remove cellular debris of bovine bone blocks, in which the decellularization in 1% Triton X-100 for 8 h, followed by a second wash in 0.1% Triton X-100 for 16 h (Gardin et al. 2015).

Additionally, ionic detergents are demonstrated as strong detergents that can completely disrupt cell membranes and fully denature proteins. Sodium dodecyl sulfate (SDS), sodium deoxycholate, and Triton X-200 are among the most commonly used ionic decellularization agents because they effectively solubilize cytoplasmic membranes, lipids, and DNA (Crapo et al. 2011; Gilbert et al. 2006). Especially, SDS is shown to be effective in solubilization of cell membranes and nuclei, therefore, is used for the decellularization of dense tissues and organ. However, the main disadvantage of SDS treatment is the high chance of eliminating growth factors. Therefore, SDS decellularization needs to be wisely adjusted in the manner of working concentration, duration, and is further applied in the combination with other reagents, including hypotonic solution and Triton-X100 (Benders et al. 2013; Kheir et al. 2011; Qiang et al. 2011; Sladkova et al. 2018). Human cancellous bone was subjected for cellular elimination in Triton X-100 and SDS treatment, while other decellularization was established according to the combination between Tris–HCl as hypotonic solution and SDS for the fabrication of porcine cartilage bone matrix.

Together with detergent treatment, enzymatic agents, including nucleases, trypsin, collagenase, lipase, dispase, thermolysin, and a-galactosidase, were reported to provide high specificity and effectiveness in the removal of cellular remnants. Enzymatic methods are mainly combined with physical and/or chemical methods to maximize the decellularization outcome while avoid any adverse effect ECM. Enzymatic protocol with trypsin was reported in decellularization of porcine femoral trabecular bone, in which 2.5 g/L trypsin was used in the combination with 100% acetone as chemical defat solution. The decellularized trabecular bone scaffolds was then coated with nanosized bioactive glass in order to increase the recellularization and neovascularization (Gerhardt et al. 2013). Porcine cancellous bone plugs from femur were decellularization by 0.05% trypsin–EDTA, 1.5% Peracetic Acid and 2.0% Triton X-100 to a novel decellularization protocol to derive a bone scaffold for orthopaedic surgery practice (Bracey et al. 2018). Nucleases, DNase and RNase, are endonucleases commonly added to detergent treatments to achieve the effective decellularization due to their specific hydrolyzation on cellular deoxyribonucleotide and ribonucleotide components. Bovine bone plugs were processed through a decellularization procedure consisting many steps of incubation in hypotonic buffer, detergent, and finally in enzymatic solution as 10 mM Tris and DNase and RNase (100 U/mL) for 6 h at 37 °C (Sladkova et al. 2018).

Due to the fact that decellularizing tissues involves using detergents may leave residual cytotoxic chemicals, as well as enzyme residues within the ECM may affect cellular response or initiate an adverse immune response (Crapo et al. 2011). An alternative method of decellularizing tissues using hydrostatic pressure (HHP, >600 MPa) has been recently developed, which was shown to shorten the processing duration and be more effective than detergents or enzymes for removing cells from dense tissues (Demazeau and Rivalain 2011; Woods and Gratzer 2005). Yoshihide Hashimoto et al. proposed the use of hydrostatic pressure to obtain decellularized porcine femur which resulted in a fine bone scaffold promoting cell infiltration with neovascularization after 6 months subcutaneous implantation in rats, and also was demonstrated to promote the osteogenic differentiation of mesenchymal stem cells in vitro (Hashimoto et al. 2011). Later then, the porcine costae were pressurized using HHP machine at 980 MPa to disrupt the cells, which presented the successful decellularization, as well as maintenance of trabecular bone structure which allowed the hMSCs to adhere and differentiate to adipocyte.

In addition of cell removal, demineralization and delipidation of bone tissues are also put on account. There is a high fat content of the bone which can interfere the decellularization due to the very poor wettability. Lipid residuals on the surface of bone ECM represent a barrier to cellular migration and may facilitate the giant cell reactions in vivo which can increase bone resorption and encapsulating fibrosis (Zhang et al. 2014). This fact indicates that delipidation should to be conducted in the first step of bone processing by using non-polar solvents such as acetone, chloroform, etc. We used chloroform in combination with methanol during delipidation procedure (Quan et al. 2014). However, the residual chloroform remained in the scaffolds after treatment causes toxicity, which should be concerned when in use. Sharing the same effect with chloroform, acetone also gives a good result in delipidation, in addition, acetone evaporates naturally at normal condition, hence, it does not leave any redundant in the scaffolds. Our previous study proposed a protocol for bone decellularization, in which fresh porcine spine bones were demineralized in 0.6 M HCl for 15 min followed by the incubation in absolute acetone for 17 h with agitation force. The protocol achieved 100% cell elimination and resulted in porous structure consisting of an interconnected network of the large pores, from 250 to 1000 μm (1 mm) in diameter, suitable for cell penetration and support vascular in growth. In this study, we want to improve protocol by decreasing acetone treating time to 12 or 6 h. Cancellous bone was derived from porcine spine bone. After removing surround tissue such as muscle, fat… and rinsing by sterilized PBS, cancellous part of spine bonc were cut into some cubes of $3 \times 3 \times 3$ mm^3 (CBC). CBCs were rinsed in Tris-NaCl for 6 h to remove all types of blood cells. Next, CBCs were immerged in 0.6 M HCl for 15 min to demineralize. Final, CBCs were shaken in absolute acetone for 6 h (Exp. 1) or 12 h (Exp. 2). Decellularization effectiveness was determined by H&E and trichrome staining. In vitro cytotoxicity test was evaluated on human fibroblasts according to ISO 10,993–5. Fibroblasts were seeded into wells of 4-well plates with concentration of 4×10^4 cells/well. Cells were incubated in DMEM/F12 supplemented with 10% FBS, in 37 °C, 5% CO_2 for 1 day. After 1 day, the medium was changed. In experimental group, acellular CBCs (aCBC) were placed on wells (4 aCBCs/4 wells). In control group, fibroblasts were maintained with fresh medium (4 wells). After 1 day, Giemsa staining and MTT assay were performed with cells to determine in vitro cytotoxicity.

After decellularizing, the cancellous blocks turned from red to white.

The result of H&E staining showed that cells were removed completely from CBCs treated by acetone for 12 h (Fig. 18.3). There are a few of cells in CBCs treated by acetone for 6 h compared to intact CBC (Figs. 18.1 and 18.2). The result of Trichrome staining showed that collagen was retained in CBCs after decellularization compared to intact CBC (Fig. 18.4). So, acetone for 12 h was used to create aCBCs for next experiments.

For scanning electron microscope (SEM) analysis, as shown in Fig. 18.5, the decellularized bone scaffold is a porous structure consisting of an interconnected network of the large pores. Pore size ranges from 250 to 1000 μm (1 mm) in diameter, which indicated that the scaffold has high porosity.

After one day of in vitro cytotoxicity test, few adherent cells died and detached from the plate. Giemsa staining showed that cell appearance in experimental group was normal, comparing to these in control group (Fig. 18.6). MTT assay showed that OD value of experimental group was 0.222 ± 0.011, and control group was 0.233 ± 0.017. The ratio is 95%. This result showed cytotoxicity level of aCBCs was 0, so the scaffold was non-toxic to human fibroblasts.

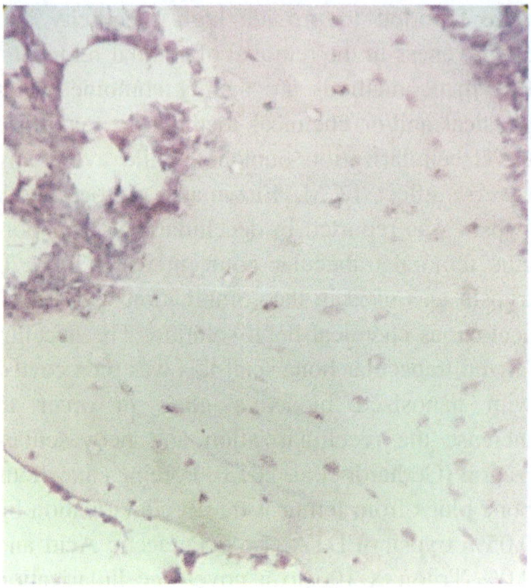

Fig. 18.1 Histology of cancellous bone by H&E staining (×400)

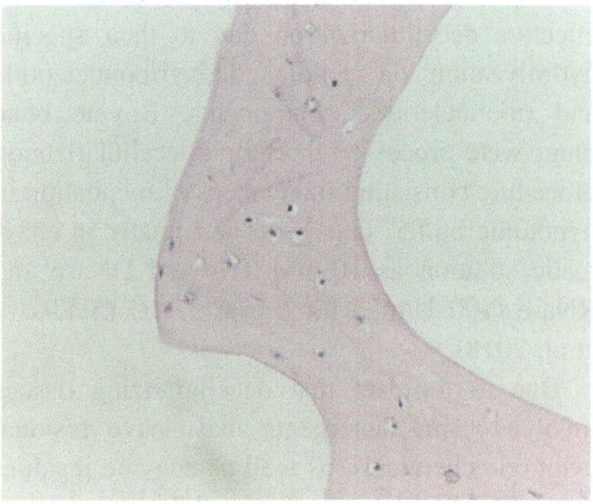

Fig. 18.2 Histology of CBC treated by acetone for 6 h by H&E staining (×400)

18.3 Cell Seeding—Recellularization

Cell seeding and preculture on scaffolds are essential procedures before in vivo implantation. Cell-seeding efficiency can allow maximized utilization of donor cells, minimal time for anchorage-dependent and shear-sensitive cells, and spatially uniform distribution of attached

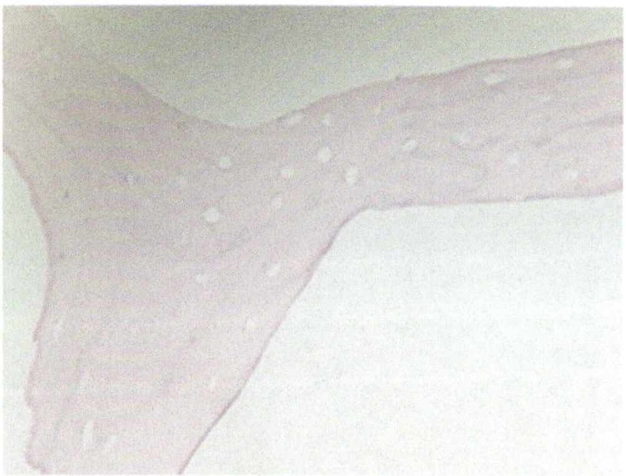

Fig. 18.3 Histology of CBC treated by acetone for 12 h by H&E staining (×400)

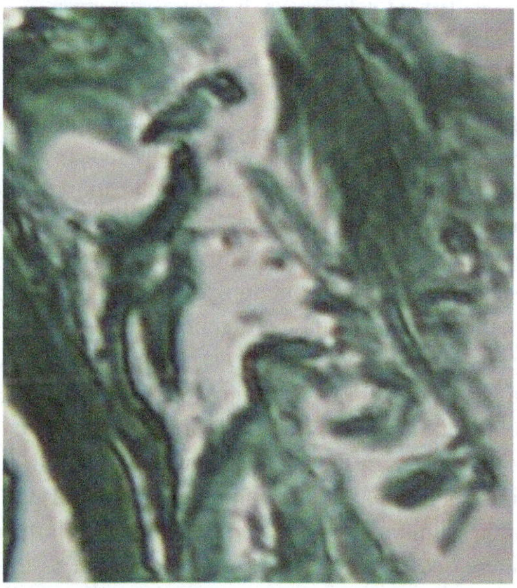

Fig. 18.4 Trichrome staining for Collagen matrix (×400)

cells. Moreover, the criteria for three-dimensional (3D) scaffolds such as pore sizes, porosity, interconnectivity, biodegradability, and mechanical integrity are explored because various scaffold properties affect efficiency of cell seeding methods; for instance, 3D scaffolds which have more regular and homogeneous pores are more accessible for the cell suspension during drop seeding because they avoid cell aggregate entrapment in the small and irregular inner pores (Chen et al. 2011; Ma et al. 2014). It

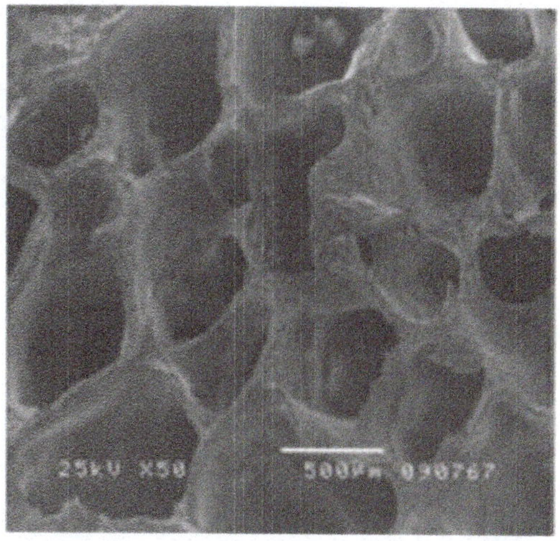

Fig. 18.5 The SEM result for identifying the spongy structure of the decellularized bone (Quan et al. 2014)

is important to investigate the relationship between pore sizes and cell seeding. The previous studies have shown that pore sizes affect to cells proliferation. The scaffolds which have pores with a size of 300–400 μm are the best for cell attachment, and the larger pores resulted in a decrease of the cell numbers in the scaffolds (Rashmi and Amarpal 2017).

Cell seeding density is known to influence the characteristics and functionality of engineered tissues. The effect of cell seeding density is dependent on the tissue type as well as the culture conditions. For example, the seeding density for bone tissue engineering is increased from 1×10^6 cells/ml to 10×10^6 cells/ml (high cell-seeding density) that induces homogenous cell distribution throughout the constructs is better (Grayson et al. 2008). At last, cell seeding volume and cell seeding time also affect to cell seeding efficiency as well as cell viability significantly (Ma et al. 2014).

Seeding cells into 3D porous scaffolds is a passive process in which the cells disperse into the scaffolds by diffusion and attach through weak molecular level driven forces such as adhesion (Weinand et al. 2009). In contrast to simple monolayer culture, seeding cells into the 3D scaffolds is difficult due to their complicated structure. Cell seeding techniques have been

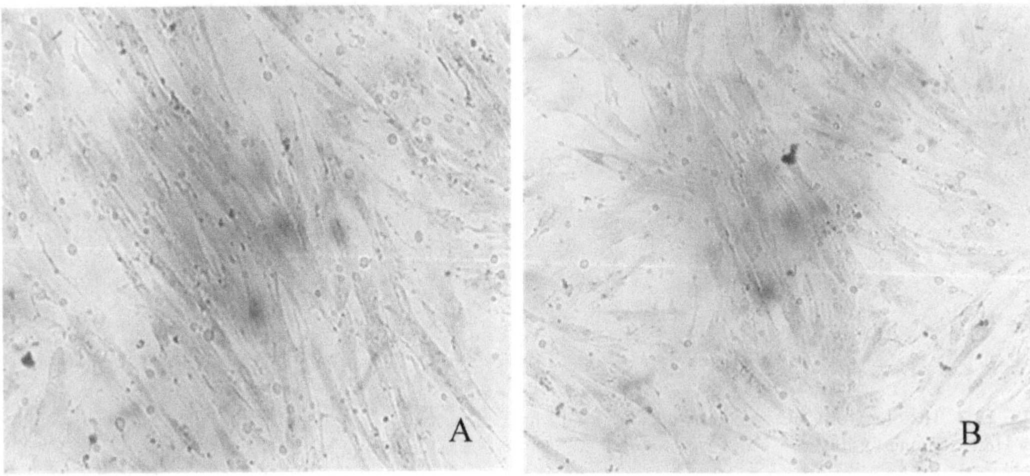

Fig. 18.6 Giemsa staining of human fibroblasts after cytotoxicity test. **a** Control group, **b** experimental group

classified into two types: direct and indirect. In direct seeding techniques, cells and scaffold are incubated concurrently in plastic dishes or a small volume of cell suspension is injected on the surface or into the center of the scaffold (Manbachi et al. 2008; Thevenot et al. 2008; Zhang et al. 2015). Indirect seeding techniques are based on the attachment of cells suspended in a dynamic environment to improve their diffusion (Zhang et al. 2015). Previous studies have demonstrated that cell binding efficiency, survival rate, distribution ability, especially in the inner parts of porous scaffolds and long-term proliferation were urgent standards for cells seeding methods (Luo et al. 2013; Roh et al. 2007; Weinand et al. 2009; Zhang et al. 2015). Multiple methods for testing of cell seeding efficiency are currently available. It can be determined by quantifying the number of cells that not only have attached to but also have penetrated the construct. This can be done by a variety of methods (hemocytometer, histology, SEM, TEM, DNA detection assay, MTT, MTS, WST-1), but can be divided into two categories: observation and manual counting, or indirect quantification (measuring a particular function of metabolically active alive cells) (Luo et al. 2013; Rashmi and Amarpal 2017; Villalona et al. 2010; Zhang et al. 2015).

Static seeding, vaccum seeding and centrifugal seeding methods are often used to transplant various cells to decellularized bone matrix scaffolds (DBMS). Static seeding is the most commonly used cell seeding technique in bone tissue engineering. This seeding technique yields seeding efficiencies of approximately 10–25%. In static seeding methods, a concentrated cell suspension is dispensed on 3D scaffold, followed by a period of rest (30–120 min) to allow the cells to enter the scaffold. With this method, the initial cell density in the scaffold can be increased by increasing the cell concentration of the suspension within a certain range, but cannot be further increased beyond a plateau level. However, this technique has several limitations such as low seeding efficiency and minimal cell penetration into scaffolds (López-Pérez et al. 2010; Luo et al. 2013; Rashmi and Amarpal 2017; Roh et al. 2007). Previous studies have demonstrated that the importance of cell distribution and proliferative behavior on 3D scaffolds provides the key to ensuring the ultimate functionality in bone tissue engineered constructs. However, seeded cells by static seeding methods tended to be concentrated on the surface and rarely infiltrated the center of 3D scaffolds (Baker and Mauck 2007; Zhang et al. 2015). Therefore, vaccum seeding methods which combine low pressure (vacuum) and vibration can help remove potential air bubbles around scaffolds and allow more cells to penetrate deeply into the scaffolds to enhance bone formation, especially for porous scaffolds compared to the static seeding methods (Hasegawa et al. 2010; Ma et al. 2014). Pressure differential systems (internal pressure or external pressure) are an extremely rapid means of cell seeding with seeding efficiencies ranging from 60 to 90%. Moreover, the simplicity of some vacuum apparatuses allows for their use as an inexpensive disposable seeding device, reducing the risk of contamination (Villalona et al. 2010).

A large number of centrifugal seeding methods have been applied to solve this problem with some achievements (Godbey et al. 2004; Ng et al. 2010; Roh et al. 2007; Zhang et al. 2015). In this technique, a scaffold is rotated (up to 2500 rpm) in a cell/medium suspension or spun along with a cell/medium suspension. Seeding efficiency under these conditions has ranged from 38 to 90%. This method has been shown to increase both seeding efficiency and penetration. Although centrifugal seeding maintains cell viability, concerns have been raised about the effect on cell morphology (Roh et al. 2007). Seeded mesenchymal stem cells (MSCs) to demineralized cancellous bone scaffold by centrifugation shows a more uniform distribution throughout the scaffold than cells are seeded by other methods. Moreover, the penetration depth in the scaffold of MSCs by centrifugation is 300–500 μm, much higher than the value of 100–300 μm by the static and injection seeding. The long-term proliferation of the MSCs in the centrifugal group is also significantly higher than that in the other groups (Weinand et al. 2009).

Recent studies find that hydrogel-assisted seeding can provide a simulated microgravity environment to allow human mesenchymal stem cells (hMSCs) to diffuse and become uniformly distributed in human demineralized cancellous bone matrix scaffolds. Hydrogels may improve the adhesion between seeded cells and the scaffold. These glue contents in mixture form an efficient cross-linking network that can support attachment and proliferation of MSCs. Therefore, higher seeding efficiency was obtained in fibrin hydrogel-assisted seeding groups (Jiang et al. 2010; Luo et al. 2013; Skardal et al. 2010; Zhao et al. 2010). Fibronectin is the most commonly described agent, but other hydrogels have been used as well, including fibrin, collagen, laminin and plasma (Villalona et al. 2010).

Currently, cells are allowed to adhere to the scaffold in a dynamic fluid flow created by a bioreactor (e.g., perfusion bioreactor). Dynamic seeding methods help facilitating adherence of cell seeding to the scaffold, increasing the speed and density of cell seeding, and improving the spatial distribution of cells in the scaffold. In addition, dynamic conditions can also support the subsequent in vitro culture of cell-scaffold constructs. A dynamic fluid flow was found to positively affect the behavior of seeded cells, such as proliferation, differentiation, and migration. Despite these advantages, however, dynamic fluid flow may result in cell detachment and damage (Jiang et al. 2010; Luo et al. 2013; Milan et al. 2009). In addition, perfusion bioreactor systems require a prolonged period of culture, so increasing the risk of bacterial and fungal contamination. Also, the use of xenogenic serum as part of the culture medium is a major obstacle for their clinical application. Moreover, the complexity of bioreactors is not suitable for more clinical applications. However, this technology has shown promising preclinical animal results regarding safety and reproducibility (Chevallier et al. 2010; Villalona et al. 2010).

We seeded bone marrow—derived mesenchymal stem cells (MSCs) to recellularize aCBCs. MSCs were trypsinized, centrifuged and adjusted to concentration of 2×10^6 cells/ml. Ten microliter of cell suspension was directly administrated to one aCBC. After 30 min of incubation, aCBCs were moved to 4-well plate (1 aCBC/well) and cultured in DMEM/F12 supplemented with 10% FBS, 37 °C, 5% CO_2 during experimental period. Growth of MSCs in aCBC is determined by MTT assay at days 1, 3, 5, 7, 9, 11 after seeding (5 aCBCs per day). Experiment was divided to 2 groups: experimental group (aCBC was seeded with MSCs) and control group (aCBC didn't contain MSCs). Results of MTT assay showed that OD value

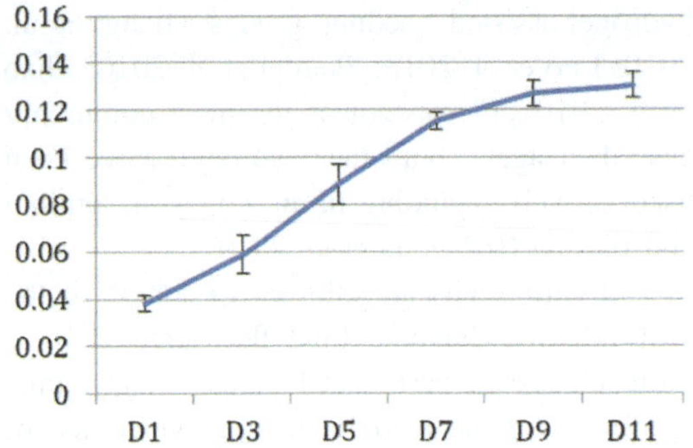

Fig. 18.7 Growth curve of MSCs in aCBC after 11 days

increased steadily from day 1–9 and stay unchanged for day 9–11 (Fig. 18.7). This result showed that cell grew in blocks. The result of H&E staining and SEM showed cells were attached in aCBC after 9 days seeding (Figs. 18.8 and 18.9).

18.4 Clinical Applications

Decellularized bone ECM is an ideal scaffold for bone tissue engineering. Some studies have shown that decellularized bone ECM consists of proteins, that are believed to change insignificantly after decellularization, such as bone morphogenetic proteins, collagen, fibronectin, laminin, hemonectin (Lee et al. 2016). This may help decellularized bone ECM to have positive influences on cells in vitro and in vivo. Furthermore, decellularized bone ECM was considered less immunogenic biological material because it was eliminated antigenic components, such as cells or DNA (Mansour et al. 2017).

Decellularized bone ECM was demonstrated to enhance adhesion, proliferation and osteogenic differentiation of many kinds of stem cells (Chen and Lv 2017). Among them, Hashimoto et al. were proven decellularized bone increased significant alkaline phosphatase activity, which promoted osteogenic differentiation of rat mesenchymal stem cells (Hashimoto et al. 2011). Bone scaffold also effected on hESC derived mesenchymal progenitor, leading the formation of large and compact bone structure (Marolt et al. 2012). Qiang Chen et al. believed that decellularized bone ECM is a suitable material for bone tissue engineering because the complex of decellularized bone ECM and modified ADSC was showed to repair bone defect (Chen et al. 2010).

Some in vivo prominent studies have been documented. The complex of native bovine decellularized bone seeded with autologous stromal/stem cells demonstrated its utility for pig facial construction. After 6 months following transplantation, this complex integrated with the host bone, and formed bone-like tissue

Fig. 18.8 H&E staining of MSC-seeded aCBC at day 9 (×100)

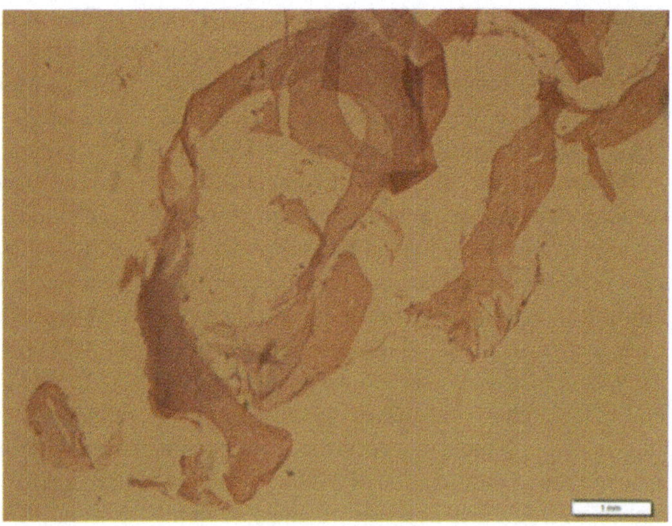

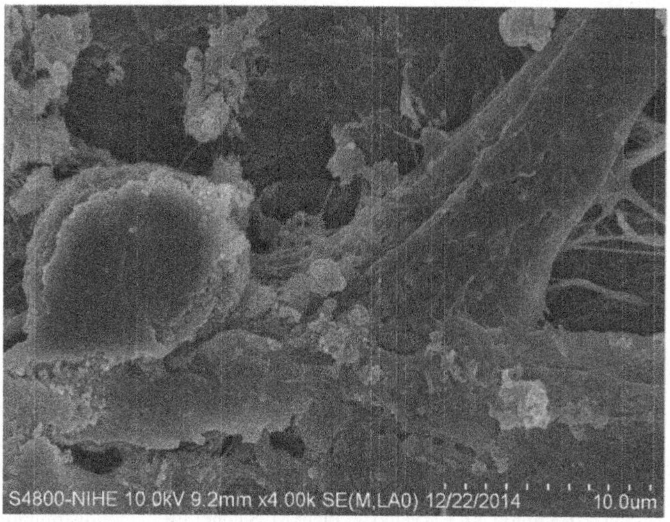

Fig. 18.9 The SEM result for hMSCs seeded aCBCs

(Bhumiratana et al. 2016). In another case, decellularized bone ECM loaded with rat mesenchymal stem cells yielded a synergic effect to stimulate bone regeneration in rat calvaria defect (Lee et al. 2016).

Notably, decellularized bone ECM has been also used in clinical. A 58-year-old woman was implanted with decellularized bovine trabecular bone seeded with autologous bone marrow cells to reconstruct 72 mm defect of distal tibia. 6 weeks after surgery, the patient could walk freely because the tibia was sufficiently stabilized (Hesse et al. 2010). Recently, decellularized bovine bone has been used to reconstruct zygomatic bone for 54-year-old female patient. Firstly, autologous parietal bone graft was used for reconstruction. However, after 3 years, autologous bone graft was resorbtion and the patient had to undergo second surgery with decellularized bovine bone graft. Luckily, after 4-year follow-up examination, face symmetry was recover (Karalashvili et al. 2017).

18.5 Conclusion

Decellularization is a process that uses different methods such as physical, chemical, enzymatic methods separately or combinedly in order to eliminate cellular remnants from original tissues or organs while minimizing any adverse effect on the structural properties, biological activity, and mechanical integrity of the remaining ECM. Decellularization of bone tissue can be performed by using of Tris-NaCl for 6 h, 0.6 M HCl for 15 min and absolute acetone for 12 h combined with shaking. Acellular cancellous bone is non-toxic and supports for human MSCs growth. Although many tests need to do to confirm safety and applicability of acellular cancellous bone, this protocol is an improved step comparing to origin and acellular cancellous bone has a huge potential to apply in regenerative medicine.

References

Alford AI, Kozloff KM, Hankenson KD (2015) Extracellular matrix networks in bone remodeling. Int J Biochem Cell Biol 65:20–31

Badylak SF (2004) Xenogeneic extracellular matrix as a scaffold for tissue reconstruction. Transpl Immunol 12 (3–4):367–377

Baker BM, Mauck RL (2007) The effect of nanofiber alignment on the maturation of engineered meniscus constructs. Biomaterials 28(11):1967–1977

Bansal M, Bhagat S, Shukla D (2009) Bovine cancellous xenograft in the treatment of tibial plateau fractures in elderly patients. Int Orthop 33(3):779–784

Barakat O, Abbasi S, Rodriguez G, Rios J, Wood RP, Ozaki C, Holley LS, Gauthier PK (2012) Use of decellularized porcine liver for engineering humanized liver organ. J Surg Res 173(1):e11–e25

Benders KE, van Weeren PR, Badylak SF, Saris DB, Dhert WJ, Malda J (2013) Extracellular matrix scaffolds for cartilage and bone regeneration. Trends Biotechnol 31(3):169–176

Bhumiratana S, Bernhard JC, Alfi D, Yeager K, Eton RE, Bova J, Gimble JM, Lopez MJ, Eisig SB, Vunjak-Novakovic G (2016) Tissue-engineered autologous grafts for facial bone reconstruction. Sci Transl Med 8 (343):343ra83–343ra83

Boskey AL (2013) Bone composition: relationship to bone fragility and antiosteoporotic drug effects. BoneKEy Rep 2

Bracey D, Seyler T, Jinnah A, Lively M, Willey J, Smith T, Van Dyke M, Whitlock P (2018) A decellularized porcine xenograft-derived bone scaffold for clinical use as a bone graft substitute: a critical evaluation of processing and structure. J Funct Biomater 9(3):45

Butler CR, Hynds RE, Crowley C, Gowers KH, Partington L, Hamilton NJ, Carvalho C, Platé M, Samuel ER, Burns AJ (2017) Vacuum-assisted decellularization: an accelerated protocol to generate tissue-engineered human tracheal scaffolds. Biomaterials 124:95–105

Calvo-Guirado JL, Gómez-Moreno G, Guardia J, Ortiz-Ruiz A, Piatelli A, Barone A, Martínez-González JM, Meseguer-Olmo L, López-Marí L, Dorado CB (2012) Biological response to porcine xenograft implants: an experimental study in rabbits. Implant Dent 21 (2):112–117

Chan LK, Leung VY, Tam V, Lu WW, Sze K, Cheung KM (2013) Decellularized bovine intervertebral disc as a natural scaffold for xenogenic cell studies. Acta Biomater 9(2):5262–5272

Chen G, Dong C, Yang L, Lv Y (2015) 3D scaffolds with different stiffness but the same microstructure for bone tissue engineering. ACS Appl Mater Interfaces 7 (29):15790–15802

Chen G, Lv Y (2017) Decellularized bone matrix scaffold for bone regeneration

Chen Q, Yang Z, Sun S, Huang H, Sun X, Wang Z, Zhang Y, Zhang B (2010) Adipose-derived stem cells modified genetically in vivo promote reconstruction of bone defects. Cytotherapy 12(6):831–840

Chen Y, Bloemen V, Impens S, Moesen M, Luyten FP, Schrooten J (2011) Characterization and optimization of cell seeding in scaffolds by factorial design: quality by design approach for skeletal tissue engineering. Tissue Eng Part C Methods 17(12):1211–1221

Chevallier N, Anagnostou F, Zilber S, Bodivit G, Maurin S, Barrault A, Bierling P, Hernigou P, Layrolle P, Rouard H (2010) Osteoblastic differentiation of human mesenchymal stem cells with platelet lysate. Biomaterials 31(2):270–278

Crapo PM, Gilbert TW, Badylak SF (2011) An overview of tissue and whole organ decellularization processes. Biomaterials 32(12):3233–3243

Datta H, Ng W, Walker J, Tuck S, Varanasi S (2008) The cell biology of bone metabolism. J Clin Pathol 61 (5):577–587

Demazeau G, Rivalain N (2011) The development of high hydrostatic pressure processes as an alternative to other pathogen reduction methods. J Appl Microbiol 110(6):1359–1369

Dimitriou R, Jones E, McGonagle D, Giannoudis PV (2011) Bone regeneration: current concepts and future directions. BMC Med 9(1):66

Downey PA, Siegel MI (2006) Bone biology and the clinical implications for osteoporosis. Phys Ther 86 (1):77–91

Dunne LW, Huang Z, Meng W, Fan X, Zhang N, Zhang Q, An Z (2014) Human decellularized adipose tissue scaffold as a model for breast cancer cell growth and drug treatments. Biomaterials 35(18):4940–4949

Elliot RR, Richards RH (2011) Failed operative treatment in two cases of pseudarthrosis of the clavicle using internal fixation and bovine cancellous xenograft (Tutobone). J Pediatr Orthopaedics B 20(5):349–353

Feng W, Fu L, Liu J, Li D (2012) The expression and distribution of xenogeneic targeted antigens on porcine bone tissue. In: Transplantation proceedings. Elsevier

Fu R-H, Wang Y-C, Liu S-P, Shih T-R, Lin H-L, Chen Y-M, Sung J-H, Lu C-H, Wei J-R, Wang Z-W (2014) Decellularization and recellularization technologies in tissue engineering. Cell Transplant 23(4–5):621–630

Gardin, C., Ricci, S., Ferroni, L., Guazzo, R., Sbricoli, L., De Benedictis, G., Finotti, L., Isola, M., Bressan, E., and Zavan, B. (2015). Decellularization and delipidation protocols of bovine bone and pericardium for bone grafting and guided bone regeneration procedures, *PloS one*, 10(7), p. e0132344.

Gerhardt LC, Widdows KL, Erol MM, Nandakumar A, Roqan IS, Ansari T, Boccaccini AR (2013) Neocellularization and neovascularization of nanosized bioactive glass-coated decellularized trabecular bone scaffolds. J Biomed Mater Res Part A 101(3):827–841

Gilbert TW (2012) Strategies for tissue and organ decellularization. J Cell Biochem 113(7):2217–2222

Gilbert TW, Sellaro TL, Badylak SF (2006) Decellularization of tissues and organs. Biomaterials 27 (19):3675–3683

Gilpin SE, Guyette JP, Gonzalez G, Ren X, Asara JM, Mathisen DJ, Vacanti JP, Ott HC (2014) Perfusion decellularization of human and porcine lungs: bringing the matrix to clinical scale. J Heart Lung Transplant 33(3):298–308

Gock H, Murray-Segal L, Salvaris E, Cowan P, d'Apice AJ (2004) Allogeneic sensitization is more effective than xenogeneic sensitization in eliciting gal-mediated skin graft rejection1. Transplantation 77(5):751–753

Godbey W, Hindy BS, Sherman ME, Atala A (2004) A novel use of centrifugal force for cell seeding into porous scaffolds. Biomaterials 25(14):2799–2805

Grayson WL, Bhumiratana S, Cannizzaro C, Chao P-HG, Lennon DP, Caplan AI, Vunjak-Novakovic G (2008) Effects of initial seeding density and fluid perfusion rate on formation of tissue-engineered bone. Tissue Eng Part A 14(11):1809–1820

Guruswamy Damodaran R, Vermette P (2018) Tissue and organ decellularization in regenerative medicine. Biotechnol Prog 34(6):1494–1505

Hasegawa T, Miwa M, Sakai Y, Niikura T, Lee S, Oe K, Iwakura T, Kurosaka M, Komori T (2010) Efficient cell-seeding into scaffolds improves bone formation. J Dent Res 89(8):854–859

Hashimoto Y, Funamoto S, Kimura T, Nam K, Fujisato T, Kishida A (2011) The effect of decellularized bone/bone marrow produced by high-hydrostatic pressurization on the osteogenic differentiation of mesenchymal stem cells. Biomaterials 32(29):7060–7067

Hesse E, Kluge G, Atfi A, Correa D, Haasper C, Berding G, Shin H-O, Viering J, Länger F, Vogt PM (2010) Repair of a segmental long bone defect in human by implantation of a novel multiple disc graft. Bone 46(5):1457–1463

Jackson D, Grood E, Wilcox P, Butler D, Simon T, Holden J (1988) The effects of processing techniques on the mechanical properties of bone-anterior cruciate ligament-bone allografts: An experimental study in goats. Am J Sports Med 16(2):101–105

Jiang T, Nukavarapu SP, Deng M, Jabbarzadeh E, Kofron MD, Doty SB, Abdel-Fattah WI, Laurencin CT (2010) Chitosan–poly (lactide-co-glycolide) microsphere-based scaffolds for bone tissue engineering: In vitro degradation and in vivo bone regeneration studies. Acta Biomater 6(9):3457–3470

Jung HJ, Vangipuram G, Fisher MB, Yang G, Hsu S, Bianchi J, Ronholdt C, Woo SLY (2011) The effects of multiple freeze–thaw cycles on the biomechanical properties of the human bone-patellar tendon-bone allograft. J Orthop Res 29(8):1193–1198

Karalashvili L, Chichua N, Menabde G, Atskvereli L, Grdzelidze T (2017) Decellularized bovine bone graft for zygomatic bone reconstruction. Med Case Rep 4 (1):52

Kheir E, Stapleton T, Shaw D, Jin Z, Fisher J, Ingham E (2011) Development and characterization of an acellular porcine cartilage bone matrix for use in tissue engineering. J Biomed Mater Res Part A 99(2):283–294

Kim SH, Shin JW, Park SA, Kim YK, Park MS, Mok JM, Yang WI, Lee JW (2004) Chemical, structural properties, and osteoconductive effectiveness of bone block derived from porcine cancellous bone. J Biomed Mater Res Part B Appl Biomater Offic J Soc Biomater Japan Soc Biomater Australian Soc Biomater Korean Soc Biomater 68(1):69–74

Ledford CK, Nunley JA, Viens NA, Lark RK (2013) Bovine xenograft failures in pediatric foot reconstructive surgery. J Pediatric Orthopaedics 33(4):458–463

Lee DJ, Diachina S, Lee YT, Zhao L, Zou R, Tang N, Han H, Chen X, Ko C-C (2016) Decellularized bone matrix grafts for calvaria regeneration. J Tissue Eng 7:2041731416680306

Liu F, Zhu Z-J, Li P, He Y-L (2013) Creation of a female rabbit model for intrauterine adhesions using mechanical and infectious injury. J Surgical Res 183 (1):296–303

López-Pérez PM, Da Silva RM, Sousa RA, Pashkuleva I, Reis RL (2010) Plasma-induced polymerization as a tool for surface functionalization of polymer scaffolds for bone tissue engineering: an in vitro study. Acta Biomater 6(9):3704–3712

Lumpkins SB, Pierre N, McFetridge PS (2008) A mechanical evaluation of three decellularization methods in the design of a xenogeneic scaffold for tissue engineering the temporomandibular joint disc. Acta Biomater 4(4):808–816

Luo F, Hou T-Y, Zhang Z-H, Xie Z, Wu X-H, Xu J-Z (2013) Effects of initial cell density and hydrodynamic culture on osteogenic activity of tissue-engineered bone grafts. PLoS One 8(1):e53697

Luo L, Eswaramoorthy R, Mulhall KJ, Kelly DJ (2016) Decellularization of porcine articular cartilage explants and their subsequent repopulation with human chondroprogenitor cells. J Mech Behav Biomed Mater 55:21–31

Ma J, Both SK, Yang F, Cui F-Z, Pan J, Meijer GJ, Jansen JA, van den Beucken JJ (2014) Concise review: cell-based strategies in bone tissue engineering and regenerative medicine. Stem Cells Transl Med 3(1):98–107

Manbachi A, Shrivastava S, Cioffi M, Chung BG, Moretti M, Demirci U, Yliperttula M, Khademhosseini A (2008) Microcirculation within grooved substrates regulates cell positioning and cell docking inside microfluidic channels. Lab Chip 8(5):747–754

Mansour A, Mezour MA, Badran Z, Tamimi F (2017) Extracellular matrices for bone regeneration: a literature review. Tissue Eng Part A 23(23–24):1436–1451

Marcos-Campos I, Marolt D, Petridis P, Bhumiratana S, Schmidt D, Vunjak-Novakovic G (2012) Bone scaffold architecture modulates the development of mineralized bone matrix by human embryonic stem cells. Biomaterials 33(33):8329–8342

Marolt D, Campos IM, Bhumiratana S, Koren A, Petridis P, Zhang G, Spitalnik PF, Grayson WL, Vunjak-Novakovic G (2012) Engineering bone tissue from human embryonic stem cells. Proc Natl Acad Sci 109(22):8705–8709

Mazza G, Rombouts K, Hall AR, Urbani L, Luong TV, Al-Akkad W, Longato L, Brown D, Maghsoudlou P, Dhillon AP (2015) Decellularized human liver as a natural 3D-scaffold for liver bioengineering and transplantation. Sci Rep 5:13079

Milan J-L, Planell JA, Lacroix D (2009) Computational modelling of the mechanical environment of osteogenesis within a polylactic acid–calcium phosphate glass scaffold. Biomaterials 30(25):4219–4226

Nakamura N, Kimura T, Kishida A (2016) Overview of the development, applications, and future perspectives of decellularized tissues and organs. ACS Biomater Sci Eng 3(7):1236–1244

Naoko Nakamura SF, Hashimoto Y, Nam K, Kimura T, Fujisato T, Iwata H, Kishida A Evaluation of decellularized bone using high-hydrostatic pressure in TERMIS

Ng R, Gurm JS, Yang ST (2010) Centrifugal seeding of mammalian cells in nonwoven fibrous matrices. Biotechnol Prog 26(1):239–245

Nonaka PN, Campillo N, Uriarte JJ, Garreta E, Melo E, de Oliveira LV, Navajas D, Farré R (2014) Effects of freezing/thawing on the mechanical properties of decellularized lungs. J Biomed Mater Res Part A Offic J Soc Biomater Japan Soc Biomater Australian Soc Biomater Korean Soc Biomater 102(2):413–419

Orlando G, Booth C, Wang Z, Totonelli G, Ross CL, Moran E, Salvatori M, Maghsoudlou P, Turmaine M, Delario G (2013) Discarded human kidneys as a source of ECM scaffold for kidney regeneration technologies. Biomaterials 34(24):5915–5925

Pathak R, Tiwari A, Kurade N, Nath A (2012) Decellularization of buffalo bone to prepare bone scaffolds for effective bone tissue engineering. J Cell Tissue Res 12 (3)

Perniconi B, Costa A, Aulino P, Teodori L, Adamo S, Coletti D (2011) The pro-myogenic environment provided by whole organ scale acellular scaffolds from skeletal muscle. Biomaterials 32(31):7870–7882

Qiang Y, Jiang P, Lu S-B, Guo Q-Y, Bin Z, Zhang L, Wang A-Y, Xu W-J, Qun X, Hu Y-C (2011) Evaluation of an extracellular matrix-derived acellular biphasic scaffold/cell construct in the repair of a large articular high-load-bearing osteochondral defect in a canine model. Chinese Med J 124(23):3930–3938

Quan TM, Vu DN, My NTN, Ha TLB (2014) Decellularization of xenogenic bone grafts for potential use as tissue engineering scaffolds. Int J Life Sci Med Res 4 (4):38

Rashmi RP, Amarpal H (2017) Evaluation of tissue-engineered bone constructs using rabbit fetal osteoblasts on acellular bovine cancellous bone matrix. Veterinary World 10(2):163

Research, G.V. (2018) Bone grafts and substitutes market size, share & trends analysis report by material type (Natural, Synthetic), by application type (Spinal Fusion, Craniomaxillofacial, Long Bone), by region, and segment forecasts, 2018–2025

Rieder E, Kasimir M-T, Silberhumer G, Seebacher G, Wolner E, Simon P, Weigel G (2004) Decellularization protocols of porcine heart valves differ importantly in efficiency of cell removal and susceptibility of the matrix to recellularization with human vascular cells. J Thorac Cardiovasc Surg 127(2):399–405

Robertson MJ, Soibam B, O'Leary JG, Sampaio LC, Taylor DA (2018) Recellularization of rat liver: an in vitro model for assessing human drug metabolism and liver biology. PloS One 13(1):e0191892

Roh JD, Nelson GN, Udelsman BV, Brennan MP, Lockhart B, Fong PM, Lopez-Soler RI, Saltzman WM, Breuer CK (2007) Centrifugal seeding increases seeding efficiency and cellular distribution of bone marrow stromal cells in porous biodegradable scaffolds. Tissue Eng 13(11):2743–2749

Sánchez PL, Fernández-Santos ME, Costanza S, Climent AM, Moscoso I, Gonzalez-Nicolas MA, Sanz-Ruiz R, Rodríguez H, Kren SM, Garrido G (2015) Acellular human heart matrix: a critical step toward whole heart grafts. Biomaterials 61:279–289

Sawkins M, Bowen W, Dhadda P, Markides H, Sidney L, Taylor A, Rose F, Badylak S, Shakesheff K, White L (2013) Hydrogels derived from demineralized and decellularized bone extracellular matrix. Acta Biomater 9(8):7865–7873

Skardal A, Sarker SF, Crabbé A, Nickerson CA, Prestwich GD (2010) The generation of 3-D tissue models based on hyaluronan hydrogel-coated microcarriers within a rotating wall vessel bioreactor. Biomaterials 31(32):8426–8435

Sladkova M, Cheng J, Palmer M, Chen S, Lin C, Xia W, Yu YE, Zhou B, Engqvist H, de Peppo GM (2018) Comparison of decellularized cow and human bone for engineering bone grafts with human induced pluripotent stem cells. Tissue Eng Part A

Sun X-J, Peng W, Yang Z-L, Ren M-L, Zhang S-C, Zhang W-G, Zhang L-Y, Xiao K, Wang Z-G, Zhang B (2011) Heparin-chitosan-coated acellular bone matrix enhances perfusion of blood and vascularization in bone tissue engineering scaffolds. Tissue Eng Part A 17(19–20):2369–2378

Tamilmahan P (2013) Development of acellular osseous xenograft for bone tissue engineering in rabbits. MV Sc. Izatnagar: Thesis Submitted to Indian Veterinary Research Institute

Thevenot P, Nair A, Dey J, Yang J, Tang L (2008) Method to analyze three-dimensional cell distribution and infiltration in degradable scaffolds. Tissue Eng Part C Methods 14(4):319–331

Villalona GA, Udelsman B, Duncan DR, McGillicuddy E, Sawh-Martinez RF, Hibino N, Painter C, Mirensky T, Erickson B, Shinoka T (2010) Cell-seeding techniques in vascular tissue engineering. Tissue Eng Part B Rev 16(3):341–350

Weinand C, Xu JW, Peretti GM, Bonassar LJ, Gill TJ (2009) Conditions affecting cell seeding onto three-dimensional scaffolds for cellular-based biodegradable implants. J Biomed Mater Res B Appl Biomater 91 (1):80–87

Woods T, Gratzer PF (2005) Effectiveness of three extraction techniques in the development of a decellularized bone–anterior cruciate ligament–bone graft. Biomaterials 26(35):7339–7349

Yin Z, Chen X, Zhu T, Hu J-J, Song H-X, Shen W-L, Jiang L-Y, Heng BC, Ji J-F, Ouyang H-W (2013) The effect of decellularized matrices on human tendon stem/progenitor cell differentiation and tendon repair. Acta Biomater 9(12):9317–9329

Zhang N, Zhou M, Zhang Y, Wang X, Ma S, Dong L, Yang T, Ma L, Li B (2014) Porcine bone grafts defatted by lipase: efficacy of defatting and assessment of cytocompatibility. Cell Tissue Banking 15(3):357–367

Zhang Z-Z, Jiang D, Wang S-J, Qi Y-S, Zhang J-Y, Yu J-K (2015) Potential of centrifugal seeding method in improving cells distribution and proliferation on demineralized cancellous bone scaffolds for tissue-engineered meniscus. ACS Appl Mater Interfaces 7 (28):15294–15302

Zhao L, Weir MD, Xu HH (2010) An injectable calcium phosphate-alginate hydrogel-umbilical cord mesenchymal stem cell paste for bone tissue engineering. Biomaterials 31(25):6502–6510

Decellularization of Nervous Tissues and Clinical Application

19

Yasuhiro Ikegami and Hiroyuki Ijima

Abstract

The nervous system is an ensemble of organs that transmit and process external information and are responsible for the adaption to the external environment and homeostasis control of the internal environment. The nervous system of vertebrates is divided into the central nervous system (CNS) and peripheral nervous system (PNS) due to its structural features. The CNS, which includes the brain and the spinal cord, processes information from external stimuli and assembles orders suitable for these stimuli. The CNS then sends signals to control other organs/tissues. On the other hand, the PNS connects the CNS to other organs/tissues and functions as a signal pathway. Therefore, the decline and loss of various functions due to injuries of the nervous system cause an impaired quality of life (QOL) and eventually the termination of life activities. Here, we report mainly on decellularized neural tissue and its application as a substrate for the regeneration of the nervous system.

Keywords

Nervous system · CNS · PNS · Decellularization

19.1 Peripheral Nervous System (PNS)

19.1.1 History

Peripheral nerve injuries are frequently caused by traffic accidents and diseases, thus leading to a decline and loss of sensory and motor function. In the United States, more than 50,000 surgical treatments are performed annually to repair peripheral nerve injuries, and the number is increasing (Hudson et al. 2004). Modern research has identified the regenerative ability of peripheral nerves through Wallerian degeneration. Therefore, providing an environment suitable for nerve regeneration was considered as a promising therapeutic approach. Preclinical studies using laboratory animals previously examined peripheral nerve regeneration by transplantation of allogenic nerve tissue. However, this approach was found to induce negative immune reactions by resident cells. Therefore, autografting (e.g., peroneal nerve) is currently performed as a standard treatment for peripheral nerve injury, although an alternative non-invasive treatment is required due to the limited donor supply and the

Y. Ikegami · H. Ijima (✉)
Department of Chemical Engineering,
Kyushu University, 744 Motooka, Nishi-ku,
Fukuoka 819-0395, Japan
e-mail: ijima@chem-eng.kyushu-u.ac.jp

A. Kajbafzadeh (ed.), *Decellularization Methods of Tissue and Whole Organ in Tissue Engineering*, Advances in Experimental Medicine and Biology 1345,
https://doi.org/10.1007/978-3-030-82735-9_19

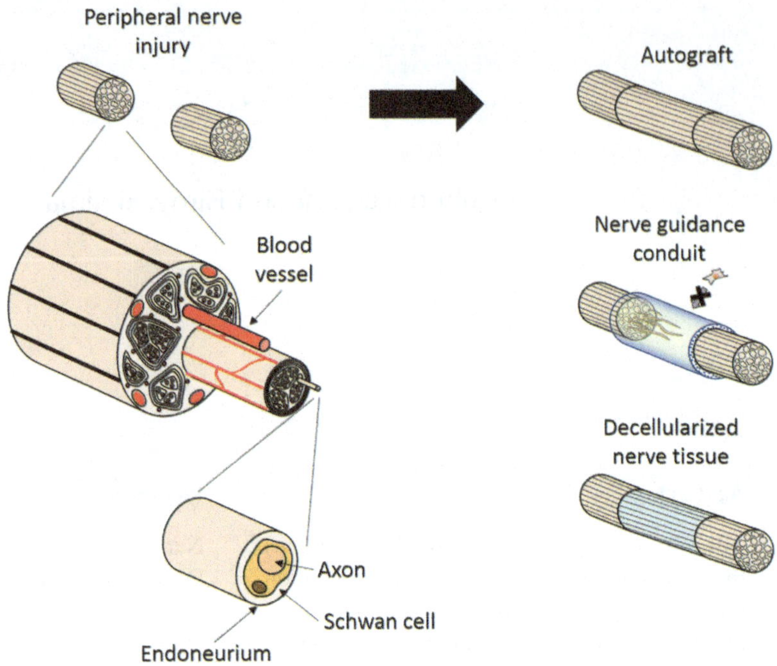

Fig. 19.1 Structure of a peripheral nerve tissue and its current therapeutic treatments

possible formation of neuromas at the site of the resected donor tissue Fig. 19.1.

Recent studies have focused on the use of nerve guidance conduits composed of natural and synthetic materials for the repair of peripheral nerves as an alternative treatment using the regenerative ability of patients. A nerve guidance conduit aims to suppress the invasion of inflammatory tissue by covering the nerve defect portion with a tubular structure and also provides an environment suitable for nerve reconstruction via the biological characteristics of the materials. Previously, nerve guidance conduits using various materials were fabricated, and their efficacy to facilitate peripheral nerve regeneration was shown by transplantation into laboratory animals such as rats and dogs (Archibald et al. 1991; Whitworth et al. 1995; Nicoli Aldini et al. 1996; Hadlock et al. 1998; Matsumoto et al. 2000; Nakamura et al. 2004). However, their recovery score did not replace that of autografting.

Alternatively, decellularized tissue has been developed as another alternative treatment for peripheral nerve injury. Decellularized tissue, fabricated by the removal of cellular components from native tissue, is associated not only with low immunological rejection, but also provides a favorable microenvironment for cells. Previous studies have shown the capabilities of decellularized tissues such as skin and heart valves, and their applications in humans have already been approved by the Food and Drug Administration (FDA) (Crapo et al. 2011). Hence, the use of decellularized tissue for peripheral nerve injury has been considered as a promising approach for nerve regeneration.

Until now, the preparation of neural tissue by decellularization has been carried out physically, chemically, and enzymatically. However, recently, chemical decellularization method using surfactants is commonly performed, focusing on the integrity of the tissue structure and on the retained ECM components (Table. 19.1) (Neubauer et al. 2007; Sondell et al. 1998; Hudson et al. 2004). The chemical decellularization method for preparing acellular neural tissue was established by Sondell et al. (Sondell et al. 1998). This method is carried out by removing cellular components using distilled water, 3% Triton X-100 in distilled water, and 4% sodium deoxycholate in distilled water and eventually storing in PBS. Although this method can effectively remove cellular components, the loss of ECM components such as laminin and the collapse of the structure has been shown to occur. Subsequently, the protocol was modified by Hudson et al., using milder surfactants such as Sulfobetaine-10 (SB-10), Triton X-200, and Sulfobetaine-16 (SB-16) (Hudson et al. 2004). This protocol succeeded in retaining the components and structure of the decellularized nervous tissue, and the remaining acellular nervous tissue promoted nerve regeneration through transplantation in a rat model. However, myelin, a structure that regulates axon function, typically remains in decellularized nerve tissue, inhibits axonal outgrowth, and negatively affects nerve regeneration. Thereafter, various researchers modified this protocol by adding some procedures for improving the removal efficiency of the cellular components using DNase/RNase and for removing the components that inhibit nerve regeneration (e.g., myelin basic protein (MBP), chondroitin sulfate proteoglycan) (Wood et al. 2014; Cai et al. 2017; Philips et al. 2018).

As a result of the above protocol changes, the decellularization method established by Hudson et al. with or without the above-mentioned minor modifications for improving its quality has been widely used due to the maintenance of the integrity of the tissue structure and of the components of the ECM.

19.1.2 Materials and Methods

Previous studies examined the transplantation of decellularized nerve tissues with various lengths ranging from 6 to 60 mm across sciatic nerve gaps. These studies reported that the recovery scores differed depending on the transplant length. For mild defects up to 10 mm, transplantation of the decellularized nerve tissue using Sondell's method showed the same therapeutic effects as autografting (Zhu and Lou 2014). However, nervous tissue capable of bridging defects 10 mm or more in length is typically required in clinical treatment. Hence, almost all of the previous studies targeted peripheral nerve defects more than 10 mm in length, and the majority of these targeted defects range from 10 to 15 mm in length. Some studies targeted defects ranging from 20 to 60 mm in length. However, the number of studies decreases as the gap length increases. With regard to defects ranging from 10 to 15 mm in length, some studies reported that the same behavioral results (i.e., sciatic functional index (SFI)) as autografting were obtained by the transplantation of acellular nervous tissue using Sondell's method (Zheng et al. 2014), whereas the recovery scores of the acellular nervous tissue were inferior to autografting in almost all other studies (Zhou et al. 2014; Zhang et al. 2014; Huang et al. 2015).

Several studies compared the recovery scores of peripheral nerve injuries by transplantation of decellularized nerve tissue prepared using Sondell's method and Hudson's method. The results showed that the acellular nerve tissues obtained using Hudson's method were superior to those obtained with Sondell's method in terms of regenerative axon density and SFI (Hudson et al. 2004; Nagao et al. 2011). Moreover, some results obtained with Hudson's method for relatively short defects were comparable to those of autografting (Nagao et al. 2011; Yan et al. 2016), though in many cases the results of relatively short defects and longer defects with Hudson's method were inferior to those of autografting (Saheb-Al-Zamani et al. 2013; Hoben et al. 2015; Marquardt et al. 2015).

19.1.3 Clinical Applications of Decellularized Peripheral Nerves

Decellularized nervous tissue obtained with Hudson's method and the chondroitinase treatment was transplanted into peripheral nerve injuries. The decellularized nerve tissue had almost the same therapeutic effects as autograft treatment due to the removal of the inhibitors of regeneration (Wood et al. 2014). The use of acellular nerve tissue has been approved for clinical trials by the FDA as a commercial product named Avance® Nerve Graft (AxiGen Inc., USA). Over 5000

Table 19.1 Previous decellularization methods of peripheral nerves (Neubauer et al. 2007; Sondell et al. 1998; Hudson et al. 2004)

Researcher	Decellularization protocol	
M. Sondell, et al.	1	Immersed in distilled water for 7 h at room temperature (RT) *replaced several times
	2	Exposed to 3% Triton X-100 in distilled water overnight at RT
	3	Agitated in 4% sodium deoxycholate in distilled water for 24 h at RT
	4	Exposed to 3% Triton X-100 in distilled water overnight at RT
	5	Agitated in 4% sodium deoxycholate In distilled water for 24 h at RT
	6	Washed with water
	7	Stored in PBS (pH 7, 2) at 4 °C
T. W. Hudson, et al.	1	Agitated in deionized distilled water for 7 h at 25 °C (*Following steps were carried out at 25 °C with constant agitation.)
	2	Replaced with a solution (SB-10 solution) containing 125 mM sulfobetaine-10 (SB-10), 10 mM phosphate, and 50 mM sodium for 15 h
	3	Rinsed In a washing solution of 50 mM phosphate and 100 mM sodium for 15 min
	4	Replaced with a solution (SB-16/Triton X-200 solution) containing 0.14% Triton X-200, 0.6 mM sulfobetaln-16 (SB-16), 10 mM phosphate, and 50 mM sodium for 24 h
	5	Rinsed with the washing solution three times (5 min per rinse)
	6	Replaced with SB-10 solution for 7 h, then washed once
	7	Replaced with SB- 16/Trlton X-200 solution for 15 h
	8	Replaced with a solution containing 10 mM phosphate and 50 mM sodium three times (15 min each)
	9	Stored in 10 mM phosphate and 50 mM sodium at 4 °C
D. Neubauer, et al.(Optional steps for removal of chondroitin sulfate proteoglycans)	1	Immersed in PBS containing 2 U/ml chondroitinase ABC for 16 h at 37 °C
	2	Rinsed twice with cold, sterile Ringer's solution (*Ringer's solution: NaCl 8.6 g, KCl 0.30 g, and CaCl2• 2H2O 0.33 g in 1000 ml)
	3	Stored on ice or at - 80 °C until use

Avance grafts have been implanted into patients worldwide, and favorable reparative effects have been obtained without any infection (Szynkaruk et al. 2013). However, the regenerative efficacy of the decellularized nerve tissue is still inferior to autografting for long peripheral nerve defects (>20 mm), and further improvements are required (Whitlock et al. 2009). Therefore, several researchers have recently attempted to facilitate nerve regeneration by combining the decellularized nerve tissue with some additives such as cells and growth factors.

19.1.4 Cell Seeding for Promotion of Peripheral Nerve Regeneration

Wallerian degeneration occurs at the distal site of a lesion after damage of peripheral nerves. There, the axons and myelin sheath constituting the peripheral nerve tissue are phagocytosed by macrophages invading the tissue, and the resident Schwann cells dedifferentiate and eventually proliferate. Subsequently, the expanding Schwann cells form a cellular column called a Büngner band and secrete neurotrophic factors (e.g., nerve growth factor (NGF)) and ECM components (e.g., laminin), leading to the re-sprouting of axons from the proximal end of the lesion. A previous study performed transplantations of degenerated isografts and decellularized nerve tissue into long rat peripheral nerve gaps 4 cm in length (Gulati 1988). Consequently, the nerve tissue including the neural cells had a higher regenerative ability compared to the acellular graft, suggesting the importance of cells in the graft during the process of nerve reconstitution. Thus, many researchers investigated the therapeutic effects of autologous Schwann cell transplantation and noted some beneficial effects. Nevertheless, there are several problems associated with this technique, such as the limited autologous Schwann cells supply from mature peripheral nerve tissue and the sacrifice of healthy tissue, thus making the use of autologous Schwann cells infeasible.

Mesenchymal stem cells (MSC) have been the recent focus of research and are considered an attractive source of cells for tissue engineering due to their ability to self-proliferate and differentiate into multiple lineages (Pittenger et al. 1999). A large number of studies showed that bone marrow-derived MSC (BMMSC) facilitate peripheral nerve regeneration by producing various neurotrophic factors such as NGF and brain-derived neurotrophic factor (BDNF) and that they may differentiate into Schwann-like cells (Chen et al. 2007). However, alternative MSC sources are urgently required because of the high invasiveness of bone marrow collection and the low content of MSC in the bone marrow (Zuk et al. 2001).

Recently, adipose-derived stem cells (ADSC), a group of mesenchymal stem cells derived from fat tissue, have received attention as an alternative to BMMSC due to their excellent characteristics. ADSCs have a similar phenotype and gene expression profile as BMMSCs, and they are enriched in fat tissue (Ugarte et al. 2003). In addition, they have a greater proliferative ability and low invasiveness compared to BMMSCs (Ugarte et al. 2003). Furthermore, ADSCs have the ability to differentiate into neural cells beyond the germ layer, and their ability to differentiate into Schwann-like cells has also been confirmed (Kingham et al. 2007). In the past, decellularized nerve tissues seeded with undifferentiated or differentiated ADSCs were transplanted to peripheral nerve defects in rats (Liu et al. 2011). Although the recovery results did not surpass those of autografting, the stimulatory effects of the seeded cells on nerve regeneration were confirmed. Presently, the mechanisms of how the seeded ADSCs affect in the body have yet been known partially, and hence, further studies are needed in the future.

19.1.5 Conclusions and Future Work on PNS Regeneration

In conclusion, the chemical decellularization method using surfactants can effectively remove the cellular components from peripheral nerve tissue. The acellular nerve tissue does not induce immunological rejection by allogeneic and xenogeneic transplantations and can retain the ECM components and structure. The acellular nerve tissue prepared with the optimized protocol had similar therapeutic effects as autografting for relatively short defects, suggesting its potential as an alternative treatment. In the case of long gaps, the reparative effects associated with the transplantation of acellular nerve grafts are presently not comparable to autografting. However, recent studies have suggested that transplantable autologous cells (e.g., ADSCs) support nerve regeneration and these studies have also confirmed the facilitating effects of combining cells with the acellular graft for nerve repair. The mechanisms via which implanted cells exert their effects are only partially known. Therefore, further studies aimed at identifying the long-term safety of the cells for the donor are necessary in the future. In other words, by understanding the mechanisms of nerve regeneration in more detail, it will be possible to more effectively utilize the combinatorial approach of decellularized nerve tissue and cells, thus potentially rendering it the novel gold standard of therapeutic treatment for peripheral nerve injury Fig. 19.2.

19.2 Central Nervous System (CNS)

19.2.1 History

In the United States, more than 1.5 million patients suffer annually from CNS diseases such as spinal cord injury (SCI) and traumatic brain injury (TBI) (McDonald and Sadowsky 2002; Thurman et al. 1999). Unlike other organs/tissues (e.g., skin, heart valve, and small intestine), the CNS has poor regenerative abilities, and damage to the CNS causes permanent motor and sensory function loss. However, the mechanisms inhibiting the regeneration of the CNS have been partially clarified, and the possibility of regenerating its components has appeared lately.

A previous study reported that central nerves regenerated approximately 30 mm in length when autologous sciatic nerve segments were implanted in a rat spinal cord injury model (David and Aguayo 1981). The results revealed that a favorable microenvironment (e.g., ECM and functional molecules) enabled the repair of the CNS. In addition, another study confirmed that the repair of SCI was caused by digestion of glial scars with a chondroitinase ABC-treatment. Glial scars are typically formed after SCI and are known to physically and biologically inhibit axonal outgrowth. Based on these results, recent studies have focused on providing a suitable environment that promotes nerve regeneration while removing and suppressing the inhibitors produced at the site of the CNS injury.

The ECM of the CNS constitutes approximately 20% of the total tissue volume and is enriched with a variety of proteoglycans, hyaluronic acids, and tenascins unlike other organs/tissues (Volpato et al. 2013; Yamaguchi 2000). Proteoglycans electrostatically bind to functional molecules such as growth factors and supply

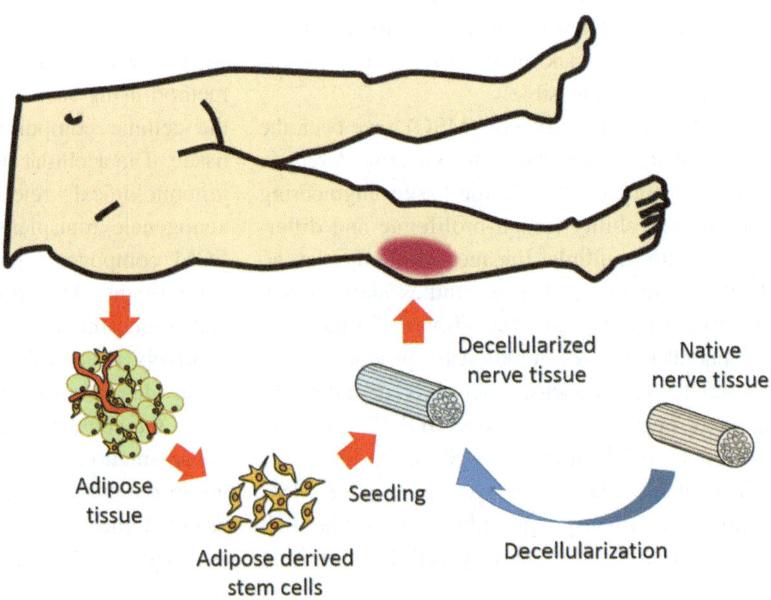

Fig. 19.2 Therapeutic scheme of peripheral nerve injuries using an acellular nerve graft and autologous cells

cells with biological signals such as proliferation and differentiation. Some reports indicated that proteoglycans themselves activated cells and controlled axonal outgrowth. Thus, the ECM has active effects on neural cells and has the potential as a biological material to construct an ideal microenvironment "niche" for central nerve regeneration.

The potential of decellularized tissue as a scaffold for tissue regeneration has already been demonstrated in other tissues (e.g., skin, heart valve, and peripheral nerve) and has been used as an FDA approved product. These scaffolds, prepared by decellularizing mammalian tissues, can suppress allogeneic and xenogeneic immune rejection by removing the cellular antigenic components. Many of the components ECM obtained by decellularization have been considered to be capable of promoting tissue regeneration. However, the decellularization method of the CNS tissue has not been sufficiently investigated compared to other organs/tissues, and there is a long way to its clinical application at present.

19.2.2 Materials and Methods

19.2.2.1 Decellularized Nerve Tissues for SCI Regeneration

Many researchers still use Sondell's method when preparing an acellular scaffold for SCI recovery because of the complexity, the use of expensive surfactants, and the presence of the residual myelin sheath in Hudson's method. The cellular components and myelin sheath of decellularized sciatic nerves are removed using Sondell's method. As a result, the main ECM components such as collagen and laminin are retained. Subsequently, decellularized sciatic nerve segments prepared by the same method were transplanted into an adult rat SCI model (2012). The results confirmed that axonal regeneration and the electrophysiological testing outcomes were improved compared to those of the SCI only condition.

Moreover, decellularization of spinal cord tissues was performed with the same method (Guo et al. 2010; Zhu et al. 2018). The resulting decellularized spinal cord tissues prepared by Sondell's method removed the cellular components and the myelin sheath, while retaining the ECM components (e.g., laminin, fibronectin, and collagen). Furthermore, it was confirmed that when the decellularized spinal cord tissues were subcutaneously transplanted to the backs of rats, the infiltration of CD4+ and CD8+ T cells was small, thus indicating that an immune response was not elicited (Guo et al. 2010). In addition, it was also confirmed that the acellular graft implanted into rat SCI models improved motor function (Zhu et al. 2018).

On the other hand, the decellularization of muscle tissue using Triton X-100 and sodium dodecyl sulfate (SDS) was performed for the preparation of regenerative scaffold for SCI (Zhang et al. 2012). The muscle tissue, like peripheral nerve tissue, has multiple channel structures and is thought to be able to guide regeneration of nerves. In addition, muscle tissue does not contain any inhibitory factor of nerve regeneration. The transplanted acellular muscle tissue was well integrated into the surrounding tissue, and axonal re-sprouting into the tissue was observed. Moreover, the sprouting axons extended inside the acellular tissue in a parallel manner, suggesting that the structure of the decellularized muscle tissue guided the axonal trajectories. As described above, it was shown that decellularized tissue also promoted remodeling after injury in the CNS.

19.2.2.2 Solubilized ECM for TBI Regeneration

Within TBI, the lesion cavities due to an ischemic stroke are irregular in size and shape, causing tissue disruption and a potential rise in the intracerebral pressure. Current therapies focus on modulating the residual tissue reactions by administering specific agents and cells capable of promoting plasticity of the remaining brain tissue to the lesion sites. However, intracerebral administration of drug or cell containing fluids causes rapid drug diffusion and death of implanted cells. Therefore, a solubilized ECM has been recently developed as an injectable hydrogel, and its potential as a minimally invasive treatment for TBI has been evaluated.

DeQuach et al. chemically decellularized a porcine brain with SDS and obtained a brain-derived ECM (B-ECM) (DeQuach et al. 2011). The B-ECM contained less intact nuclei than native porcine brain tissue and preserved the main ECM components such as collagen, sulfated glycosaminoglycans (sGAG), and laminin. Moreover, an enzymatically solubilized B-ECM promoted better growth and maturation of human iPS cell-derived neurons in vitro compared to Matrigel. Furthermore, the solubilized B-ECM spontaneously formed a gel at the injected site in vivo. These results suggested that the solubilized B-ECM may be able to fill randomly shaped lesion cavities in the brain and may function as a potential carrier of additional factors (e.g., cells and functional molecules) to promote regeneration after TBI.

In addition, Medberry et al. produced spinal cord-derived ECM (SC-ECM) and B-ECM from enzymatically solubilized porcine spinal cords and brains, respectively (Medberry et al. 2013). Both of these solubilized ECMs were formed into a gel but had different rheological properties. In addition, when neural cells (N1E-115) were cultured in solubilized ECM dilutions, both dilutions showed a dose-dependent increase in the number of cells with neurite extension. However, only the B-ECM dilution showed a dose-dependent increase in neurite length with increasing concentrations of ECM. These results imply that both solubilized CNS-derived ECMs have the potential to facilitate nerve regeneration, but the effects are different depending on the ECM sources. This difference may be due to the differing composition of the ECM based on its origin.

Additionally, Wu et al. injected the porcine B-ECM into a mouse TBI model (Wu et al. 2017). Their results showed that the ECM treatment reduced neurodegeneration in the hippocampus and white matter following injury and improved neurobehavioral function in comparison with the phosphate-buffered saline (PBS)-treated condition. Furthermore, the ECM treatment improved the TBI-induced gliosis and the microglial pro-inflammatory responses. As mentioned above, the solubilized B-ECM has a high potential as a therapeutic treatment for TBI; however, proper selection of the ECM source based on the effects on nerve defects is required.

19.2.3 Cell Transplantations for CNS Injuries

Cell transplantation has been used as another therapeutic strategy for CNS injuries such as SCI and TBI. Several studies transplanted native peripheral nerve tissue into optic nerve injuries as counterparts of acellular grafts (Berry et al. 1988, 1988). In both conditions, it was observed that the regenerated axons sprouted at the interface of the recipient tissue and the donor tissue, while the regenerated axons infiltrated and extended only into the cellular tissue. These results indicated that the presence of viable neural cells (e.g., Schwann cells) was important for the extension of the optic nerve into the decellularized nerve tissue, suggesting that the neurotrophic factors produced by the resident cells were effective for nerve regeneration in the acellular graft.

For the repair of SCI, a wide range of cell types have been transplanted, and their efficacy evaluated, such as Schwann cells, bone marrow stromal cells, MSCs, and amniotic epithelial cells (AECs). Liu et al. bridged the spinal cord lesion gap by transplantation of decellularized spinal cord tissue alone and seeded with human umbilical cord blood-derived MSCs (hUCB-MSCs) (Liu et al. 2013). Behavioral analysis showed that there was a significant recovery of locomotor activity not only in the cell-seeded condition but also in the acellular graft alone condition as compared with the SCI only condition. However, there was no statistical difference in the functional recovery with the presence of cells. Xenogenically implanted hUCB-MSCs were expected to induce immune rejection, leading to the presence of an insufficient number of cells to recognize the effects of hUCB-MSC's, such as the production of neurotrophic factors.

Moreover, Xue et al. seeded rat AECs in acellular muscle tissue and evaluated the combinatorial effects when transplanted into rat SCI models as compared to an acellular graft only (Xue et al. 2013). AECs have some advantages such as low immunogenicity and anti-inflammatory properties and retain stem cell characteristics. In addition, AECs produce neurotrophic factors such as BDNF and neurotrophin (NT)-3. When the acellular muscle tissues, with or without AECs, were implanted in the SCI, the histological, behavioral, and electrophysiological outcomes showed a significant improvement compared to the SCI only condition. In addition, compared with the acellular tissue alone, the co-implantation of acellular muscle tissue and AECs further promoted the re-myelination of nerve fibers and functional recovery. It was also found that the transplantation of viable cells was effective in promoting nerve regeneration.

Furthermore, Yin et al. transplanted decellularized spinal cord tissue injected with rat ADSCs into rat spinal cord lesion sites (Yin et al. 2018). Although all the regenerative mechanisms were not revealed, the results implied that the acellular graft seeded with rat ADSCs facilitated the functional recovery of SCI to a greater extent than the acellular graft alone because of the promotion of axon outgrowth and the reduction of gliosis. This result indicates that transplantation of allogeneic acellular grafts seeded with autologous cells collected from the donors themselves is effective for not only PNS injuries but also CNS injuries. Therefore, these treatments are considered highly feasible approaches for nerve regeneration.

On the other hand, injections of solubilized ECM containing cell suspensions into CNS injuries have also been performed. In recent studies, bioactive scaffolds are no longer being used to merely to physically protect transplanted stem cells and to support cell survival. Previous studies showed that urinary bladder-derived ECM (UB-ECM, non-neuronal ECM) had anti-inflammatory properties and degraded rapidly in vivo. Moreover, injected UB-ECM reduced the lesion volume as well as myelin disruption in TBI and improved vestibulomotor function (Zhang et al. 2013). However, no significant differences in cognitive recovery between the UB-ECM- and PBS-treated groups were confirmed. Another research group injected the UB-ECM into a TBI combined with mouse neural stem cells (NSCs) (Wang et al. 2013). The results showed that only UB-ECM containing NSCs attenuated memory and cognitive impairments, suggesting that both the cells and the ECM were essential for the functional recovery of TBI.

Previous studies reported that the effects of UB-ECM (non-neural ECM) on neural cells differed from those of B-ECM (neural ECM) (Crapo et al. 2014), a finding which may be attributable to the difference in the composition of the ECM. Consequently, the efficacy of cell transplantation with neural ECM for TBI will need to be evaluated and compared with that of non-neural ECM in the future.

19.2.4 Conclusions and Future Works for CNS Regeneration

Chemically decellularized nerve tissue has a low immunological antigenicity and retains the main ECM components including collagen and laminin. Transplantation of the acellular nerve tissue promoted recovery and regeneration after SCI. Decellularized muscle tissue (non-neural ECM) also promoted regeneration after SCI due to the retained ECM components and structure. Moreover, solubilized acellular nerve tissue formed a gel in vivo after its injection into the cavity of the lesion caused by TBI. Injection of ECM scaffolds in lesions of the brain with smaller volumes promoted recovery of motor function and improvement of the inflammatory responses. In addition, co-implantation of the ECM scaffold and cells enhanced its efficacy in both of SCI and TBI. As the mechanisms involved in CNS regeneration are poorly understood, the effects of various variables, including ECM composition and seeded cell types, should be examined in the future Fig. 19.3.

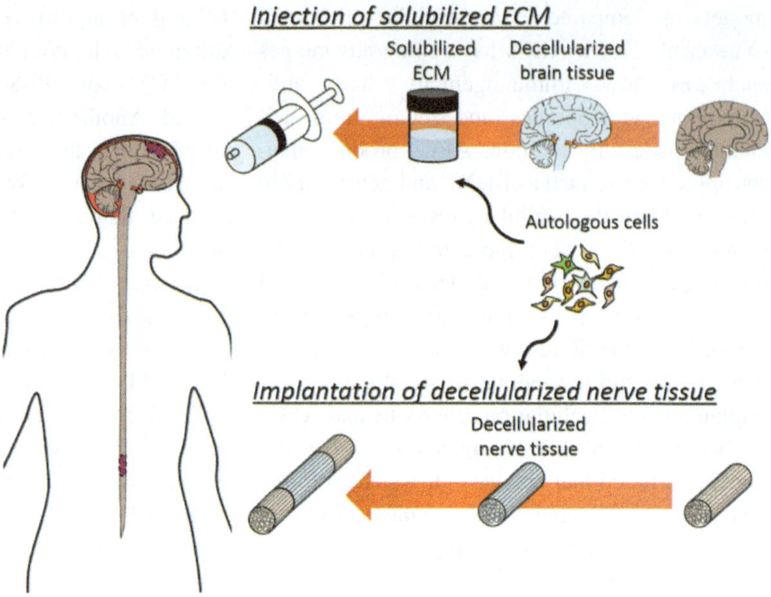

Fig. 19.3 Therapeutic scheme of central nerve injuries using tissue-engineered neural ECM scaffolds and autologous cell

19.3 Overall Summary and Future Works

In this section, we reported the efficacy of decellularized nerve tissue for nerve regeneration and its applications. Acellular nerve tissue had regenerative effects not only in the PNS but also in the CNS. There is currently no ideal protocol for selectively removing the factors that inhibit regeneration from the ECM and selectively retaining the regeneration promoting factors. Moreover, not all of the mechanisms of nerve regeneration via decellularized nerve tissue have been clarified. However, further examinations involving combinations of additional factors (e.g., cells and functional molecules) and the improvement of the decellularization method may lead to the achievement of optimal nerve regeneration in the future.

References

Archibald SJ, Krarup C, Shefner J, Li ST, Madison RD (1991) A collagen-based nerve guide conduit for peripheral nerve repair: an electrophysiological study of nerve regeneration in rodents and nonhuman primates. J Comp Neurol 306(4):685–696

Berry M, Hall S, Follows R, Rees L, Gregson N, Sievers J (1988) Response of axons and glia at the site of anastomosis between the optic nerve and cellular or acellular sciatic nerve grafts. J Neurocytol 17(6):727–744

Berry M, Rees L, Hall S, Yiu P, Sievers J (1988) Optic axons regenerate into sciatic nerve isografts only in the presence of Schwann cells. Brain Res Bull 20 (2):223–231

Cai M, Huang T, Hou B, Guo Y (2017) Role of demyelination efficiency within acellular nerve scaffolds during nerve regeneration across peripheral defects. Biomed Res Int 2017:1–10

Chen CJ, Ou YC, Liao SL, Chen WY, Chen SY, Wu CW, Wang CC, Wang WY, Huang YS, Hsu SH (2007) Transplantation of bone marrow stromal cells for peripheral nerve repair. Exp Neurol 204(1):443–453

Crapo PM, Gilbert TW, Badylak SF (2011) An overview of tissue and whole organ decellularization processes. Biomaterials 32(12):3233–3243

Crapo PM, Tottey S, Slivka PF, Badylak SF (2014) Effects of biologic scaffolds on human stem cells and implications for CNS tissue engineering. Tissue Eng Part A 20(1–2):313–323

David S, Aguayo AJ (1981) Axonal elongation into peripheral nervous system "bridges" after central nervous system injury in adult rats. Science 214 (4523):931–933

De Ugarte DA, Morizono K, Elbarbary A, Alfonso Z, Zuk PA, Zhu M, Dragoo JL, Ashjian P, Thomas B, Benhaim P, Chen I, Fraser J, Hedrick MH (2003) Comparison of multi-lineage cells from human adipose tissue and bone marrow. Cells Tissues Organs 174(3):101–109

De Ugarte DA, Alfonso Z, Zuk PA, Elbarbary A, Zhu M, Ashjian P, Benhaim P, Hedrick MH, Fraser JK (2003) Differential expression of stem cell mobilization-associated molecules on multi-lineage cells from adipose tissue and bone marrow. Immunol Lett 89 (2–3):267–270

DeQuach JA, Yuan SH, Goldstein LS, Christman KL (2011) Decellularized porcine brain matrix for cell culture and tissue engineering scaffolds. Tissue Eng Part A 17(21–22):2583–2592

Gulati AK (1988) Evaluation of acellular and cellular nerve grafts in repair of rat peripheral nerve. J Neurosurg 68(1):117–123

Guo SZ, Ren XJ, Wu B, Jiang T (2010) Preparation of the acellular scaffold of the spinal cord and the study of biocompatibility. Spinal Cord 48(7):576–581

Hadlock T, Elisseeff J, Langer R, Vacanti J, Cheney M (1998) A tissue-engineered conduit for peripheral nerve repair. Arch Otolaryngol Head Neck Surg 124 (10):1081–1086

Hoben G, Yan Y, Iyer N, Newton P, Hunter DA, Moore AM, Sakiyama-Elbert SE, Wood MD, Mackinnon SE (2015) Comparison of acellular nerve allograft modification with Schwann cells or VEGF. Hand (n y) 10(3):396–402

Huang H, Xiao H, Liu H, Niu Y, Yan R, Hu M (2015) A comparative study of acellular nerve xenografts and allografts in repairing rat facial nerve defects. Mol Med Rep 12(4):6330–6336

Hudson TW, Zawko S, Deister C, Lundy S, Hu CY, Lee K, Schmidt CE (2004) Optimized acellular nerve graft is immunologically tolerated and supports regeneration. Tissue Eng 10(11–12):1641–1651

Hudson TW, Liu SY, Schmidt CE (2004) Engineering an improved acellular nerve graft via optimized chemical processing. Tissue Eng 10(10–12):1346–1358

Kingham PJ, Kalbermatten DF, Mahay D, Armstrong SJ, Wiberg M, Terenghi G (2007) Adipose-derived stem cells differentiate into a Schwann cell phenotype and promote neurite outgrowth in vitro. Exp Neurol 207 (2):267–274

Li C, Zhang X, Cao R, Yu B, Liang H, Zhou M, Li D, Wang Y, Liu E (2012) Allografts of the acellular sciatic nerve and brain-derived neurotrophic factor repair spinal cord injury in adult rats. PLoS One 7(8): e42813

Liu G, Cheng Y, Guo S, Feng Y, Li Q, Jia H, Wang Y, Tong L, Tong X (2011) Transplantation of adipose-derived stem cells for peripheral nerve repair. Int J Mol Med 28(4):565–572

Liu J, Chen J, Liu B, Yang C, Xie D, Zheng X, Xu S, Chen T, Wang L, Zhang Z, Bai X, Jin D (2013) Acellular spinal cord scaffold seeded with mesenchymal stem cells promotes long-distance axon regeneration and functional recovery in spinal cord injured rats. J Neurol Sci 325(1–2):127–136

Marquardt LM, Ee X, Iyer N, Hunter D, Mackinnon SE, Wood MD, Sakiyama-Elbert SE (2015) Finely tuned temporal and spatial delivery of GDNF promotes enhanced nerve regeneration in a long nerve defect model. Tissue Eng Part A 21(23–24):2852–2864

Matsumoto K, Ohnishi K, Kiyotani T, Sekine T, Ueda H, Nakamura T, Endo K, Shimizu Y (2000) Peripheral nerve regeneration across an 80-mm gap bridged by a polyglycolic acid (PGA)-collagen tube filled with laminin-coated collagen fibers: a histological and electrophysiological evaluation of regenerated nerves. Brain Res 868(2):315–328

McDonald JW, Sadowsky C (2002) Spinal-cord injury. Lancet 359(9304):417–425

Medberry CJ, Crapo PM, Siu BF, Carruthers CA, Wolf MT, Nagarkar SP, Agrawal V, Jones KE, Kelly J, Johnson SA, Velankar SS, Watkins SC, Modo M, Badylak SF (2013) Hydrogels derived from central nervous system extracellular matrix. Biomaterials 34(4):1033–1040

Nagao RJ, Lundy S, Khaing ZZ, Schmidt CE (2011) Functional characterization of optimized acellular peripheral nerve graft in a rat sciatic nerve injury model. Neurol Res 33(6):600–608

Nakamura T, Inada Y, Fukuda S, Yoshitani M, Nakada A, Itoi S, Kanemaru S, Endo K, Shimizu Y (2004) Experimental study on the regeneration of peripheral nerve gaps through a polyglycolic acid-collagen (PGA-collagen) tube. Brain Res 1027(1–2):18–29

Neubauer D, Graham JB, Muir D (2007) Chondroitinase treatment increases the effective length of acellular nerve grafts. Exp Neurol 207(1):163–170

Nicoli Aldini N, Perego G, Cella GD, Maltarello MC, Fini M, Rocca M, Giardino R (1996) Effectiveness of a bioabsorbable conduit in the repair of peripheral nerves. Biomaterials 17(10):959–962

Philips C, Campos F, Roosens A, Sánchez-Quevedo MDC, Declercq H, Carriel V (2018) Qualitative and quantitative evaluation of a novel detergent-based method for decellularization of peripheral nerves. Ann Biomed Eng 46(11):1921–1937

Pittenger MF, Mackay AM, Beck SC, Jaiswal RK, Douglas R, Mosca JD, Moorman MA, Simonetti DW, Craig S, Marshak DR (1999) Multilineage potential of adult human mesenchymal stem cells. Science 284 (5411):143–147

Saheb-Al-Zamani M, Yan Y, Farber SJ, Hunter DA, Newton P, Wood MD, Stewart SA, Johnson PJ, Mackinnon SE (2013) Limited regeneration in long acellular nerve allografts is associated with increased Schwann cell senescence. Exp Neurol 247:165–177

Sondell M, Lundborg G, Kanje M (1998) Regeneration of the rat sciatic nerve into allografts made acellular through chemical extraction. Brain Res 795(1–2):44–54

Szynkaruk M, Kemp SW, Wood MD, Gordon T, Borschel GH (2013) Experimental and clinical evidence for use of decellularized nerve allografts in peripheral nerve gap reconstruction. Tissue Eng Part B Rev 19(1):83–96

Thurman DJ, Alverson C, Dunn KA, Guerrero J, Sniezek JE (1999) Traumatic brain injury in the

United States: a public health perspective. J Head Trauma Rehabil 14(6):602–615
Volpato FZ, Führmann T, Migliaresi C, Hutmacher DW, Dalton PD (2013) Using extracellular matrix for regenerative medicine in the spinal cord. Biomaterials 34(21):4945–4955
Wang JY, Liou A, Ren ZH, Zhang L, Brown BN, Cui XT, Badylak SF, Cai YN, Guan YQ, Leak RK, Chen J, Ji X, Chen L (2013) Neurorestorative effect of urinary bladder matrix-mediated neural stem cell transplantation following traumatic brain injury in rats. CNS Neurol Disord Drug Targets 12(3):413–425
Whitlock EL, Tuffaha SH, Luciano JP, Yan Y, Hunter DA, Magill CK, Moore AM, Tong AY, Mackinnon SE, Borschel GH (2009) Processed allografts and type I collagen conduits for repair of peripheral nerve gaps. Muscle Nerve 39(6):787–799
Whitworth IH, Brown RA, Doré C, Green CJ, Terenghi G (1995) Orientated mats of fibronectin as a conduit material for use in peripheral nerve repair. J Hand Surg Br 20(4):429–436
Wood MD, Kemp SW, Liu EH, Szynkaruk M, Gordon T, Borschel GH (2014) Rat-derived processed nerve allografts support more axon regeneration in rat than human-derived processed nerve xenografts. J Biomed Mater Res A 102(4):1085–1091
Wu Y, Wang J, Shi Y, Pu H, Leak RK, Liou AKF, Badylak SF, Liu Z, Zhang J, Chen J, Chen L (2017) Implantation of brain-derived extracellular matrix enhances neurological recovery after traumatic brain injury. Cell TransplAnt 26(7):1224–1234
Xue H, Zhang XY, Liu JM, Song Y, Li YF, Chen D (2013) Development of a chemically extracted acellular muscle scaffold seeded with amniotic epithelial cells to promote spinal cord repair. J Biomed Mater Res A 101(1):145–156
Yamaguchi Y (2000) Lecticans: organizers of the brain extracellular matrix. Cell Mol Life Sci 57(2):276–289
Yan Y, Wood MD, Hunter DA, Ee X, Mackinnon SE, Moore AM (2016) The effect of short nerve grafts in series on axonal regeneration across isografts or acellular nerve allografts. J Hand Surg Am 41(6): e113–e121
Yin H, Jiang T, Deng X, Yu M, Xing H, Ren X (2018) A cellular spinal cord scaffold seeded with rat adipose-derived stem cells facilitates functional recovery via enhancing axon regeneration in spinal cord injured rats. Mol Med Rep 17(2):2998–3004
Zhang XY, Xue H, Liu JM, Chen D (2012) Chemically extracted acellular muscle: a new potential scaffold for spinal cord injury repair. J Biomed Mater Res A 100 (3):578–587
Zhang L, Zhang F, Weng Z, Brown BN, Yan H, Ma XM, Vosler PS, Badylak SF, Dixon CE, Cui XT, Chen J (2013) Effect of an inductive hydrogel composed of urinary bladder matrix upon functional recovery following traumatic brain injury. Tissue Eng Part A 19(17–18):1909–1918
Zhang Y, Zhang H, Katiella K, Huang W (2014) Chemically extracted acellular allogeneic nerve graft combined with ciliary neurotrophic factor promotes sciatic nerve repair. Neural Regen Res 9(14):1358–1364
Zheng C, Zhu Q, Liu X, Huang X, He C, Jiang L, Quan D (2014) Improved peripheral nerve regeneration using acellular nerve allografts loaded with platelet-rich plasma. Tissue Eng Part A 20(23–24):3228–3240
Zhou X, He B, Zhu Z, He X, Zheng C, Xu J, Jiang L, Gu L, Zhu J, Zhu Q, Liu X (2014) Etifoxine provides benefits in nerve repair with acellular nerve grafts. Muscle Nerve 50(2):235–243
Zhu G, Lou W (2014) Regeneration of facial nerve defects with xenogeneic acellular nerve grafts in a rat model. Head Neck 36(4):481–486
Zhu J, Lu Y, Yu F, Zhou L, Shi J, Chen Q, Ding W, Wen X, Ding YQ, Mei J, Wang J (2018) Effect of decellularized spinal scaffolds on spinal axon regeneration in rats. J Biomed Mater Res A 106(3):698–705
Zuk PA, Zhu M, Mizuno H, Huang J, Futrell JW, Katz AJ, Benhaim P, Lorenz HP, Hedrick MH (2001) Multilineage cells from human adipose tissue: implications for cell-based therapies. Tissue Eng 7(2):211–228

The Three-Dimensional Structure of Tissues and Organs: A Scanning Electron Microscopic Atlas for Research and Education

20

Shabnam Sabetkish
and Abdol-Mohammad Kajbafzadeh

In this atlas we explain what does scanning electron microscope (SEM) means, focus on discussing the SEM method, its consumption, principle, development, samples preparation, and applications in materials science. This chapter also represents a variety of SEM images of more than 20 organs such as urologic, connective, nervous, skeletal, and muscle tissues, blood, bone marrow, glands, and other organs. The current atlas offers an exclusive and wide-ranging look at the structure and function of normal and decellularized tissues/organs at the subcellular and molecular level, an important perspective in understanding the field of tissue engineering and regenerative medicine.

- Presents the major systems of the tissues and organs through SEM images.
- Has images prepared almost exclusively from human/animal tissues.
- Contains sets of 3D images in most chapters.

SEM is a great luxury image system that is infrequently used for routine work at medical schools. This method is employed to show the arrangement of the collagen fibrillar network of various tissues. The SEM was first introduced in 1938 (Von Ardenne), and in 1965 the first SEM instrument came into the market (von Ardenne 1985). We do suppose that SEM examination is complementary to histological tests for the best and most accurate interpretation of biological function as well as evaluation of the structure of fairly complex biological structures. By the application of SEM, it is easier to comprehend the three-dimensional images of the tissues and organs. Evidence demonstrated that enhancing our understanding and knowledge of the structure and function of normal and decellularized tissues at the subcellular and molecular levels advance our chances of finding treatments for various diseases.

For better comprehension of the readers, this indispensable tool produces images of a sample by scanning the surface phenomena with a focused beam of high-energy electrons (Akhtar et al. 2018). These electrons generate several signals at the surface of solid specimens. The signals that are obtained from electron-sample interactions show data about the specimen regarding external morphology (shape and size), composition (elements and compounds that constitute the material), crystallography (the arrangement of atoms in the materials), smoothness or roughness, and orientation of materials of the sample (Zhou et al. 2006). In the next step, information is gathered over a chosen region of

S. Sabetkish · A.-M. Kajbafzadeh (✉)
Pediatric Urology and Regenerative Medicine Research Center, Section of Tissue Engineering and Stem Cells Therapy, Children's Medical Center, Tehran University of Medical Sciences,
No. 62, Dr. Gharib's Street, Keshavarz Boulevard, 1419433151 Tehran, Iran
e-mail: kajbafzd@sina.tums.ac.ir

A. Kajbafzadeh (ed.), *Decellularization Methods of Tissue and Whole Organ in Tissue Engineering*, Advances in Experimental Medicine and Biology 1345,
https://doi.org/10.1007/978-3-030-82735-9_20

the surface of the sample, and a 2-dimensional picture is produced that exhibits spatial variations in these possessions. Approximately 1 cm–5 μ in width can be imaged as areas ranges in a scanning mode with conventional SEM methods (Ul-Hamid 2018). A wide range of magnifications is possible, from 20× to approximately 30,000×, with a spatial resolution of 50–100 nm (Srinivasulu et al. 2017). The size of the electron spot is critical for the spatial resolution of the SEM, which depends on both the wavelength of the electrons and the electron-optical system.

Without any doubt, this versatile advanced instrument is vital in all fields that necessitate classification of solid materials and there is no other tool with the extent of applications in the study of solid materials that compare with the SEM. It is also important to note that most of the SEM's are effortless to function, with user-friendly "instinctive" interfaces (He et al. 2019). This electron microscopy technique needs minimal and trouble-free sample preparation. Additionally, data achievement is fast as less than 5 min. Modern SEMs also produce statistics in digital formats, which are extremely portable (Fandrich et al. 2007).

Sample preparation depends on the character of the samples and the data required. Preparation comprises attainment of a sample that will fit into the SEM chamber and some accommodation to avoid charge build-up on electrically insulating samples (Zhou et al. 2006). For conventional imaging in the SEM, specimens must be electrically conductive, to prevent scanning faults and other image artifacts, and also they are needed to be electrically grounded to avert the accumulation of electrostatic charge (Assumpção and Ferri 2017). Non-conducting substances are generally coated with an ultrathin coating of electrically conducting material, by low-vacuum sputter coating or by high-vacuum evaporation (Echlin 2011). Conductive materials use I gold, gold/palladium alloy, platinum, iridium, tungsten, chromium, osmium, and graphite for specimen coating (Heu et al. 2019).

Although SEM is applied to envisage exterior images of a tissue/organ without any given internal information (McMahon 2008), it is still considered as a dominant appliance used in characterizing crystallographic, magnetic, and electrical features of the sample. This technique also is able to determine if any morphological modifications of the particle have occurred after changing the sample surface with other molecules or even after the decellularization process (Goldstein 2012).

In our center, several images from more than 25 organs have been taken from natural and decellularized tissues in order to evaluate the efficacy of the decellularization process for cell removal and assure the maintenance of the extra cellular matrix (ECM) structure on SEM as a multipurpose state-of-the-art instrument. For this purpose, we fix the specimens using2.5% glutaraldehyde and wash them in PBS three times, and left them for 24 h at 3 °C. For complete drying, we dehydrate the specimens in graded ethanol-water series (30, 50, 70, 90, and 100%; and use critical point drying in CO_2. The specimens are then placed under ultraviolet light for the next 10 h. Finally, to improve the electron conductivity, the dehydrated specimens are coated with a thin layer (approximately 2 nm thick) of gold/palladium (Au/Pd) by the application of a Gatan ion beam coater. Samples are then examined using field emission SEM (FE-SEM; JSM-6340F, JEOL, Tokyo, Japan). The coated samples are evaluated with an acceleration voltage of 10 kV and a working distance of 8 mm is used (S3500N; Hitachi High Technologies America).

Several papers state the main steps leading to the technique of SEM of vascular corrosion casts (Lametschwandtner et al. 1990; Giuvărăşteanu 2007; Vrbova et al. 2020; Lametschwandtner and Minnich 2020; Gorczyca et al. 2017). In our lab, we also decellularized rat lungs and preserved the inner microarchitecture, then seeded cells onto the stable bioscaffold. Casting of the decellularized scaffold and SEM of casts were performed to confirm preservation of geometrical properties for further cell seeding. Foam wound dressing (Cavi-Care; Smith & Nephew, Victoria, Australia) was applied for this process. The two liquid components were mixed and injected through the trachea into the lung tissue to form white foam that conformed to the contours of the lung microstructure. Both lungs were remained at room temperature for 24 h to let the injected material dry meticulously. In the next step, the lungs were placed in 90% acetic acid for 10 min to digest the tissue. SEM was used to verify the detailed internal structure of terminal alveoli in casts of natural and decellularized lung tissues. The results demonstrated that SEM examination of molds showed the preservation of terminal alveolar architecture after the decellularization process. No obvious difference was detected between the two molds (Kajbafzadeh et al. 2015).

We believe that this atlas would be beneficial for those who can afford a scanning electron microscope in their labs, for the student of functional anatomy as for the serious research workers.

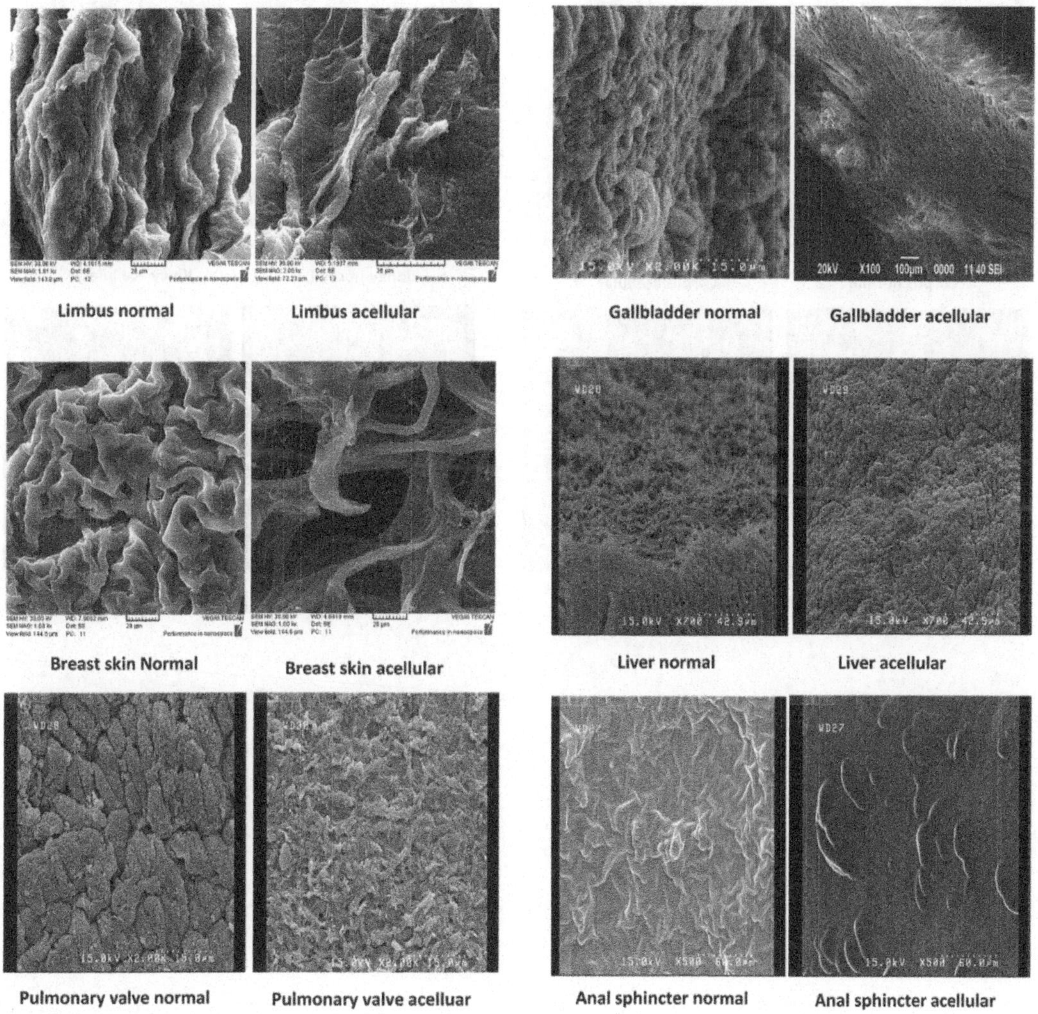

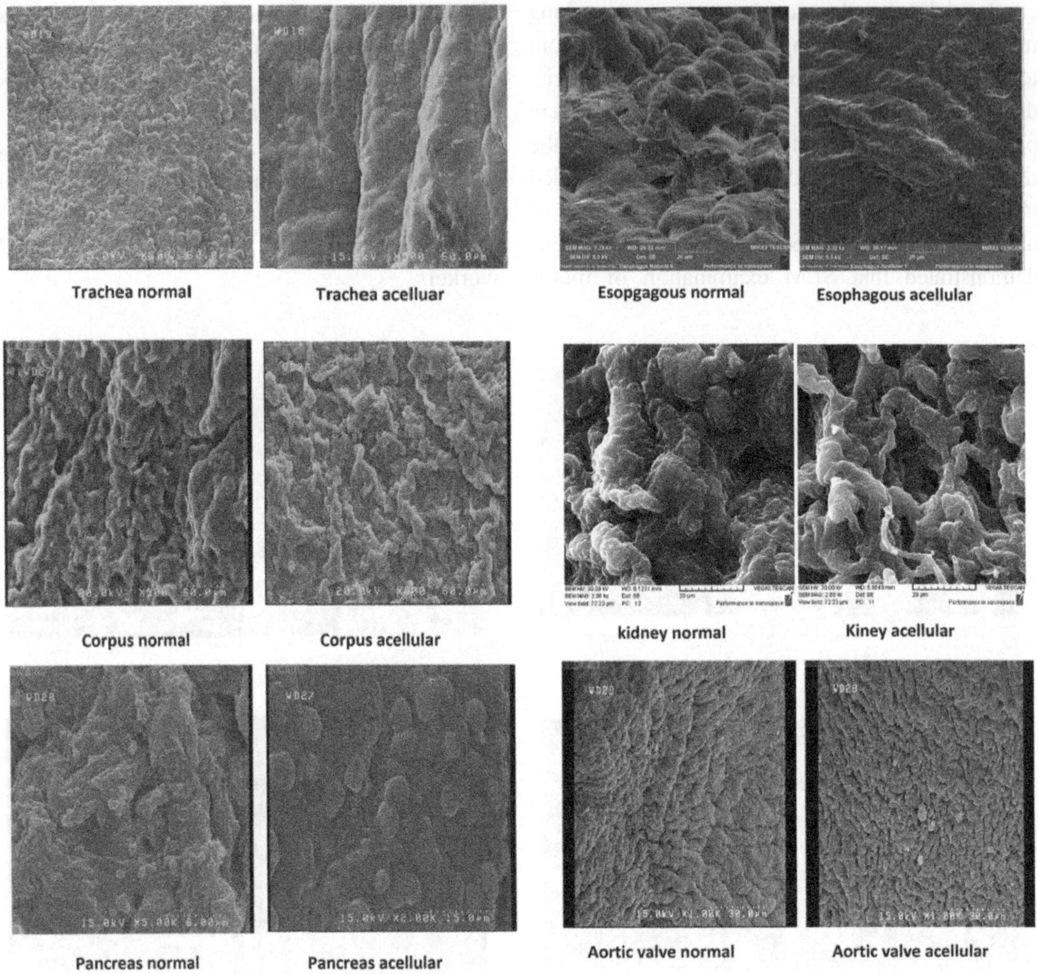
Trachea normal
Trachea acelluar
Esopgagous normal
Esophagous acellular
Corpus normal
Corpus acellular
kidney normal
Kiney acellular
Pancreas normal
Pancreas acellular
Aortic valve normal
Aortic valve acellular

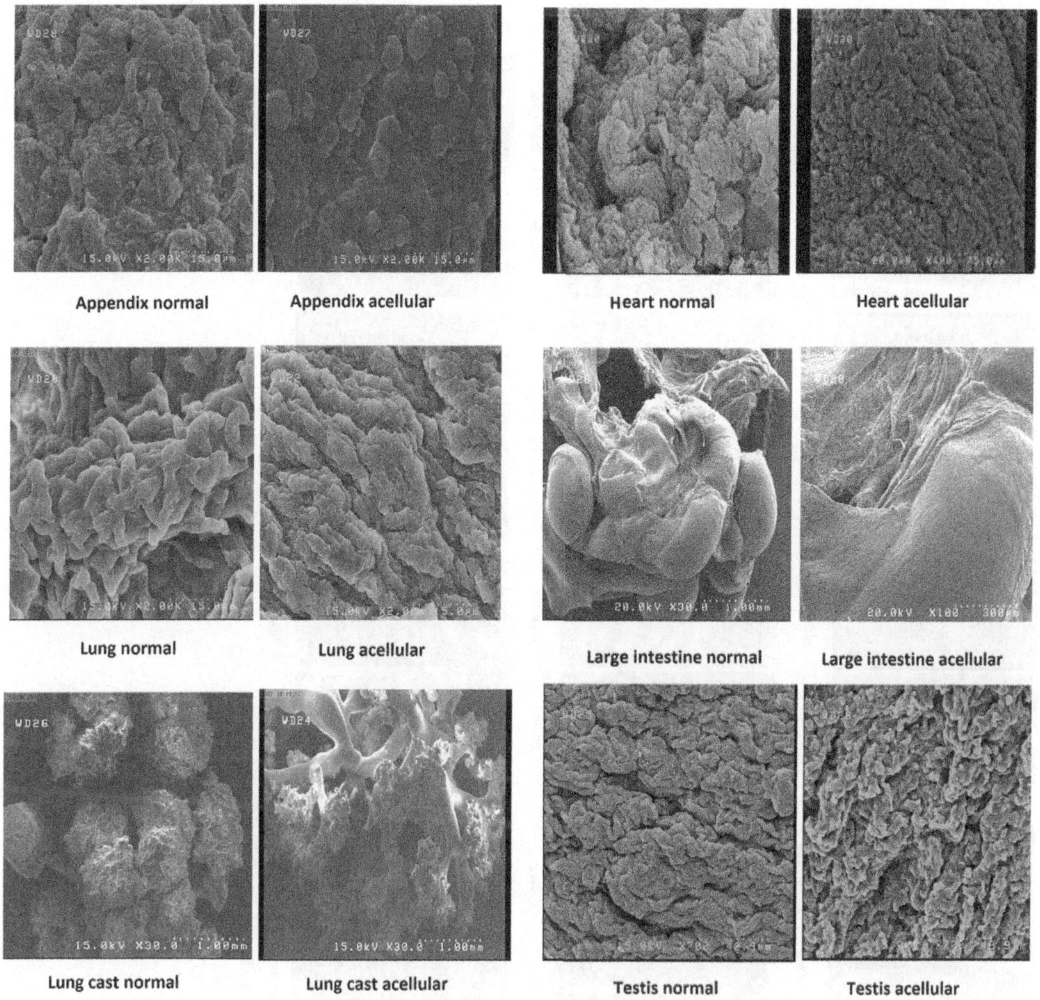
Appendix normal
Appendix acellular
Heart normal
Heart acellular
Lung normal
Lung acellular
Large intestine normal
Large intestine acellular
Lung cast normal
Lung cast acellular
Testis normal
Testis acellular

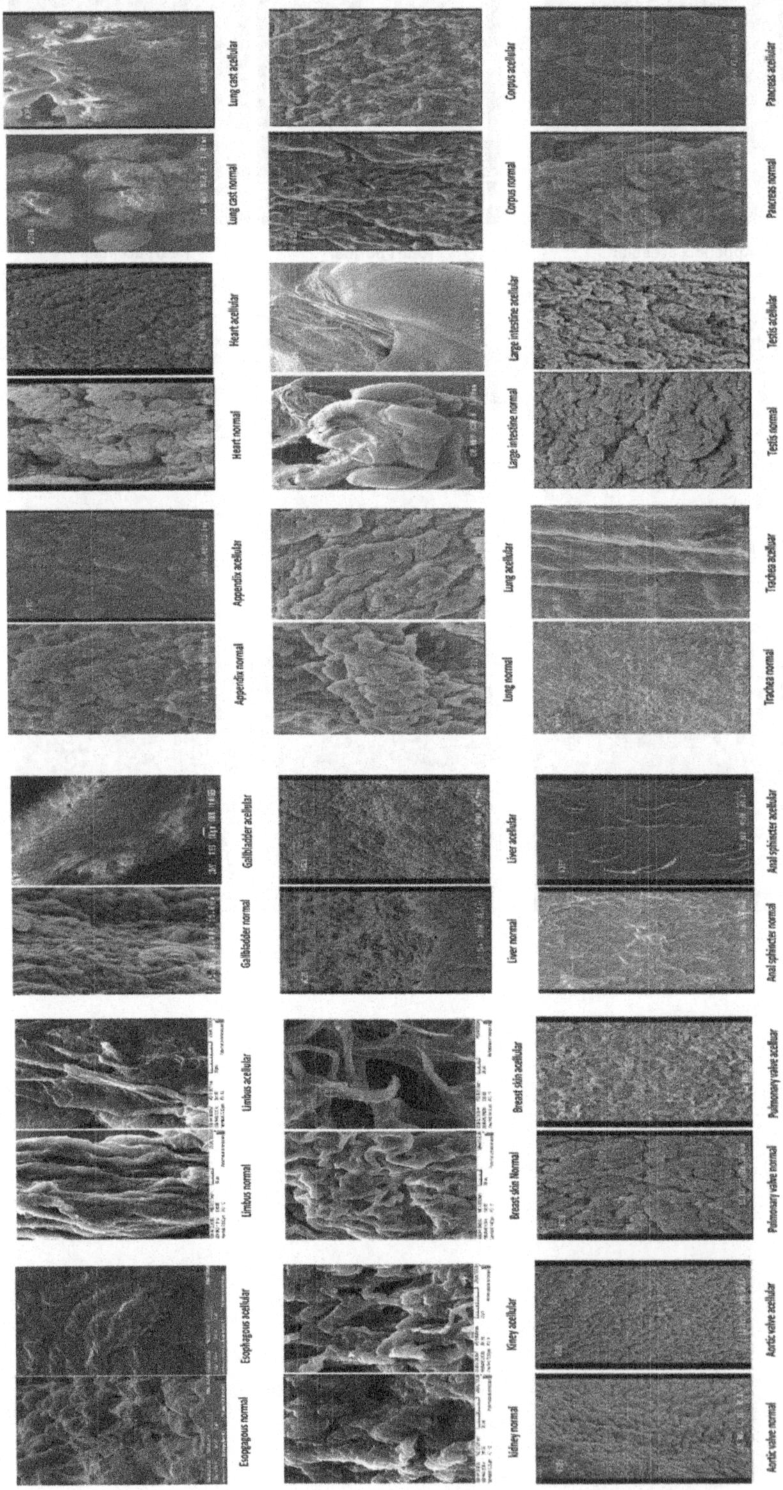
Lung cast normal
Lung cast acellular
Heart normal
Heart acellular
Appendix normal
Appendix acellular
Corpus normal
Corpus acellular
Large intestine normal
Large intestine acellular
Lung normal
Lung acellular
Pancreas normal
Pancreas acellular
Testis normal
Testis acellular
Trachea normal
Trachea acellular
Esopgagous normal
Esophagus acellular
Limbus normal
Limbus acellular
Gallbladder normal
Gallbladder acellular
kidney normal
Kiney acellular
Breast skin Normal
Breast skin acellular
Liver normal
Liver acellular
Aortic valve normal
Aortic valve acellular
Pulmonary valve normal
Pulmonary valve acellular
Anal sphincter normal
Anal sphincter acellular

References

Akhtar K, Khan SA, Khan SB, Asiri AM (2018) Scanning electron microscopy: principle and applications in nanomaterials characterization. In: Handbook of materials characterization. Springer, pp 113–45

de Assumpção Pereira-da-Silva M, Ferri FA (2017) Scanning electron microscopy. In: Nanocharacterization techniques. Elsevier, pp 1–35

Echlin P (2011) Handbook of sample preparation for scanning electron microscopy and X-ray microanalysis. Springer Science & Business Media

Fandrich R, Gu Y, Burrows D, Moeller K (2007) Modern SEM-based mineral liberation analysis. Int J Miner Process 84(1–4):310–320

Giuvărăşteanu I (2007) Scanning electron microscopy of vascular corrosion casts–standard method for studying microvessels. Rom J Morphol Embryol 48(3):257–261

Goldstein J (2012) Practical scanning electron microscopy: electron and ion microprobe analysis. Springer Science & Business Media

Gorczyca J, Tomaszewski KA, Henry BM, Pekala PA, Pasternak A, Mizia E et al (2017) The vascular microarchitecture of the human fetal pancreas: a corrosion casting and scanning electron microscopy study. Pancreas 46(1):124–130

He T, Sun Z, Shi Q, Zhu M, Anaya DV, Xu M et al (2019) Self-powered glove-based intuitive interface for diversified control applications in real/cyber space. Nano Energy 58:641–651

Heu R, Shahbazmohamadi S, Yorston J, Capeder P (2019) Target material selection for sputter coating of SEM samples. Microscopy Today 27(4):32–36

Kajbafzadeh A, Sabetkish N, Sabetkish S, Tavangar S, Beigi RH, Talebi M et al (2015) Lung tissue engineering and preservation of alveolar microstructure using a novel casting method. Biotech Histochem 90(2):111–123

Lametschwandtner A, Minnich B (2020) Renal microvasculature in the adult pipid frog, Xenopus laevis: a scanning electron microscope study of vascular corrosion casts. J Morphol

Lametschwandtner A, Lametschwandtner U, Weiger T (1990) Scanning electron microscopy of vascular corrosion casts—technique and applications: updated review. Scanning microscopy 4(4):889–940; discussion 1

McMahon G (2008) Analytical instrumentation: a guide to laboratory, portable and miniaturized instruments. Wiley & Sons

Srinivasulu B, Dev PB, Murthy P (2017) Chemical Characterization of Samaguna Balijarita Kajjali (Black Sulphide of Mercury). Int J Pharm Phytopharmacol Res 2(1):16–19

Ul-Hamid A (2018) A beginners' guide to scanning electron microscopy. Springer

von Ardenne M (1985) On the history of scanning electron microscopy, of the electron microprobe, and of early contributions to transmission electron microscopy. In: The beginnings of electron microscopy. Academic Press Orlando, pp 1–21

Vrbova R, Bradna P, Bartos M, Roubickova A (2020) The effect of disinfectants on the accuracy, quality and surface structure of impression materials and gypsum casts: a comparative study using light microscopy, scanning electron microscopy and micro computed tomography. Dental Mater J 2019-065

Zhou W, Apkarian R, Wang ZL, Joy D (2006) Fundamentals of scanning electron microscopy (SEM). In: Scanning microscopy for nanotechnology. Springer, pp 1–40

The manufacturer's authorised representative in the EU is Springer Nature Customer Service Centre GmbH, Europaplatz 3, 69115 Heidelberg, Germany. If you have any concerns regarding our products, please contact ProductSafety@springernature.com

Printed and bound by CPI Group (UK) Ltd, Croydon, CR0 4YY
28/09/2026
02224091-0020